Springer Series in Geomechanics and Geoengineering

Editors: Wei Wu · Ronaldo I. Borja

Jacek Tejchman

Shear Localization in Granular Bodies with Micro-Polar Hypoplasticity

Springer

Professor Wei Wu, Institut für Geotechnik, Universität für Bodenkultur, Feistmantelstraße 4, 1180 Vienna, Austria, E-mail: wei.wu@boku.ac.at

Professor Ronaldo I. Borja, Department of Civil and Environmental Engineering, Stanford University, Stanford, CA 94305-4020, USA, E-mail: borja@stanford.edu

Author

Professor Dr. Jacek Tejchman
University of Gdansk
Department of Civil and Environmental Engineering
Narutowicza 11/12
80-952 Gdansk
Poland
E-Mail: tejchmk@pg.gda.pl

ISBN 978-3-642-08948-0 e-ISBN 978-3-540-70555-0

DOI 10.1007/978-3-540-86654-1

Springer Series in Geomechanics and Geoengineering ISSN 1866-8755

Cover Design: Erich Kirchner Heidelberg, Germany.

Printed in acid-free paper

5 4 3 2 1 0

springer.com

Preface

This book includes a numerical investigation of shear localization in granular materials within micro-polar hypoplasticity, which was carried out during my long research stay at the Institute of Soil and Rock Mechanics at Karlsruhe University from 1985 to 1996.

I dedicate my book to Prof. Gerd Gudehus from Germany, the former head of the Institute of Rock and Soil Mechanics at Karlsruhe University and the supervisor of my scientific research during my stay in Karlsruhe, who encouraged me to deal with shear localization in granular bodies within micro-polar hypoplasticity. I greatly appreciate his profound knowledge, kind help constructive discussions, and collegial attitude to his co-workers.

I am thankful to the both series editors: Prof. Wei Wu from Universität für Bodenkultur in Austria and Prof. Ronaldo Borja from Stanford University in USA for their helpful suggestions with respect to the contents and structure of the book.

I am also grateful to Dr. Thomas Ditzinger and Mrs. Heather King from the Springer Publishing Company and SPS data processing team for their help in editing this book.

Gdansk, June 2008

Jacek Tejchman

Contents

1 Introduction

This chapter describes shortly the phenomenon of shear localization in dry and cohesionless granular materials. It presents the main problems in granular bodies related to this phenomenon. In addition, it summarizes different continuum approaches capable to properly describe shear localization using a finite element method and enhanced constitutive models. A micro-polar hypoplastic constitutive model is briefly described to numerically investigate shear localization in dry and cohesionless granular materials during mainly monotonic deformation paths. The differences between hypoplastic and conventional elasto-plastic continuum models are stressed. The outline of the book is given.

Localization of deformation in the form of narrow zones of intense shearing is a fundamental phenomenon in granular materials (Vardoulakis 1980, Gudehus 1986, Yoshida et al. 1994, Tejchman 1989, 1997, Harris et al. 1995, Desrues et al. 1996, Alshibli and Sture 2000, Leśniewska and Mróz 2001, Lade 2002, Desrues and Viggiani 2004, Gudehus and Nübel 2004). Thus, it is of a primary importance to take it into account while modeling their behaviour. Localization under shear occurs either in the interior domain in the form of a spontaneous shear zone as a single shear zone, a multiple or a regular pattern of zones (Han and Vardoulakis 1991, Harris et al, 1995, Desrues et al. 1996). It can be also created at interfaces in the form of an induced single shear zone where structural members are interacting and stresses are transferred from one member to the other (Uesugi et al. 1988, Tejchman 1989, Hassan 1995). The localized shear zones inside of the material are closely related to an unstable behaviour of the entire earth structure. An understanding of the mechanism of the formation of shear zones is important since first they act as a precursor to ultimate soils failure and second for a realistic estimation of forces transferred from the surrounding granular body to the structure, e.g. in the problems of foundations, slopes, silos, piles and earth retaining walls. The multiple patterns of shear zones are not usually taken into account in engineering calculations.

Within shear zones, pronounced grain rotations (Oda et al. 1982, Uesugi et al. 1988, Tejchman 1989) and curvatures connected to couple stresses (Oda 1993), large strain gradients (Vardoulakis 1980), high void ratios together with material softening (negative second-order work) (Desrues et al. 1996) and void ratio fluctuations (Löffelmann 1989) are observed. Non-coaxiality (understood as a coincidence of the directions of the principal stresses and principal plastic strain increments) takes place

J. Tejchman: Shear Localiz. in Gran. Bodies, SSGG, pp. 1–10, 2008.
springerlink.com

(Gutierrez and Vardoulakis 2007). The thickness of shear localization depends on many various factors such as: the mean grain diameter (Vardoulakis 1980, Tejchman 1989, Tatsuoka et al. 1991, 1994, 1997), pressure level (Tatsuoka et al. 1991, Desrues and Hammad 1989), initial void ratio (Tejchman 1989, Desrues and Hammad 1989), direction of deformation (Tatsuoka et al. 1994), shear velocity (Löffelmann 1989), grain roughness and grain size distribution (Tejchman 1989, Yoshida et al. 1994, Viggiani et al. 2001, Desrues and Viggiani 2004).

Extensive experimental studies have been conducted on shear localization in granular materials (Vardoulakis 1980, Tejchman 1989, Yoshida et al. 1994, Desrues et al. 1996, Vardoulakis et al. 1995, Tatsuoka et al. 1990, 1997, Finno et al. 1996, Alshibli and Sture 2000, Lade 2002, Rechenmacher 2006). They investigated various aspects of shear localization, such as shear resistance, localization criteria and analytical expressions for shear zone orientations, thickness of shear zones and distribution of void ratio.

Localization was treated analytically as a bifurcation problem by Mandel (1966), Rudnicki and Rice (1975) and Vardoulakis (1980). Realistic plastic numerical solutions of geotechnical problems can be only found with constitutive models which are able to describe the formation of shear zones with a certain thickness and spacing, i.e. the constitutive model has to be endowed with a characteristic length of micro-structure. There are several approaches to capture spontaneous and induced shear localization in a quasi-static regime: e.g. the second-gradient (Vardoulakis and Aifantis 1991, de Borst et al. 1992, Sluys 1992, de Borst and Mühlhaus 1992, Pamin 1994, 2004, Vardoulakis and Sulem 1995, Oka et al. 2001, Chambon et al. 2001, Zervos et al. 2001, di Prisco et al. 2002, Maier 2002, Shi and Chang 2003, Tejchman 2004), non-local (Bazant et al. 1987, Brinkgreve 1994, Schanz 1998, Marcher and Vermeer 2001, di Prisco et al. 2002, Maier 2002, Tejchman 2004), micro-polar (Bogdanova-Bontscheva and Lippmann 1975, Becker and Lippmann 1977, Mühlhaus 1987, 1989, Mühlhaus and Vardoulakis 1989, Tejchman 1989, 1997, Gudehus and Tejchman 1991, de Borst 1991, Slyus 1992, Tejchman and Wu 1993, 1995, 1997, Tejchman and Bauer 1996, Tejchman et al. 1999, Tejchman and Gudehus 2001, Ehlers and Volk 1998, Yoshida et al. 1997, Maier 2002, Huang and Bauer 2003, Manzari 2004, Gudehus and Nübel 2004, Arslan et al. 2008) and viscous ones (Loret and Prevost 1991, Sluys 1992, Sluys and de Borst 1994, Belytschko et al. 1994, Lodygowski and Perzyna 1997, Ehlers and Volk 1998, Ehlers and Graf 2003). The approaches regularize the ill-posedness (i.e. preserve the well-posedness) of the underlying incremental boundary value problem (Benallal et al. 1987, de Borst et al. 1992) caused by strain-softening and localization (the differential equations of motion in a regularized problem remain elliptic for quasi-static problems and hyperbolic during dynamic calculations) and prevent pathological discretization sensitivity. Thus, objective and properly convergent numerical solutions for localized deformation (mesh-insensitive load-displacement diagram and mesh-insensitive deformation pattern) are achieved (Sluys 1992, Brinkgreve 1994, Tejchman 1997, Maier 2002). Otherwise, FE results are completely controlled by the size and orientation of the mesh and thus produce unreliable results, i.e. the shear zones become narrower upon mesh refinement (element size becomes the characteristic length) and computed force-displacement curves change considerably depending on the width of the calculated shear zone (Tejchman 1989, Brinkgreve 1994, Maier 2002). In addition, a premature divergence of incremental FE calculations is often met. The presence of a characteristic length allows also to take into account microscopic inhomogeneities triggering shear localization (e.g. grain size, size and spacing of micro-defects) and to capture a deterministic size effect of a specimen (dependence of strength and other mechanical properties on the size of the specimen) observed experimentally on softening granular and brittle specimens. This is made possible since the ratio *l/L* governs the response of the model (*l* – characteristic length of micro-structure, *L* – size of the structure). Since the shear zone thickness depends on many different factors, it is of a primary importance to use a constitutive model taking them into account.

Other numerical technique which enables us to remedy the drawbacks of standard FE-methods and to obtain mesh-independent results during the description of the formation of shear zones is the so-called strong discontinuity approach which allows a finite element with a displacement discontinuity (Larsson and Larsson 2000, Regueiro and Borja 2001, Lai et al. 2003, Vermeer et al. 2004, Simone and Sluys 2004). However, in this case, the patterning of intersecting shear zones inside of the material and wall shear zones have not been obtained yet. Moreover, it does not take into account a characteristic length of micro-structure.

The formation of shear zones inside of granular materials has been also numerically investigated with discrete element models in order to gain some insight into the microscopic mechanism (Oda and Kazama 1998, Oda and Iwashita 2000, Thornton and Zhang 2003, Thornton 2004, Tykhoniuk et al. 2004, Alonso-Marroquin et al. 2004, Pena et al. 2007, Ord et al. 2007, Tordesillas 2007, David et al. 2007).

To better describe shear localization, the following numerical techniques have been additionally used such as: remeshing (Pastor and Peraire 1989, Hicks 1998, Ehlers and Volk 1998, Hicks et al. 2001, Ehlers et al. 2001, Ehlers and Graf 2003), multi-scaling (Gitman 2006) which are very useful in larger geotechnical problems and an element-free Galerkin concept (Belytschko et al. 1996, Pamin et al. 2003) wherein a high order of continuity for shape functions is provided. To avoid large mesh distortions, an Arbitrary Lagrangian-Eulerian approach can be used (Stoker 1999, Wójcik and Tejchman 2007) wherein the material flows through the mesh.

The intention of this book is to present the capability of a hypoplastic constitutive law enriched by a characteristic length of microstructure in the form of a mean grain diameter to capture shear localization in cohesionless granular bodies on the basis of some FE analyses for plane strain and axisymmetric cases during monotonic and cyclic deformation paths.

Granular materials consist of grains in contact, and of voids. Their micromechanical behaviour is inherently discontinuous, heterogeneous and non-linear. Despite the discrete nature of granular materials, their mechanical behaviour can be reasonably described by continuum models, in particular elasto-plastic (Vermeer 1982, Lade 1997, Pestana and Whittle 1999, Al Hattamleh et al. 2005) and hypoplastic ones (Darve 1995, Gudehus 1996, Kolymbas, 2000). A hypoplastic constitutive model (which was formulated at Karlsruhe University in Germany, Gudehus 1996, 2006, 2007, Bauer 1996, 2000, von Wolffersdorff 1996) is an alternative to elasto-plastic formulations for continuum modelling of granular materials. It describes the evolution of effective stress components with the evolution of strain components by a differential equation including isotropic linear and non-linear tensorial functions. In contrast to elasto-plastic models, the decomposition of deformation components into elastic and plastic parts, yield surface, plastic potential, flow rule and hardening rule are not needed. Moreover, both the coaxiality and stress-dilatancy rule are not assumed in advance. The constitutive law takes into account the influence of density, pressure and direction of deformation. Hypoplastic constitutive models without a characteristic length can describe realistically the onset of shear localization, but not its further evolution. A characteristic length can be introduced into hypoplasticity by means e.g. of micro-polar, non-local or second-gradient theories (Maier 2002, Tejchman 2004, 2005). In this paper, a micro-polar continuum was adopted. A micro-polar continuum which is a continuous collection of particles behaving like rigid bodies combines two kinds of deformations at two different levels, viz: micro-rotation at the particle level and macro-deformation at the structural level. For the case of plane strain, each material point has three degrees of freedom: two translations and one independent rotation. The gradients of the rotation are related to curvatures, which are associated with couple stresses. The presence of couple stresses gives rise to a non-symmetry of the stress tensor and to a characteristic length. The micro-polar model makes use of rotations and couple stresses which have clear physical meaning for granular materials. The rotations can be observed during shearing, but remain negligible during homogeneous deformations. Pasternak and Mühlhaus (2005) have demonstrated that the additional rotational degree of freedom of a micro-polar continuum arises naturally by mathematical homogenization of an originally discrete system of spherical grains with both contact forces and contact moments.

The potential of a micro-polar hypoplastic constitutive law to describe shear localization in granular bodies was demonstrated with several FE solutions of various boundary value problems. The FE calculations with a micro-polar hypoplastic model were carried for plane strain compression, monotonic and symmetric cyclic shearing of an infinite layer, biaxial test, strip foundation, earth pressure, direct and simple shear test, wall direct shear test, contractant soil and direct cyclic shearing under constant normal stiffness conditions. In addition, the FE investigations of wall boundary conditions, a deterministic and statistical size effect and textural anisotropy were performed.

The book includes 6 Sections and is organized as follows. After a short introduction in Section 1, Section 2 includes a summary of experimental results of shear localization in dry cohesionless granular bodies. In Section 3, a micro-polar hypoplastic constitutive law is described with a calibration procedure of material constants. Later, numerical results with a finite element method on the basis of a micro-polar constitutive law are demonstrated for different boundary value problems involving shear localization (Sections 4 and 5). Finally, general conclusions from the research are enclosed (Section 6).

Throughout the book, both matrix-vector and tensor notations are used. A single underscore denotes a vector and a double underscore denotes a matrix. For summation of vector and matrix components, the Einstein's rule is applied. A superposed circle indicates objective time derivation and a superposed dot indicates material time derivation of a particular quantity. Compressive stress and shortening strain are taken as negative (thus, dilatancy is positive and contractancy is negative).

References

Al Hattamleh, O., Muhunthan, B., Zbib, H.M.: Stress distribution in granular heaps using multi-slip formulation. Int. J. Numer. Analyt. Meth. Geomech. 29, 713–727 (2005)

Alonso-Marroquin, F., Herrmann, H.J., Vardoulakis, I.: Micromechanical aspects of soil plasticity; an investigation using a discrete model of polygonal particles. In: Vermeer, P.A., Ehlers, W., Herrmann, H.J., Ramm, E. (eds.) Modelling of Cohesive-Frictional Materials, Balkema, pp. 45–69 (2004)

Alshibli, K.A., Sture, S.: Shear band formation in plane strain experiments of sand. J. Geotech. Geoenviron. Engng. ASCE 126(6), 495–503 (2000)

Arslan, H., Sture, S., Willam, K.: Finite element simulation of localization in granular material by micro-polar continuum approach. Comp. Geotech. (in press, 2008)

Bauer, E.: Calibration of a comprehensive hypoplastic model for granular materials. Soils Found. 36(1), 13–26 (1996)

Bauer, E.: Conditions of embedding Casagrande's critical states into hypoplasticity. Mech. Coh.-Frict. Mater. 5, 125–148 (2000)

Bazant, Z., Lin, F., Pijaudier-Cabot, G.: Yield limit degradation: non-local continuum model with local strain. In: Owen, R., et al. (eds.) Proc. Int. Conf. Computational Plasticity, Barcelona, pp. 1757–1780 (1987)

Becker, M., Lippmann, H.: Plane plastic flow of granular model material. Arch. Mech. 29, 829–846 (1977)

Belytschko, T., Chiang, H., Plaskacz, E.: High resolution two dimensional shear band computations: imperfections and mesh dependence. Com. Meth. Appl. Mech. Engng. 119, 1–15 (1994)

Beltyschko, T., Krongauz, Y., Organ, D., Fleming, M., Krysl, P.: Meshless methods: an overview and recent developments. Comp. Meth. Appl. Mech. Engrg. 139, 3–47 (1996)

Benallal, A., Billardon, R., Geymonat, G.: Localization phenomena at the boundaries and interfaces of solids. In: Desai, C.S., et al. (eds.) Proc. of the 3rd Int. Conf. Constitutive Laws for Engineering Materials: Theory and Applications, Tucson, Arizona, pp. 387–390 (1987)

Bogdanova-Bontscheva, N., Lippmann, H.: Rotations symmetrisches ebenes Fliessen eines granularen Modellmaterials. Acta Mech. 21, 93–113 (1975)

de Borst, R.: Simulation of strain localization: a reappraisal of the Cosserat continuum. Engng. Comput. 8, 317–332 (1991)

de Borst, R., Mühlhaus, H.B., Pamin, J., Sluys, L.Y.: Computational modelling of localization of deformation. In: Owen, D.R.J., Onate, E., Hinton, E. (eds.) Proc. of the 3rd Int. Conf. Comp. Plasticity, pp. 483–508. Pineridge Press, Swansea (1992)

de Borst, R., Mühlhaus, H.B.: Gradient dependent plasticity: formulation and algorithmic aspects. Int. J. Numer. Methods Engng. 35, 521–539 (1992)

Brinkgreve, R.: Geomaterial models and numerical analysis of softening. PhD thesis, Delft University, 1-153 (1994)

Chambon, R., Matsushima, T., Caillerie, D.: Plastic continuum with microstructure, local second gradient theories for geomaterials: localisation studies. Int. J. Solids Structures 38, 8503–8527 (2001)

Darve, F., Flavigny, E., Megachou, M.: Yield surfaces and principle of superposition revisited by incrementally non-linear constitutive relations. Int. J. Plasticity 11(8), 927–948 (1995)

David, C.T., Garcia-Rojo, R., Herrmann, H.J., Luding, S.: Powder Flow Testing with 2D and 3D biaxial and triaxial simulations. J. Particle and Particle System Charact. 24(1), 29–33 (2007)

Desrues, J., Hammad, W.: Shear banding dependency on mean pressure level in sand. In: Dembicki, E., et al. (eds.) Proc. Int. Workshop on Numerical Methods for Localization and Bifurcation of Granular Bodies, Gdansk, Poland, pp. 57–67 (1989)

Desrues, J., Chambon, R., Mokni, M., Mazerolle, F.: Void ratio evolution inside shear bands in triaxial sand specimens studied by computed tomography. Géotechnique 46(3), 529–546 (1996)

Desreues, J., Viggiani, G.: Strain localization in sand: overview of the experiments in Grenoble using stereo photogrammetry. Int. J. Numer. Anal. Meth. Geomech. 28(4), 279–321 (2004)

Ehlers, W., Volk, W.: On theoretical and numerical methods in the theory of porous media based on polar and non-polar elasto-plastic solid materials. Int. J. Solids Structures 35, 4597–4616 (1998)

Ehlers, W., Ellsiepen, P., Ammann, M.: Time- and space-adaptive computations of localization phenomena in standard and micropolar frictional porous materials. In: Mühlhaus, H.B., Dyskin, A., Pasternak, E. (eds.) Bifurcation Localisation Theory in Geomechanics, Balkema, pp. 139–147 (2001)

Ehlers, W., Graf, T.: Adaptive computation of localization phenomena in geotechnical applications. In: Labuz, J., Drescher, A. (eds.) Bifurcations and Instabilities in Geomechanics, Swets, Zeitlinger, pp. 247–262 (2003)

Finno, R.J., Harris, W., Mooney, M., Viggiani, G.: Strain localization and undrained steady state of sand. Journal of Geotechnical Engineering ASCE 122(6), 462–473 (1996)

Gitman, I.: Representative volumes and multi-scale modeling of quasi-brittle materials. PhD Thesis, University of Delft (2006)

Gudehus, G.: Einige Beiträge der Bodenmechanik zur Entstehung und Auswirkung von Diskontinuitäten. Felsbau 4, 190–195 (1986)

Gudehus, G.: A comprehensive constitutive equation for granular materials. Soils Found. 36(1), 1–12 (1996)

Gudehus, G., Nübel, K.: Evolution of shear bands in sand. Geotechnique 54(3), 187–201 (2004)

Gudehus, G.: Physical background of hypoplasticity. In: Wu, W., Yu, H.S. (eds.) Modern Trends in Geomechanics, pp. 17–37. Springer, Heidelberg (2006)

Gudehus, G.: Physical Soil Mechanics. Springer, Heidelberg (2008)

Gutierrez, M., Vardoulakis, I.: Energy dissipation and post-bifurcation behaviour of granular soils. Int. J. Num. Anal. Meth. Geomech. 31, 435–455 (2007)

Han, C., Vardoulakis, I.: Plane strain compression experiments on water saturated fine-grained sand. Geotechnique 41, 49–78 (1991)

Harris, W.W., Viggiani, G., Mooney, M.A., Finno, R.J.: Use of stereo photogrammetry to analyze the development of shear bands in sand. Geotech. Test. J. 18(4), 405–420 (1995)

Hassan, A.H.: Etude experimentale et numerique du comportement local et global d'une interface sol granulaire. PHD thesis, Grenoble University (1995)

Hicks, M.A., Yap, T.Y., Bakar, A.A.: Adaptive and fixed mesh study of localisation in a strain-softening soil. In: Mühlhaus, H.B., Dyskin, A., Pasternak, E. (eds.) Bifurcation and Localisation Theory in Geomechanics, Balkema, pp. 147–155 (2001)

Huang, W., Bauer, E.: Numerical investigations of shear localization in a micro-polar hypoplastic material. J. Num. Anal. Meth. Geomech. 27, 325–352 (2003)

Kolymbas, D.: Introduction to hypoplasticity. Advances in Geotechnical Engineering and Tunneling, Balkema, Rotterdam, Brookfield (2000)

Lade, P.V.: Elasto-plastic stress-strain theory for cohesionless soil with curved yield surfaces. Int. J. Solid Structures 13, 1019–1035 (1977)

Lade, P.V.: Instability, shear banding and failure in granular materials. I. J. Solids Structures 39, 3337–3357 (2002)

Lai, T.Y., Borja, R.I., Duvernay, B.G., Meehan, R.L.: Capturing strain localization behind a geosynthetic-reinforced soil wall. I. J. Num. Anal. Meths. Geomech. 27, 425–451 (2003)

Larsson, J., Larsson, R.: Finite-element analysis of localization of deformation and fluid pressure in elastoplastic porous medium. Int. J. Solids Structures 37, 7231–7257 (2000)

Leśniewska, D., Mróz, Z.: Study of evolution of shear band systems in sand retained by flexible wall. Int. J. Numer. Anal. Meth. Geomech. 25, 909–932 (2001)

Lodygowski, T., Perzyna, P.: Numerical modelling of localized fracture of inelastic solids in dynamic loading process. Int. J. Num. Meth. Eng. 40(22), 4137–4158 (1997)

Loret, B., Prevost, J.H.: Dynamic strain localization in fluid-saturated porous media. J. Engng. Mech. ASCE 117, 907–922 (1991)

Löffelmann, F.: Theoretische und experimentelle Untersuchungen zur Schüttgut-Wand-Wechselwirkung und zum Mischen und Entmischen von Granulaten. Publication Series of the Institut für Mechanische Verfahrenstechnik, Universität Karlsruhe (1989)

Maier, T.: Numerische Modellierung der Entfestigung im Rahmen der Hypoplastizität. PhD Thesis, Dortmund University, Germany (2002)

Mandel, J.: Conditions de stabilite et postulat de Drucker. In: Proceedings IUTAM Symposium Rheology and Soil Mechanics, pp. 58–68. Springer, Berlin (1966)

Manzari, M.T.: Application of micropolar plasticity to post failure analysis in geomechanics. Int. J. Numer. Anal. Meths. Geomech. 28(10), 1011–1032 (2004)

Marcher, T., Vermeer, P.A.: Macromodelling of softening in non-cohesive soils. In: Vermeer, P.A., et al. (eds.) Continuous and Discontinuous Modelling of Cohesive-Frictional Materials, pp. 89–110. Springer, Heidelberg (2001)

Mühlhaus, H.B.: Berücksichtigung von Inhomogenitäten im Gebirge im Rahmen einer Kontinuumstheorie. Veröffentlichung des Instituts für Boden- und Felsmechanik, Universität Karlsruhe, Heft 106, 1-65 (1987)

Mühlhaus, H.B.: Application of Cosserat theory in numerical solutions of limit load problems. Ing. Arch. 59, 124–137 (1989)

Mühlhaus, H.B., Vardoulakis, I.: The thickness of shear bands in granular materials. Geotechnique 37, 271–283 (1989)

Mühlhaus, H.B.: Continuum models for layered and blocky rock. In: Hudson, J.A., Fairhurst, C. (eds.) Comprehensive Rock Engineering, vol. 2, pp. 209–231. Pergamon Press, Oxford (1990)

Oda, M., Konishi, J., Nemat-Nasser, S.: Experimental micromechanical evaluation of strength of granular materials, effects of particle rolling. Mech. Mater. 1, 269–283 (1982)

Oda, M.: Micro-fabric and couple stress in shear bands of granular materials. In: Thornton, C. (ed.) Powders and Grains, Rotterdam, Balkema, pp. 161–167 (1993)

Oda, M., Kazama, H.: Microstructure of shear bands and its relation to the mechanisms of dilatancy and failure of dense granular soils. Geotechnique 48, 465–481 (1998)

Oda, M., Iwashita, K.: Study on couple stresses and shear band development in granular media based on numerical simulation analyses. Int. J. of Enging. Science 38, 1713–1740 (2000)

Oka, F., Jing, M., Higo, Y.: Effect of transport of pore water on strain localisation analysis of fluid-saturated strain gradient dependent viscoplastic geomaterial. In: Mühlhaus, H.-B., et al. (eds.) Bifurcation and Localisation Theory in Geomechanics, Swets, Zeitlinger, Lisse, pp. 77–83 (2001)

Ord, A., Hobbs, B., Regenauer-Lieb, K.: Shear band emergence in granular materials – a numerical study. Int. J. Num. Anal. Meth. Geomech. 31, 373–393 (2007)

Pamin, J.: Gradient-dependent plasticity in numerical simulation of localisation phenomena. PhD Thesis, Delft University, 1-134 (1994)

Pamin, J., Askes, H., de Borst, R.: Two gradient plasticity theories discretized with the element-free Galerkin method. Comp. Meth. Appl. Mech. Enging. 192, 2377–2403 (2003)

Pamin, J.: Gradient-enhanced continuum models: formulations, discretization and applications. Monograph 301, 1–174 (2004)

Pasternak, E., Mühlhaus, H.B.: Cosserat continuum modelling of granulate materials. In: Valliappan, S., Khalili, N. (eds.) Computational Mechanics – New Frontiers for New Millennium, pp. 1189–1194. Elsevier Science, Amsterdam (2001)

Pastor, M., Peraire, J.: Capturing shear bands via adaptive remeshing techniques. Euromech. 248 (1989); Non-linear soil-structure interaction

Pena, A.A., García-Rojo, R., Herrmann, H.J.: Influence of particle shape on sheared dense granular media. Granular Matter 3-4, 279–292 (2007)

Pestana, J.M., Whittle, A.J.: Formulation of a unified constitutive model for clays and sands. Int. J. Num. Anal. Meth. Geomech. 23, 1215–1243 (1999)

Rechenmacher, A.L.: Grain-scale processes governing shear band initiation and evolution in sands. J. Mech. Physics Solids 54, 22–45 (2006)

Regueiro, R.A., Borja, R.I.: Plane strain finite element analysis of pressure sensitive plasticity with strong discontinuity. Int. J. Solids Structures 38(21), 3647–3672 (2001)

Rudnicki, J.W., Rice, J.R.: Conditions of the localization of deformation in pressure-sensitive dilatant materials. J. Mech. Physics Solids 23, 371–394 (1975)

Schanz, T.: A constitutive model for cemented sands. In: Adachi, T., Oka, F., Yashima, A. (eds.) Localisation and Bifurcation Theory for Soils and Rocks, Balkema, Rotterdam, pp. 165–172 (1998)

Shi, Q., Chang, C.S.: Numerical analysis for the effect of heterogeneity on shear band formation. In: Proc. 16th ASCE Engineering Mechanics Conference, University of Washington, Seattle, pp. 1–11 (2003)

Simone, A., Sluys, L.J.: Continous-discountinous modeling of mode-I and mode-II failure. In: Vermeer, P.A., Ehlers, W., Herrmann, H.J., Ramm, E. (eds.) Modelling of Cohesive-Frictional Materials, Balkema, pp. 323–337 (2004)

Sluys, L.J.: Wave propagation, localisation and dispersion in softening solids. PhD Thesis, Delft University of Technology (1992)

Sluys, L.J., de Borst, R.: Dispersive properties of gradient and rate-dependent media. Mech. Mater. 183, 131–149 (1994)

Stoker, C.: Developments of the Arbitrary Lagrangian-Eulerian method in non-linear solid mechanics. PhD Thesis, University of Delft (1999)

Tatsuoka, F., Nakamura, S., Huang, C.C., Tani, K.: Strength anisotropy and shear band direction In plane strain test of sand. Soils and Foundations 30(1), 35–54 (1990)

Tatsuoka, F., Okahara, M., Tanaka, T., Tani, K., Morimoto, T., Siddiquee, M.S.A.: Progressive failure and particle size effect in bearing capacity of footing on sand. In: Proc. of the ASCE Geotechnical Engineering Congress, vol. 27(2), pp. 788–802 (1991)

Tatsuoka, F., Siddiquee, M.S.A., Yoshida, T., Park, C.S., Kamegai, Y., Goto, S., Kohata, Y.: Testing methods and results of element tests and testing conditions of plane strain model bearing capacity tests using air-dried dense Silver Buzzard Sand. Internal Report of the University of Tokyo, pp. 1–129 (1994)

Tatsuoka, F., Goto, S., Tanaka, T., Tani, K., Kimura, Y.: Particle size effects on bearing capacity of footing on granular material. In: Asaoka, A., Adachi, T., Oka, F. (eds.) Deformation and Progressive Failure in Geomechanics, pp. 133–138. Pergamon, Oxford (1997)

Tejchman, J.: Scherzonenbildung und Verspannungseffekte in Granulaten unter Berücksichtigung von Korndrehungen. Publication Series of the Institute of Soil and Rock Mechanics, University Karlsruhe, vol. 117, pp. 1-236 (1989)
Tejchman, J., Wu, W.: Numerical study on shear band patterning in a Cosserat continuum. Acta Mech. 99, 61–74 (1993)
Tejchman, J., Gudehus, G.: Silo-music and silo-quake, experiments and a numerical Cosserat approach. Powder Technology 76(2), 201–212 (1993)
Tejchman, J., Wu, W.: Experimental and numerical study of sand-steel interfaces. Int. J. Num. Anal. Meth. Geomech. 19(8), 513–537 (1995)
Tejchman, J., Bauer, E.: Numerical simulation of shear band formation with a polar hypoplastic model. Comp. Geotech. 19(3), 221–244 (1996)
Tejchman, J.: Modelling of shear localisation and autogeneous dynamic effects in granular bodies. Publication Series of the Institute of Soil and Rock Mechanics, University Karlsruhe, vol. 140, pp. 1-353 (1997)
Tejchman, J., Herle, I., Wehr, J.: FE-studies on the influence of initial void ratio, pressure level and mean grain diameter on shear localisation. Int. J. Num. Anal. Meth. Geomech. 23, 2045–2074 (1999)
Tejchman, J., Gudehus, G.: Shearing of a narrow granular strip with polar quantities. J. Num. Anal. Methods in Geomechanics 25, 1–28 (2001)
Tejchman, J.: Influence of a characteristic length on shear zone thickness in hypoplasticity with different enhancements. Comp. Geotech. 31(8), 595–611 (2004)
Tejchman, J.: Finite element modeling of shear localization in granular bodies in hypoplasticity with enhancements. Gdansk University of Technology, Gdańsk (2005)
Thornton, C., Zhang, L.: Numerical simulations of the direct shear test. Chem. Eng. Technol. 26(2), 1–4 (2003)
Thornton, C., Ciomocos, M.T., Adams, M.J.: Numerical simulations of diametrical compression tests on agglomerates. Powder Technology 140, 258–267 (2004)
Tordesillas, A.: Force chain buckling, unjamming transitions and shear banding in dense granular assemblies. Philos. Mag. J. (in press, 2007)
Tykhoniuk, R., Luding, S., Tomas, J.: Simulation der Scherdynamik kohäsiver Pulver. Chem.-Ing.-Technik 76, 59–62 (2004)
Uesugi, M., Kishida, H., Tsubakihara, Y.: Behaviour of sand particles in sand-steel friction. Soils Found. 28(1), 107–118 (1988)
von Wolffersdorff, P.A.: A hypoplastic relation for granular materials with a predefined limit state surface. Mech. Coh.-Frict. Mater. 1, 251–271 (1996)
Wojcik, M., Tejchman, J.: Numerical simulations of granular material flow in silos with and without insert. Arch. Civil Enng. LIII 2, 293–322 (2007)
Vardoulakis, I.: Shear band inclination and shear modulus in biaxial tests. Int. J. Num. Anal. Meth. Geomech. 4, 103–119 (1980)
Vardoulakis, I., Aifantis, E.C.: A gradient flow theory of plasticity for granular materials. Acta Mech. 87, 197–217 (1991)
Vardoulakis, I., Sulem, J.: Bifurcation Analysis in Geomechanics. Blackie Academic and Professional, Glasgow (1995)
Vermeer, P.: A five-constant model unifying well-established concepts. In: Gudehus, G., Darve, F., Vardoulakis, I. (eds.) Proc. Int. Workshop on Constitutive Relations for Soils, Balkema, pp. 175–197 (1982)
Vardoulakis, I., Goldschneider, M., Gudehus, G.: Formation of shear bands in sand bodies as a bifurcation problem. International Journal of Numerical and Anal. Methods in Geomechanics 2, 99–128 (1995)

Viggiani, G., Kuntz, M., Desrues, J.: An experimental investigation of the relationship between grain size distribution and shear banding in granular materials. In: Vermeer, P.A., et al. (eds.) Continuous and Discontinuous Modelling of Cohesive Frictional Materials, pp. 111–127. Springer, Berlin (2001)

Yoshida, T., Tatsuoka, F., Siddiquee, M.S.A.: Shear banding in sands observed in plane strain compression. In: Chambon, R., Desrues, J., Vardoulakis, I. (eds.) Localisation and Bifurcation Theory for Soils and Rocks, Balkema, Rotterdam, pp. 165–181 (1994)

Yoshida, N., Arai, T., Onishi, K.: Elasto-plastic Cosserat finite element analysis of ground deformation under footing load. In: Asaoka, A., Adachi, T., Oka, F. (eds.) Deformation and Progressive Failure in Geomechanics, pp. 395–401. Pergamon, Oxford (1997)

Zervos, A., Papanastasiou, P., Vardoulakis, I.: A finite element displacement formulation for gradient plasticity. Int. J. Numer. Meth. Engng. 50, 1369–1388 (2001)

2 Literature Overview on Experiments

Abstract. This chapter describes several laboratory experiments on shear localization in dry and cohesionless granular materials during different boundary values problems. These experiments include plane strain compression, triaxial compression, wall friction tests, biaxial compression, earth pressures on a retaining walls, strip foundation, pull-out tests and silo flow. The experiments were carried out with different initial densities and mean grain diameters of sand, and wall roughness. During tests, attention was paid to load-displacements diagrams, shear zone thickness and shear zone spacing.

Different techniques were used to visualize shear zones: colored layers and markers (Tejchman 1989, Yoshida et al. 1994), x-rays (Vardoulakis 1977, Michalowski 1984, Tejchman 1989), gamma-rays (Tan and Fwa 1991), photogrammetry and stereo-photogrammetry (Desrues 1984, Desrues and Viggiani 2004), tomography (Mokni 1992, Desrues et al. 1996, Jaworski and Dyakowski 2001, Niedostatkiewicz and Tejchman 2007, Marashdeh et al. 2008), digital image correlation (DIC) and particle image velocimetry PIV (Nübel 2002, Michalowski and Shi 2003, Slominski 2003, Rechenmacher Finno 2004, Slominski et al. 2007, Kozicki and Tejchman 2007, Niedostatkiewicz and Tejchman 2007).

Below, results of some laboratory tests with dry granular specimens are shortly described. The experimental data on water-saturated granular materials can be found in papers by Han and Vardoulakis (1991), Harris et al. (1995) and Mokni and Desrues (1998).

Plane strain compression

Plane strain compression tests under a constant lateral pressure were carried out by Vardoulakis (1977, 1980) and Vardoulakis et al. (1978) at Karlsruhe University in the apparatus shown in Fig.2.1, wherein a sample 4×8×14 cm^3 was wrapped into a rubber mould of 0.3 mm thickness. A so-called Karlsruhe dry sand was used. The index properties of sand were: mean grain diameter d_{50}=0.45-0.50 mm, grain size among 0.08 mm and 1.8 mm, uniformity coefficient U=2, maximum specific weight γ_d^{max}=17.4 kN/m^3, minimum void ratio e_{min}=0.53, minimum specific weight γ_d^{min}=14.6 kN/m^3 and

J. Tejchman: Shear Localiz. in Gran. Bodies, SSGG, pp. 11–46, 2008.
springerlink.com

maximum void ratio e_{max}=0.84. Two side plates served for the condition of plane strain; they carried polished stainless steel plate and a silicone grease to prevent boundary friction. The base and top plates were also polished and lubricated and carried 10 mm diameter porous stone to keep the sample centered. The base plate was placed on a movable roller bearing. The experiments were carried out either with a roller bearing (VV) or without roller bearing (XV) or with a clamped piston (XV1, VV1) or a hinged piston (XV2, VV2) (Tab.2.1). In some tests, the specimen did not include any artificial imperfection. In some other tests, a small artificial initial imperfection (side notch or loose sand inclusion) was included. Some experimental results are shown in Figs.2.2 and 2.3 and in Tab.2.1.

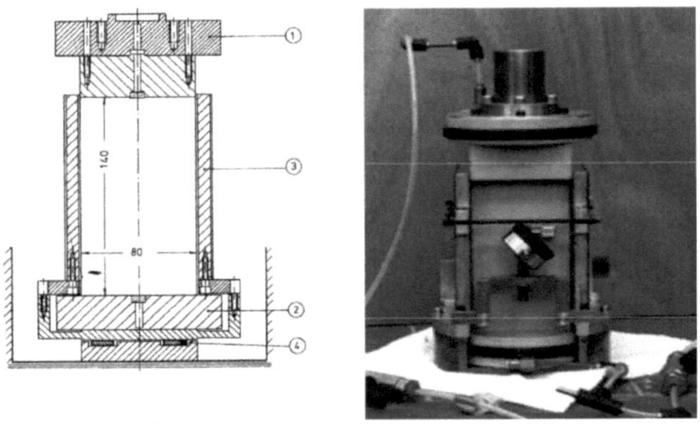

a) b)

Fig. 2.1. Plane strain compression apparatus: a) system, b) photograph: 1. top plate, 2. base plate, 3. side plates, 4. roller bearing (Vardoulakis 1977)

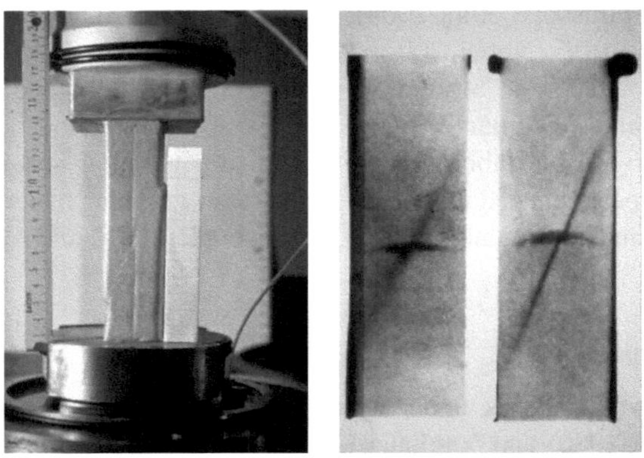

a) b)

Fig. 2.2. Formation of shear zone in dense specimen without initial (a) and with initial imperfection in the form of loose sand inclusion on the basis of x-rays radiographs (b)

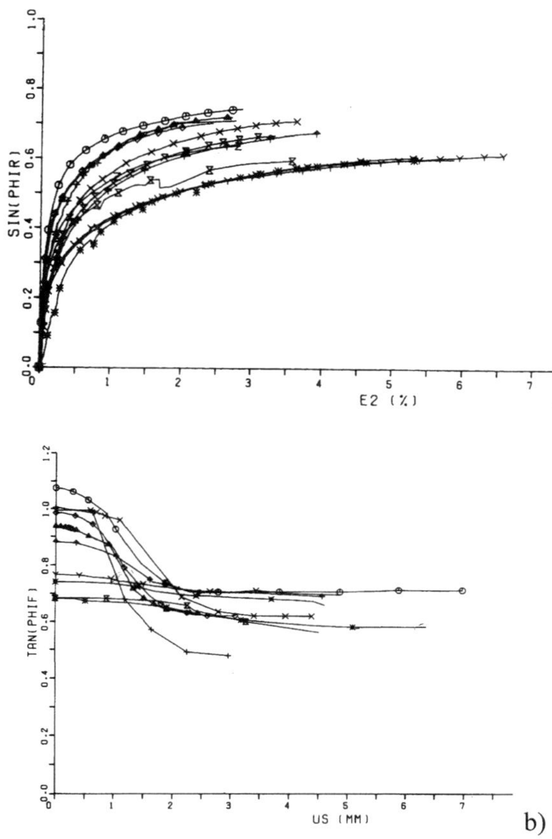

a)

b)

Fig. 2.3. Pre-failure stress-strain behaviour (a) and post-failure stress-displacement (b); $\sin\phi_r = (\sigma_1 - \sigma_2)/(\sigma_1 + \sigma_2)$, $\varepsilon_2 = -\ln(h/h_o)$, σ_1 – vertical principal stress, σ_2 – lateral pressure, h_o – initial height, $\tan\phi_f = \tau_f / \sigma_f$, τ_f, σ_f – shear and normal stress in the shear zone, *us* – relative displacement of two rigid bodies (Vardoulakis et al. 1978)

The tests showed that an internal shear zone was spontaneously formed at the peak of the stress-strain curve. The thickness of the shear zone was about 3-5 mm, i.e. (10-15)×d_{50}, and the inclination of the shear zone to the horizontal was approximately θ =52-67°. The inclination was in accordance with the formula by Arthur et al. (1977): $\theta \cong 45° + (\phi_p + \upsilon_p)/4$ (ϕ_p – the angle of internal friction at peak, υ_p – the dilatancy angle at peak). The shear zones were less steep in initially looser samples than in dense samples. The rotating top plate decreased the shear zone inclination. The shape of the shear zone was influenced by the type of the imperfection (shear zones with a notch were slightly curved). Looser sands were more sensitive than dense ones. The maximum angle of internal friction decreased almost linearly with increasing void ratio.

Table 2.1. Details of some plane strain compression tests (Vardoulakis et al. 1978)

Test	n_o	$(\varepsilon_2)_p$ [%]	ϕ_p [°]	θ [°]	ϕ_r	σ_2 [kPa]
VV1.08	0.355	2.8	47.6	63-65	35.4	120
XV2.02	0.364	2.7	46.1	57-58	33.9	100
XV1.20	0.372	2.8	45.6	67	27.1	55
XV1.06	0.358	2.4	44.8	67	32.5	80
VV1.09	0.381	3.8	44.3	64-65	31.3	230
XV1.22	0.385	3.9	42.2	60	34.7	180
XV1.09	0.364	3.3	41.6	63	-	200
XV2.01	0.388	2.8	39.9	55-58	-	100
VV1.13	0.429	6.5	38.3	59-60	35.3	89
XV1.21	0.424	5.3	37.2	56-60	33.9	200
XV2.03	0.419	5.9	37.1	52	30.0	150
VV2.02	0.395	3.6	37.1	52-54	29.8	75

n_o – initial porosity, $\varepsilon_2 = -\ln(h/h_o)$, ϕ - internal friction angle, θ - shear zone inclination, σ_2 – lateral pressure, '*p*' – peak state, '*r*' - residual state

The plane strain compression tests were also carried out at Grenoble University in the apparatus developed by Desrues (1984) and later modified by Hammad (1991) (Fig.2.4). The height and width of the specimen varied in the range of 75-350 mm and 80-175 mm, respectively. The side walls were 50 mm thick glass plates. All surfaces in contact with the specimen were lubricated with silicon grease to minimize friction. During tests, the bottom and top plates could be either locked, prevented from rotating and translating in the horizontal direction, or allowed to rotate without translating, or the top plate could be free to translate in the horizontal direction but not to rotate (while the bottom plate was locked). The strain-controlled axial loading was applied through a screw jack at the top of the device. The tests were carried out on Hostun sand with different initial densities, mean grain size, specimen size and slenderness, lateral pressure and imperfection type (Desrues 1984, Hammad 1991, Mokni 1992). In some other tests, the top and bottom plates were not lubricated. The shear zones were detected using stereo-photogrammetry. The index properties of sand were: mean grain diameter d_{50}=0.32 mm, grain size among 0.08-1.8 mm, uniformity coefficient U=1.7, maximum specific weight γ_d^{max}=15.99 kN/m^3 and minimum specific weight γ_d^{min}=13.24 kN/m^3.

In Tab.2.2 the observed different types of shear localization (depending strongly on boundary conditions) are schematically depicted from the experiments by Desrues (1984) (γ_o – initial unit weight, e_0 – initial void ratio, σ'_3 – lateral pressure, H_o – initial height, L_o – initial width). Figs.2.5-2.7 present the evolution of shear strain from different tests. Fig.2.8 depicts the stress-strain and volumetric responses from tests on dense sand.

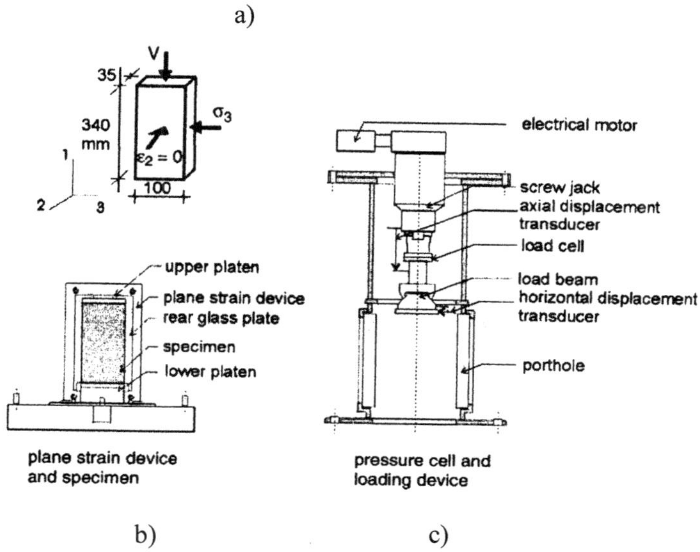

Fig. 2.4. Schematic diagram of a plane strain apparatus at the University of Grenoble: a) specimen geometry, b) plane strain device and specimen, c) pressure cell and loading device (Desrues and Viggiani 2004)

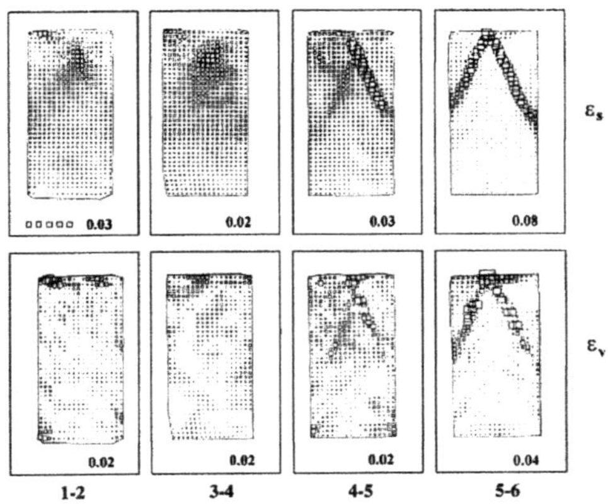

Fig. 2.5. Evolution of shear strain and volumetric strain intensity in time from stereo-photogrammetry (test 'shf00' of Tab.2.2) (Desrues and Viggiani 2004)

The results showed that depending on boundary conditions and slenderness of the specimen, various patterns of shear zones were observed including even parallel and crossing shear zones. The onset of shear localization took place slightly before the peak of the stress ratio. The shear zone reflection at rigid boundaries was a typical

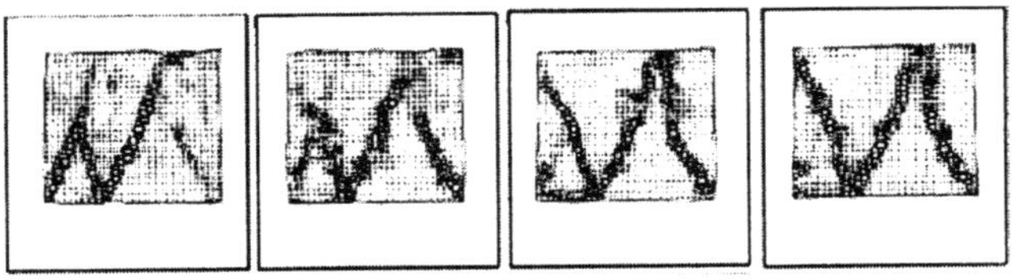

Fig. 2.6. Evolution of shear strain intensity in time from stereo-photogrammetry (dense stout sand specimen, slenderness ratio H0/Lo=0.95) (Mokni 1992)

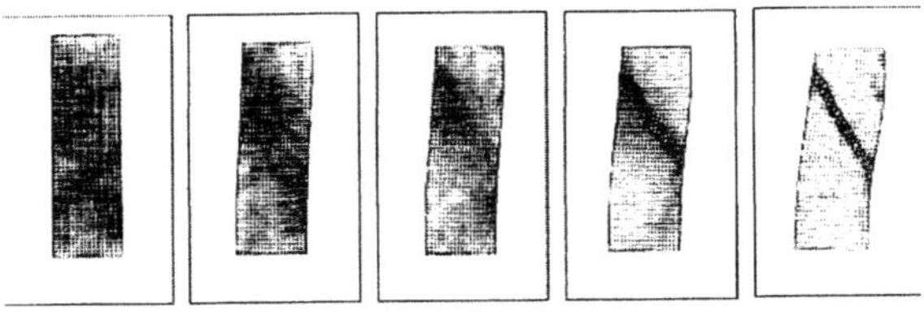

Fig. 2.7. Evolution of shear strain intensity in time from stereo-photogrammetry (loose sand, slenderness ratio H_0/L_o=3.35, locked bottom plate) (Hammad 1991)

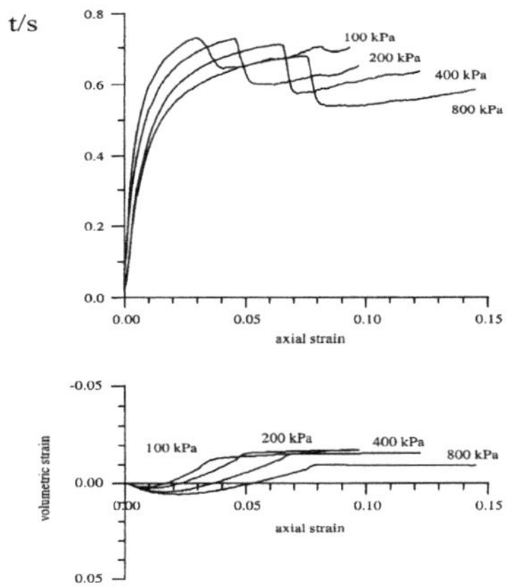

Fig. 2.8. Stress-strain responses versus global axial strain from tests on dense sand: a) stress ratio t/s ($t=(\sigma_1-\sigma_3)/2$, $s=(\sigma_1+\sigma_3)/2$), b) global volumetric strain (σ_i – principle stresses) (Desrues and Viggiani 2004)

mode of propagation in stout specimens. Shear zones were steeper in dense specimens than in loose ones. The shear zone inclination decreased with increasing pressure. The shear zone thickness decreased as the confining stress and initial density increased. By reducing the specimen size or its slenderness, the onset of strain localization was retarded, the steepness of the shear zone was reduced and its width was increased. For

Table 2.2. Observed shear localization (schematically) in experiments by Desrues (1984) from Desrues and Viggiani (2004)

Test	γ_0 (kN/m^3)	e_0	σ_3' (kPa)	H_0/L_0	Lubrication (Y/N)	End plates (L/R/TT)*	Observed localization	'imperfections'
SH00	15.75	0.650	80	2.00	N	L		
shf00	14.93	0.741	90	2.00	Y	L		
shf01	14.93	0.741	90	2.00	Y	L		
shf02	13.68	0.900	85	2.20	N	R		
shf03	14.05	0.850	90	2.20	Y	L		
shf04	13.68	0.900	80	3.60	N	L		
shf05	15.65	0.661	80	2.00	Y	L		
shf06	15.65	0.661	80	2.00	Y	R		left tilting 
shf07	15.65	0.661	80	2.00	Y	L		right tilting 
shf08	15.65	0.661	80	2.00	Y	R		
shf09	15.65	0.661	80	2.00	Y	L		
shf10	15.65	0.661	80	2.00	Y	L		soft inclusion
shf11	15.47	0.680	80	2.00	Y	L		hard inclusion
shf12	15.65	0.661	80	2.00	Y	TT		lateral load
shf13	15.65	0.661	80	2.00	Y	L		
shf14	15.65	0.661	80	3.55	Y	L		
shf15	15.65	0.661	80	3.55	Y	L		
shf16	15.65	0.661	80	3.55	N	L		
shf17	15.65	0.661	80	2.00	N	L		
shf18	15.65	0.661	80	3.55	N	L		
shf19	15.65	0.661	80	3.55	Y	R + TT		
shf20	13.97	0.860	80	2.00	Y	R + TT		

*L: both platens locked, i.e. prevented from rotating and translating in the horizontal direction; R: both platens allowed to rotate without translating; TT: top platen free to translate in the horizontal direction, but not to rotate, while the bottom platen is locked in place

higher slenderness ratios, other bifurcation modes (e.g. buckling) were more likely to occur. The shear zone thickness increased with increasing particle size; its inclination was not affected by the mean grain size and non-uniformity of sand grading. The imperfection dictated the location of the shear zone and acted as a trigger for the onset of shear localization.

In turn, the granular specimens in laboratory tests at Tokyo University of Yoshida et al. (1994) were 20 cm high, 16 cm long and 8 cm wide (in the direction of the minor principal stress), Fig.2.9. The specimens were covered with a 0.3 mm thick latex rubber membrane. The top and bottom surfaces were in contact with well

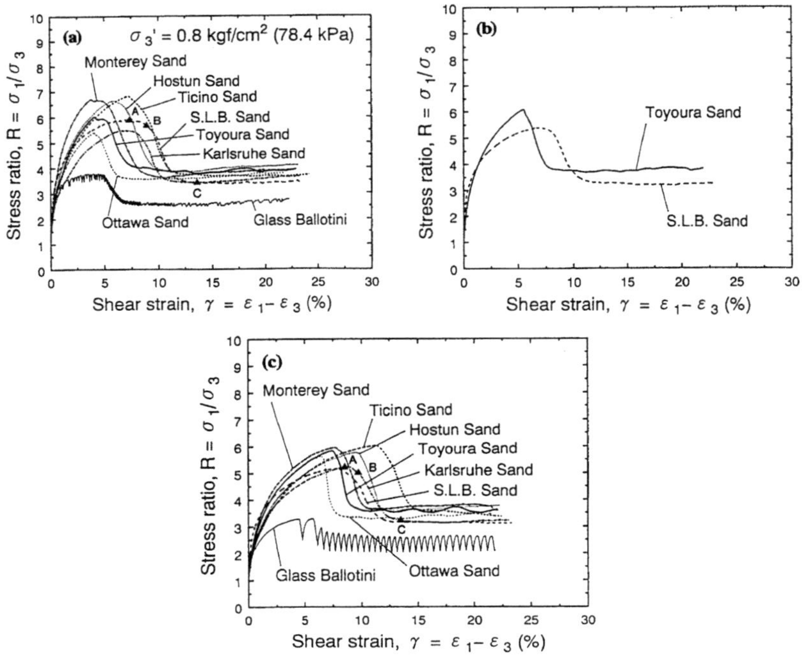

Fig. 2.9. Relationships between the stress ratio σ_1/σ_3 and the average shear strain $\gamma=\varepsilon_1$-ε_3: a) σ_1=80 kPa, b) σ_1=200 kPa, c) σ_1=400 kPa (σ_1 , σ_3 – vertical and horizontal principal stress, ε_1, ε_3 – vertical and horizontal principle strain) (Yoshida et al. 1994)

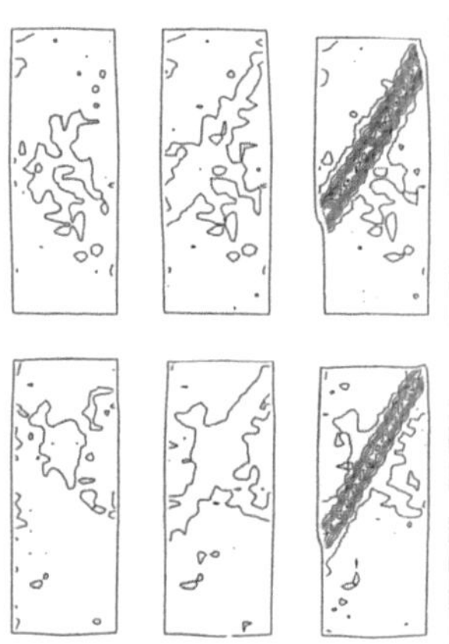

Fig. 2.10. Evolution of a shear zone in plane strain compression on the basis of contours of shear strain in SLB sand with lateral pressure of σ_3=80 kPa (upper row) and in Karlsruhe sand with σ_3=400 kPa (lower row) (Yoshida et al. 1994)

Table 2.3. Experimental data from plane strain compression tests (Yoshida et al. 1994)

Material	σ_3 [kPa]	e_o	ϕ_{peak} [°]	ϕ_{res} [°]	θ [°]	t_s/d_{50}	γ_{peak} [%]
Hostun sand	80	0.616	47.6	35.9	63	20	6.0
	400	0.648	44.8	34.2	58	9.3	9.5
Toyoura sand	80	0.649	45.7	35.5	66	22	4.1
	200	0.660	45.9	35.0	65	19	5.5
	400	0.661	45.0	33.7	66	15	7.4
Ticino sand	80	0.657	48.1	34.8	61	10	7.2
	400	0.679	45.7	34.5	60	7.2	11.1
Monterey sand	80	0.604	47,8	34.4	66	12	3.8
	400	0.643	45.5	34.7	59	8.2	7.7
SLB sand	80	0.549	44.7	32.7	59	9.8	7.4
	200	0.548	43.4	31.3	62	9.4	6.9
	400	0.547	42.5	30.8	61	8.9	7.6
Karlsruhe sand	80	0.621	43.8	33.0	59	10	7.1
	400	0.636	42.8	31.0	58	9.3	8.9
Ottawa sand	80	0.598	43.4	34.2	70	20	4.1
	400	0.608	44.3	32.2	65	20	6.4
Glass ballotini	80	0.573	35.7	26.5	54	19	4.0
	400	0.621	32.3	26.2	53	18	5.8

polished stainless steel boundaries of the cap and pedestal. They were lubricated by means of a latex rubber sheet smeared with silicon grease.

Fig.2.9 shows some results for different sands and lateral pressures. The information about the lateral pressure σ_3, initial void ratio e_o, internal friction angle at peak ϕ_{peak} and at residual state ϕ_{res}, shear zone inclination to the horizontal θ, ratio between the shear zone thickness t_s and mean grain diameter d_{50} and shear strain at peak γ_{peak} is given in Tab.2.3. The results showed that shear localization started immediately before the peak in the form of multiple shear bands. A single well-defined shear zone was attained only after the peak (Fig.2.10). The most important material factor controlling the shear deformation was the particle size. The rate of the increase in the shear zone thickness was the largest near the peak state, and it decreased monotonically with shear deformation towards nearly zero at the residual state. The shear zone thickness, (8-22)$\times d_{50}$, decreased, in general, with increasing pressure. The shear zone inclination (relative to the horizontal direction) decreased with increasing particle size.

The plane strain compression tests by Pradhan (1997) indicated that the internal friction angle at peak decreased with increasing direction of the bedding plane. The shear zone thickness depended on confining pressure and mean grain size. The thickness decreased with increasing confining pressure and decreasing particle size. In turn, the shear zone inclination depended on confining pressure, anisotropy and mean grain size. The inclination to the horizontal decreased as the mean grain diameter and pressure level increased.

Wall friction tests

Own wall friction tests (Tejchman 1989, Tejchman and Wu 1995) were carried out in a plane strain apparatus by Vardoulakis (1977) according to the principle shown in Fig.2.11. Fig.2.12 presents the measured wall friction coefficients and volume changes for dense and loose Karlsruhe sand with the different wall roughness. In addition, the results of the wall friction in a parallelly-guided direct shear box are shown (Tejchman 1989, Tejchman and Wu 1995). The radiographs of density changes in sand for a rough and very rough wall are shown in Fig.2.13.

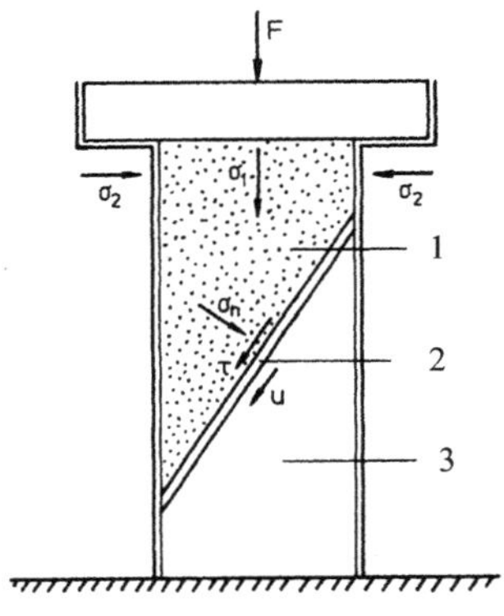

Fig. 2.11. Measurement of wall friction in a plane strain apparatus: 1: sand specimen, 2. steel wall, 3. wooden wedge (τ, σ_n – shear and normal stress in the plane of shearing, u – sand displacement along the wall, σ_1 – vertical principle stress, σ_3 – horizontal principle stress) (Tejchman 1989)

The thickness of the shear zone formed along the wall and inclined to the bottom under the angle of α=65° ($\alpha = 45^o + \phi^p / 2$) was approximately 1 mm (2×d_{50}) for a rough wall and 3 mm (6×d_{50}) for a very rough wall. The peak wall friction angle increased with increasing wall roughness and initial density. It was higher during direct wall shearing.

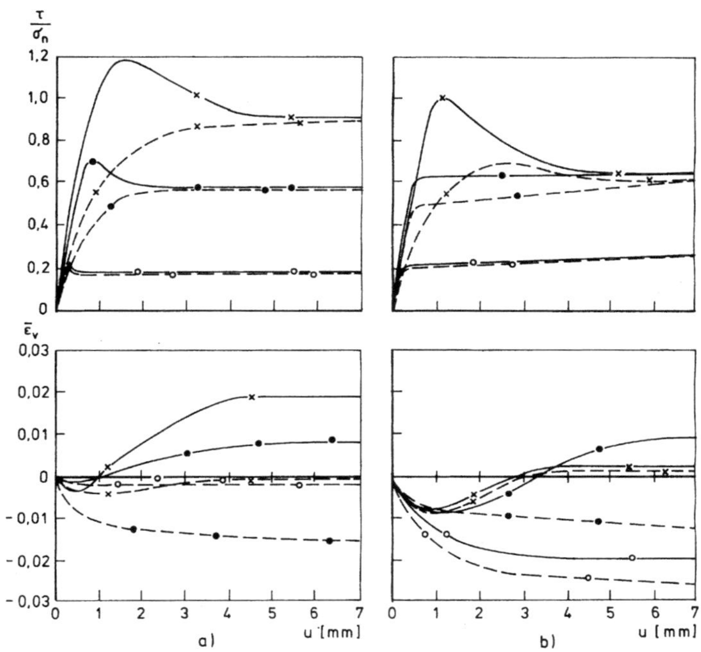

Fig. 2.12. Results of wall friction: a) parallelly-guided direct shear box and (b) plane strain apparatus at the normal stress σ_n=100-400 kPa (τ, σ_n – shear and normal stress in the plane of shearing, u – sand displacement along the wall, $\bar{\varepsilon}_v$ - mean volume change, × - very rough, • - rough, o – smooth, — dense, --- loose) (Tejchman 1989)

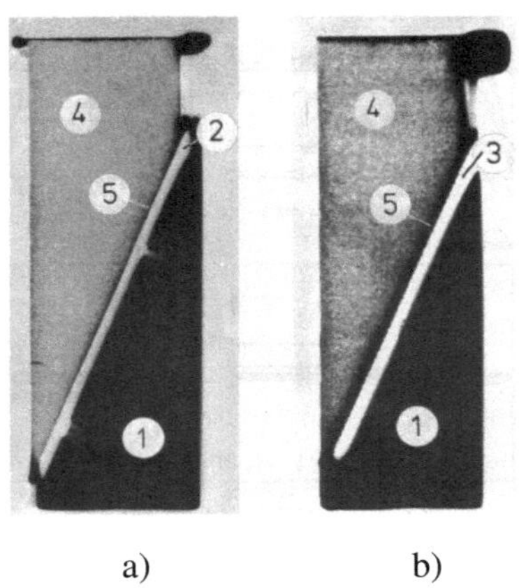

Fig. 2.13. Formation of a wall shear zone along: a) rough surface and b) very rough surface in a plane strain compression apparatus: 1. wooden wedge, 2. rough wall, 3. very rough wall, 4. sand specimen, 5. shear zone (Tejchman 1989)

Triaxial compression

Axisymmetric triaxial tests were carried out by Desrues et al. (1996). The diameter of the sand specimen was 100 mm and the height was 100 or 200 mm. The axial load and axial displacement were measured. A special anti-friction system was used. The shear zones were detected using X-ray tomography. The tests were carried out on dense and loose sand with lubricated and non-lubricated specimens.

In initially dense specimens, localization of deformation was observed to depend greatly on test conditions. In a dense specimen with non-lubricated ends, one slightly curved shear zone was created (Fig.2.14). When the specimen was short, the localized deformation was organized with a single rigid cone attached only to one of the platens (the other platen did not generate any cone). Outside of the cone, a complex pattern was observed in the form of plane strain shear mechanisms associated in pairs of lines (Fig.2.15).

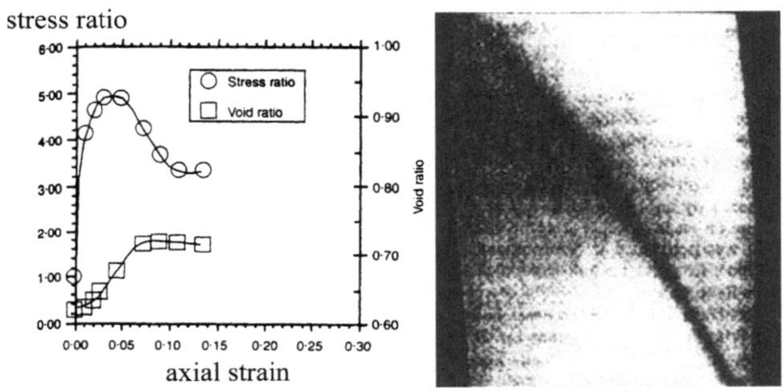

a) b)

Fig. 2.14. Evolution of the stress ratio and void ratio (a) and formation of a shear zone (b) during the experiment with a high dense specimen and non-lubricated ends (Desrues et al. 1996)

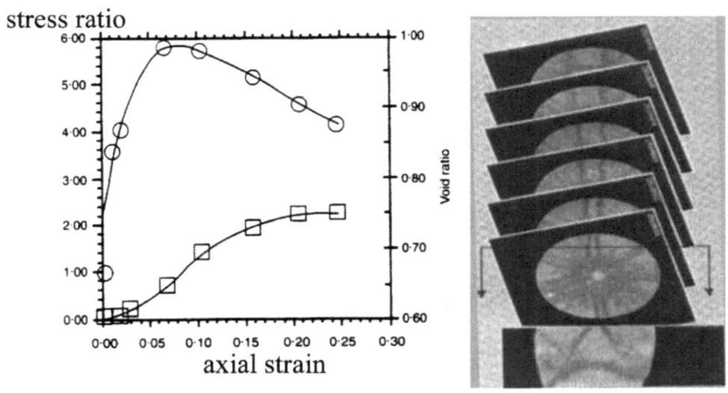

a) b)

Fig. 2.15. Evolution of the stress ratio and void ratio (a) and formation of shear zones (b) during the experiment with a low dense specimen and lubricated ends (Desrues et al. 1996)

Biaxial compression

Tests were carried out with Karlsruhe sand in a biaxial apparatus at Karlsruhe University (Kohse 2003) (with movable horizontal and vertical rigid walls). The dimensions of the sand specimen were: 80×113×50 mm^3. A constant deformation velocity of 0.011 mm/min was prescribed to both horizontal walls, and a constant pressure of 20 kPa was prescribed to both vertical walls.

Fig.2.16 shows the stress-strain curve and evolution of shear zones on the basis of volume changes obtained with the PIV-method (Nübel 2002). During compression, a pattern of shear zones was created; shear zone was reflected from rigid walls.

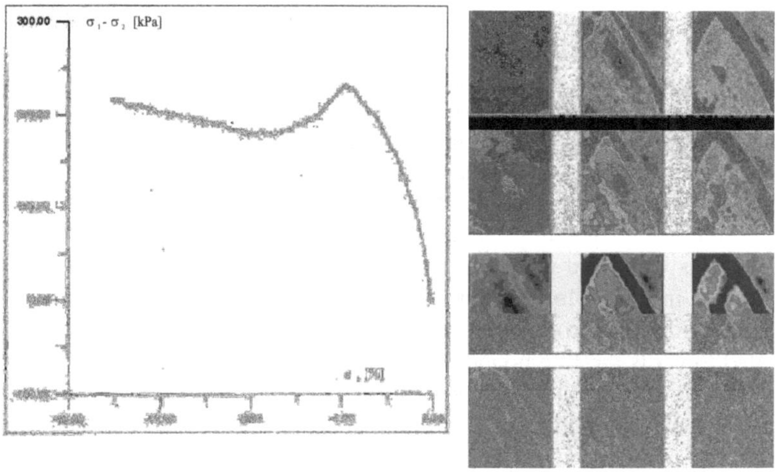

Fig. 2.16. Biaxial compression: stress-strain curve and evolution of volume changes in time (Kohse 2003)

Tests with earth pressures on a retaining wall

Comprehensive experimental studies on earth pressure in sand were carried out at Cambridge University between 1962 and 1974. During this period, a number of researchers (Arthur 1962, James 1965, Lucia 1966, May 1967, Bransby 1968, Adeosun 1968, Lord 1969, Smith 1972 and Milligan 1974) carried out experiments on the active and passive failure of a mass of dry sand deforming under plane strain conditions. The type of the wall movement: passive wall translation (Lucia 1966), active wall rotation about its top (Lord 1969), passive wall rotation about its top (Arthur 1962, James 1965, Lord 1969), active wall rotation about its toe (Smith 1962, Milligan 1974), passive wall rotation about its toe (Bransby 1968, Adeosun 1968), wall height (0.152 m and 0.33 m), wall roughness: smooth and rough (James 1965, Milligan 1974)), wall flexibility: rigid wall and flexible wall (Milligan 1974), initial density of sand: dense and loose (James 1965, Lord 1969) and surcharge were varied.

In case of a small apparatus (Arthur 1962), the wall was 152 mm high and 152 mm wide. In the remaining cases (using a large earth pressure apparatus), the retaining wall was 330 mm high and 190 mm wide. The dimensions of the sand specimen behind the wall were: 346 mm (height), 382 mm (length) and 200 mm (width) in a small apparatus, and 1500 mm (height) 1420 mm (length) and 195 mm (width) in a large apparatus. Sand was poured at a different height from a moving hopper when the wall was fixed. In addition, dredged and backfilled experiments were carried out (Milligan 1974). In the first case, sand on one side of the wall was successfully removed. In the second case, the wall was partially embedded in sand and filling was continued on one side. Moreover, tests on soil cutting with an inclined wall were carried out (Bransby 1968, Adeosun 1968).

The sand used was a rounded coarse quartz Leighton Buzzard sand (maximum void ratio 0.70, minimum void ratio 0.51, grain size between 0.6-1.2 mm, mean grain diameter 0.6 mm). The evolution of shear localization in sand was recognized using a radiographic technique which was able to detect density changes.

Different modes of shear zones have been observed during passive and active earth pressure tests depending mainly on the type of the wall motion and surcharge. In passive tests with rigid walls rotating about the top, one or two curved shear zones were obtained in sand. Multiple curved shear zones of a similar shape were observed during tests with a wall rotating about the bottom. They occurred at the wall top and propagated towards the free boundary. During tests with a translating rigid wall, one slightly curved shear zone starting to form from the wall bottom, and secondary radial shear zones beginning at the wall top appeared. In active tests with rigid walls, nearly parallel straight zones or a mesh of intersecting parallel zones close to the wall (wall rotating around the bottom) or a single curved zone (wall rotating around the top) were observed. The details of the tests at the Cambridge University and radiographs from Cambridge Archive of Radiographs were given by Leśniewska (2000). Figs.2.17 and 2.18 show the formation of shear zones during a passive and an active state.

Experimental studies of passive earth pressure on a retaining wall in sand were also performed at the Karlsruhe University by Gudehus (1986), Gudehus and Schwing (1986), Schwing (1991). In these experiments,, the wall height, h, was 0.15 m or 0.20 m. In the case of h=0.20 m, the dimensions of the sand specimen were: 570 mm (height), 630 mm (length) and 200 mm (width). In turn, in the case of h=0.15 m, the dimensions of the sand specimen were: 200 mm (height), 400 mm (length) and 200 mm (width). The material used was the Karlsruhe sand. During a passive wall translation, one observed in dense sand a pattern of shear zones consisting of one major slightly curved shear zone starting to form at the wall toe and propagating towards the free boundary, radial zones linking the wall top and the curved shear zone, and one shear zone parallel to the bottom of the sand body (Fig.2.19).

The experiments on earth pressures were also carried out by Nübel (2002) with Karlsruhe sand (Figs.2.20 and 2.21). The dimensions of the sand specimen were 15×18×9 cm^3. In the active case, a shear zone was created with a thickness of (11-15)×d_{50} and an inclination against the bottom of 68°. The thickness of a shear zone in the passive case was 20×d_{50}.

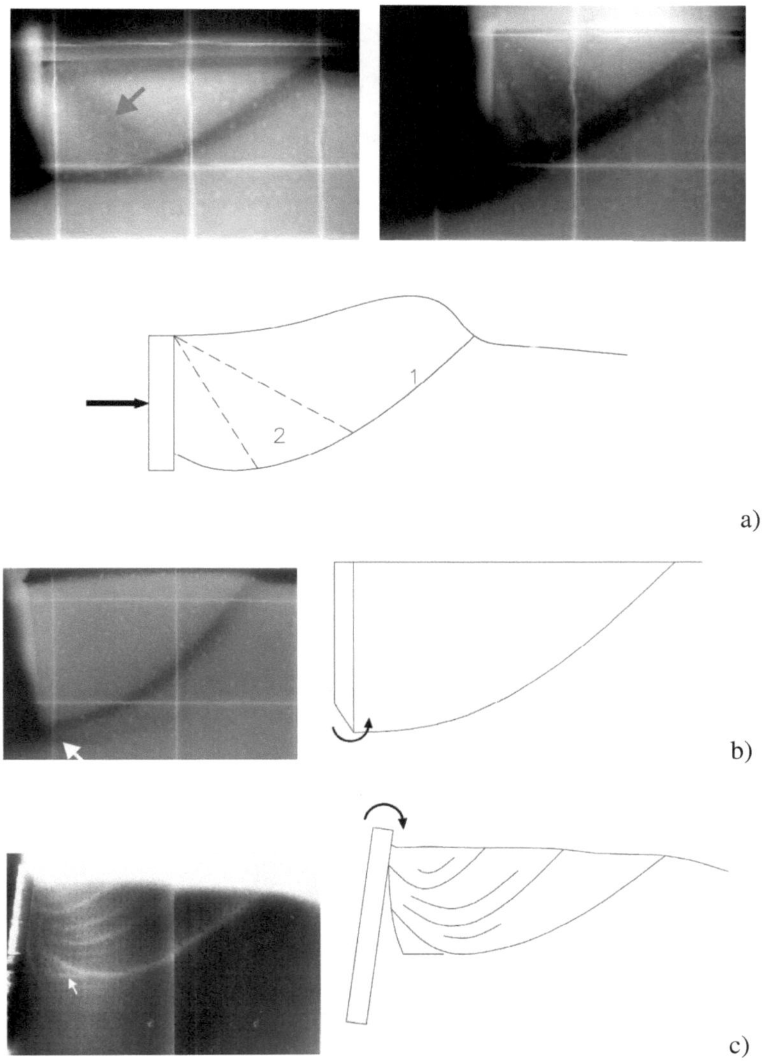

Fig. 2.17. Shear zones observed in experiments (radiograhs and schematically): a) during passive wall translation (Lucia 1966), b) during passive wall rotation around the top (Arthur 1962) and c) during passive wall rotation around the bottom (Bransby 1968) from Leśniewska (2000)

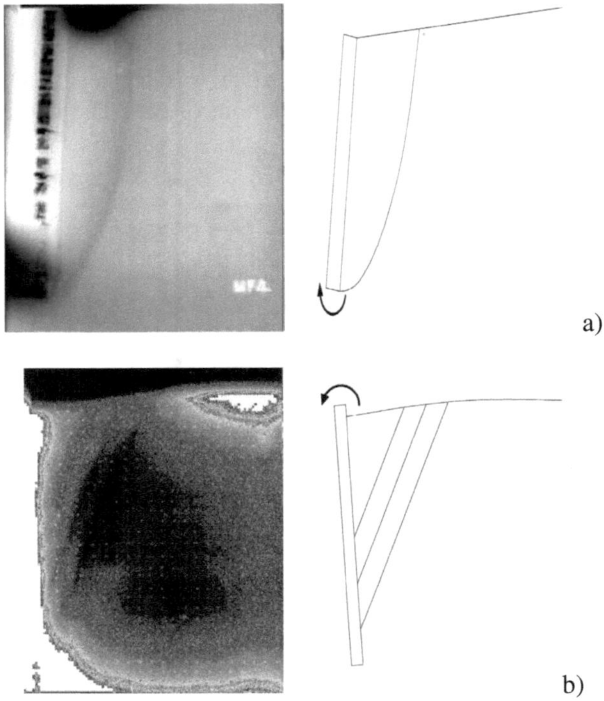

Fig. 2.18. Shear zones observed in experiments (radiograhs and schematically): a) during active wall rotation around the top (Lord 1969) and b) during active wall rotation around the bottom (Smith 1972) from Leśniewska (2000)

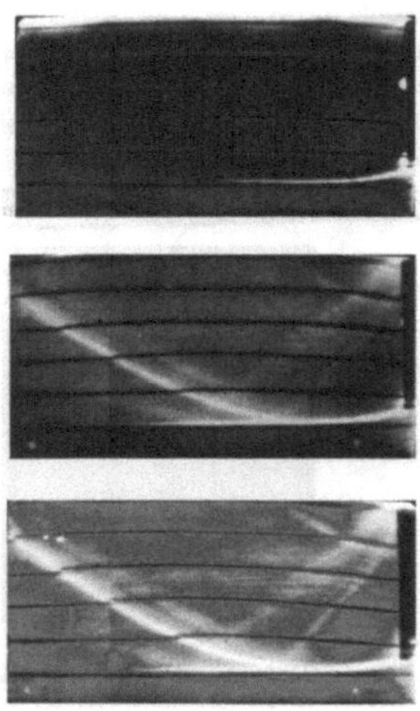

Fig. 2.19. Formation of shear zones in the passive case from X-radiographs (Gudehus and Schwing 1986)

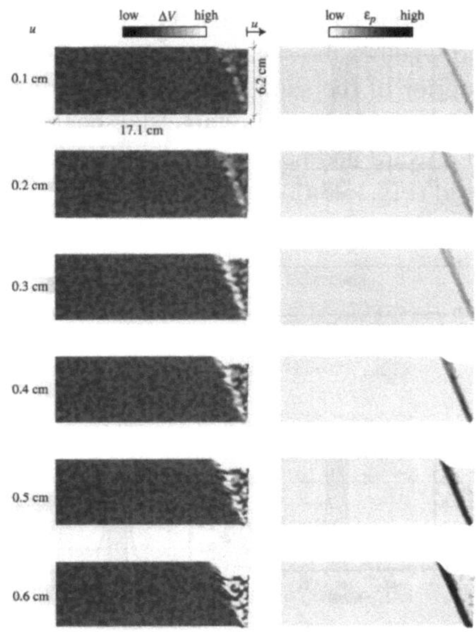

a) b)

Fig. 2.20. Formation of a shear zone in the active case on the basis of the PIV method: a) volume change, b) deviatoric deformation (Nübel 2002)

In Fig.2.22, the results of the author's experiments on the formation of shear zones in passive and active cases are shown (Tejchman 1997). The experimental set-up consisted of a steel movable piston, a rubber membrane and a perspex box containing sand specimen (0.2×0.6×0.1 m^3). The rubber membrane was placed on the box bottom and connected at its ends to the piston and box wall. The rubber membrane was very rough (it was achieved by sticking sand particles to its surface). During a piston displacement, the deformation was first generated in the rubber and was next transferred to the sand body. Due to imperfections along the rubber membrane, several shear zones were created.

Trap-door tests

The tests on the evolution of shear zones during an active and passive mode of the trap-door in sand were carried out by Vardoulakis et al. (1981) and Graf (1984). The dimensions of the sand specimen were 100×15×50 cm^3. The width of the outlet varied. The failure patterns were observed by using thin horizontal coloured sand layers placed in the sand body or by means of x-rays (Fig.2.23).

A very small upward displacement of the trap-door (passive case) yielded two almost symmetrical shear zones proceeded from the edges of the trap-door (Fig.2.23b). The shear zones propagated into the earth body until reaching finally the free surface. The inclination of shear zones in initially dense sand was about 70-75°

and in initially loose sand about 90°. It depended on the ratio between the outlet width and specimen height and whether the shear zone tip reached the free surface or came to a dead stop in the interior of the sand body. During a downward displacement of the trap-door (active case), two almost symmetric shear zone appeared above the outlet. They propagated upward and next crossed with each other. Afterwards a flock of shear zones was created (Fig.2.23a).

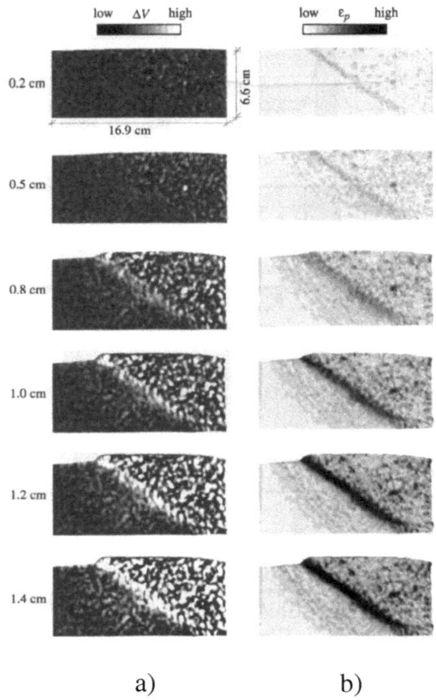

Fig. 2.21. Formation of the shear zone in the passive case on the basis of the PIV method: a) volume change, b) deviatoric deformation (Nübel (2002)

Fig. 2.22. Formation of shear zones in the passive (a) and active case (b) (Tejchman 1997)

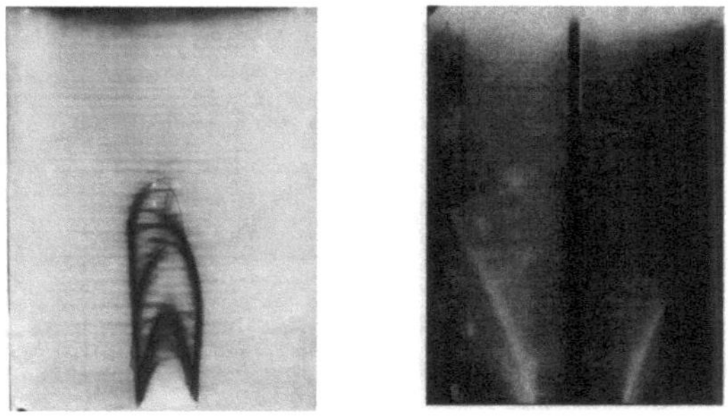

a) b)

Fig. 2.23. Formation of a shear zone in the active (a) and passive trap-door problem (b) on the basis of x-ray radiographs (Graf 1984)

Strip foundation tests

The experiments with different widths of the strip foundation on dense SLB sand were performed by Tatsuoka et al. (1997). To determine a scale effect due to the pressure level and grain size, the 1g and centrifuge tests were performed. In the 1g tests, a sand box had the dimensions 25×60×30 cm^3 (for model footings with a width of B_o=0.5, 1.0, 2.5 cm) and 40×183×60 cm^3 (for a model footing with a width of B_o=10 cm). In the centrifuge tests, a 10 cm wide × 50 cm long × 30 cm deep sand box was used with B_o=2 cm and 3 cm. The footing base was made rough by gluing a sheet of sand paper. The loading of footings was central. Figs.2.24 and 2.25 show the relationships between the normalized vertical force and normalized vertical displacement from 1g and centrifuge tests. The progressive failure on the basis of local shear strain contours is presented in Fig.2.26. The normalized vertical force increased with increasing ratio d_{50}/B_o (except of the case with B_o=0.5 cm) and decreasing pressure level. The thickness of the shear zone was about 6 mm (10×d_{50}).

Wall pull-out experiments

A wall pull-out experiment was carried out by Slominski (2003). A very rough vertical wall was pulled out from dense Karlsruhe sand. The volume changes were observed by means of the PIV-method. Fig.2.27 presents the evolution of localization of deformation at the wall. A large dilatant region occurred behind the wall.

Tests with confined granular flow in silos

Model tests were performed with dry sand in a plane strain model silo with parallel (bin) and convergent walls (hopper) and a slowly moveable bottom (Tejchman 1989). The dimensions of the bin were: 0.5 m (height), 0.6 m (length) and 0.10-0.30 m (width). In the case of a hopper, the height was 0.5 m, length 0.6 m, width at the

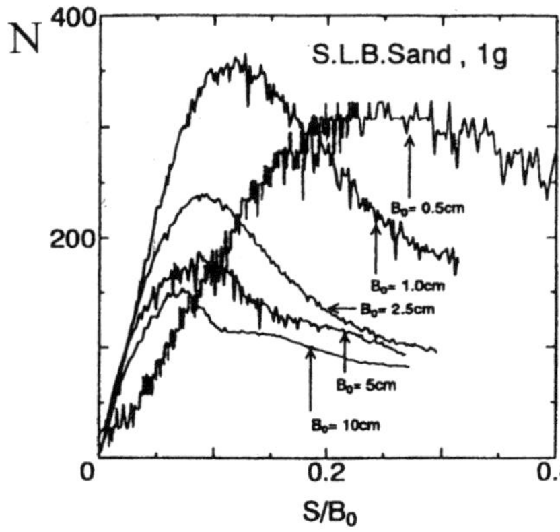

Fig. 2.24. Relationship between the normalised vertical force N=2q/($\gamma_d B_o$) and the normalised vertical displacement S/B_o from 1g tests on SLB sand (γ_d – initial unit weight, B_o – footing width, q – mean vertical normal stress, s – vertical displacement) (Tatsuoka et al. 1997)

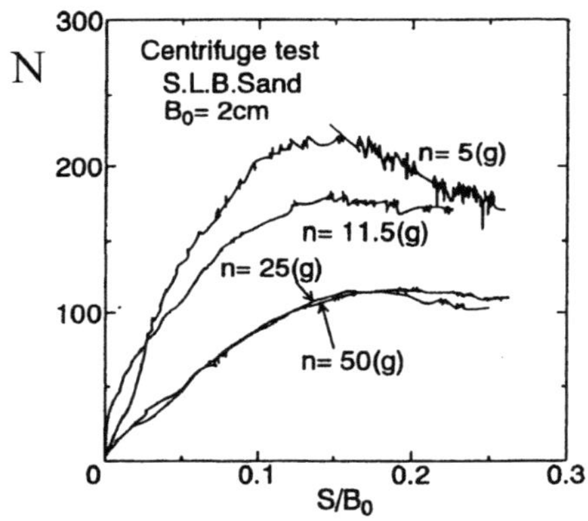

Fig. 2.25. Relationship between the normalised vertical force N=2q/($\gamma_d B_o$) and the normalised vertical displacement S/B_o from centrifuge tests on SLB sand (γ_d - unit weight, B_o – footing width, q – mean vertical normal stress, s – vertical displacement) (Tatsuoka et al. 1997)

bottom 0.10 m, width at the top 0.20-0.30 m and wall inclination to the vertical α=5.6°-11.4°. A discharge was induced by lowering the bottom plate at a constant velocity of 5 mm/h (quasi-static flow). The tests were carried out with dense and loose Karlsruhe sand (d_{50}=0.45 mm). The walls were smooth, rough or very rough.

In Fig.2.28, the evolution of the resultant vertical force on the bottom in a bin and hopper is presented. Figs.2.29 and 2.30 show displacement profiles obtained with thin colored sand layers. The flow in a bin was of a plug type except for a narrow shear zone adjacent to the wall. The thickness of the shear zone was approximately 5 mm (11×d_{50}) at the smooth wall, 20 mm (45×d_{50}) at the very rough wall with

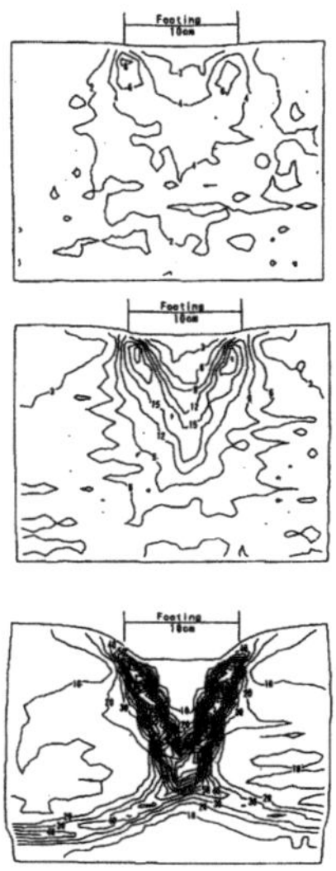

Fig. 2.26. Evolution of local shear strain contours from 1g test with footing of width B_o=10 cm on SLB sand (Tatsuoka et al. 1997)

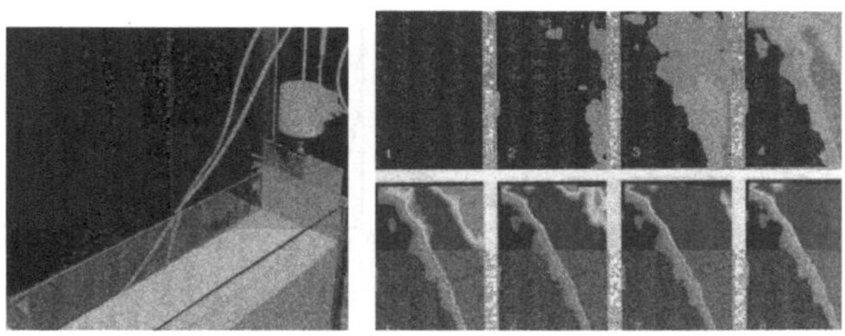

a) b)

Fig. 2.27. Experiment with a pull-out wall: set-up (a) and evolution of silatant region behind the wall for the increasing vertical wall displacement (1-8) (b) on the basis of the PIV method (Slominski 2003)

loose Karlsruhe sand and 15-20 mm ((33-45)×d_{50}) at the very rough wall with dense Karlsruhe sand. In the last case, secondary shear zones appeared also inside of the flowing material. In the case of a hopper, the shear zones occurred along the walls as well and only inside of the moving dense solid. Due to that, the flow was non-symmetric. The flow of initially loose sand was always symmetric. The thickness of the wall shear zone with coarse Karlsruhe sand (d_{50}=1.0 mm) was 10 mm (10×d_{50}) at the smooth wall, 25 mm (25×d_{50}) at the very rough wall with initially loose sand and 22-25 mm ((22-25)×d_{50}) at the very rough wall with initially dense sand.

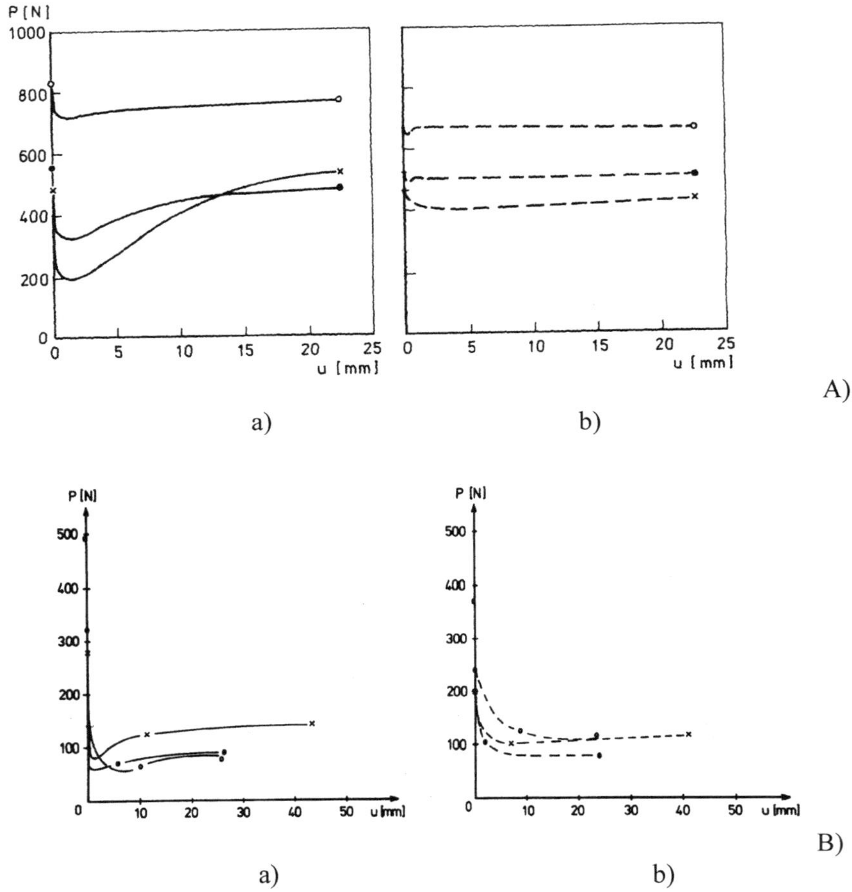

Fig. 2.28. Evolution of the resultant vertical force on the bottom P after bottom displacement u for dense sand and loose sand: A) bin, B) hopper (α=11.4°), a) dense sand, b) loose sand, o - smooth walls, • - rough walls, × - very rough walls (Tejchman 1989)

In silo model tests by Nedderman and Laohakul (1980) and Ananda et al (2007), the thickness of a wall shear zone increased with bin width and particle diameter. Takahashi and Yanai (1973) reported that the wall shear zone thickness in a bin increased also with flow velocity.

In addition, measurements of porosity changes in bulk solids during granular flow in model silos using two different non-invasive methods, namely Electrical Capacitance Tomography (ECT) and Particle Image Velocimetry (PIV) (Slominski et al. 2007, Niedostatkiewicz and Tejchman 2007) were carried out. The measurements with a ECT method were performed with a cylindrical perspex silo model (height h=2.0 m, diameter d=0.2 m, wall thickness 5 mm) supported by a steel frame and emptied gravitationally through the outlet with a diameter d_o smaller than d

(d_o=0.07-0.12 m). During the emptying process, strong dynamic effects in the form of regular almost harmonic pulsations occurred in the upper part of the silo during mass flow (i.e. 1.0-2.0 m above the outlet) due to a dynamic interaction between the silo fill and silo structure (Tejchman and Gudehus 1993, Tejchman 1998). The frequency of self-excited dynamic effects in sand at the bottom was equal approximately to the natural frequencies of the entire silo structure. These strong dynamic effects were observed only in the upper part of the silo during the mass flow of the bulk solid. When funnel flow occurred, the effects were dampened by the non-moving material at wall. The mode shapes of the silo during flow were very similar to the modes of bells (Wilde et al. 2008). The role of the clapper was taken on by radial forces generated by the silo fill hitting the hopper at the outlet. The dominant frequency of the sound signal during flow was equal to 100 Hz and corresponded to the 1st ovalling silo mode shape. This mode was very similar to the first bell mode. However, the music of bells is only due to free vibrations of the shell and in the case of the model silo there was a dynamic interaction between the oscillating silo and the falling and oscillating sand column (Wilde et al. 2008).

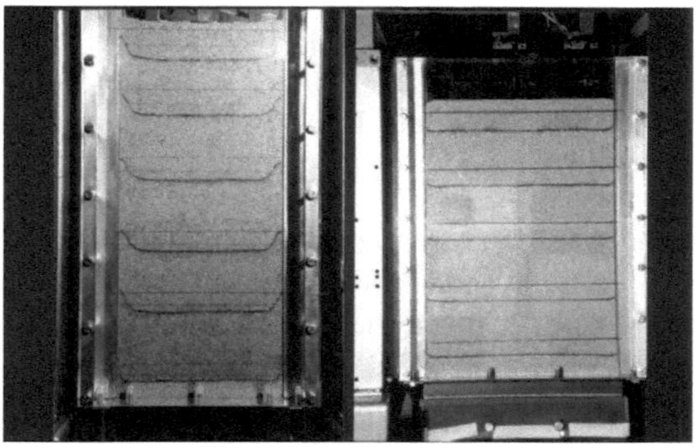

a) b)

Fig. 2.29. Displacements in a bin: a) very rough walls and loose sand, b)smooth walls, dense sand (Tejchman 1989)

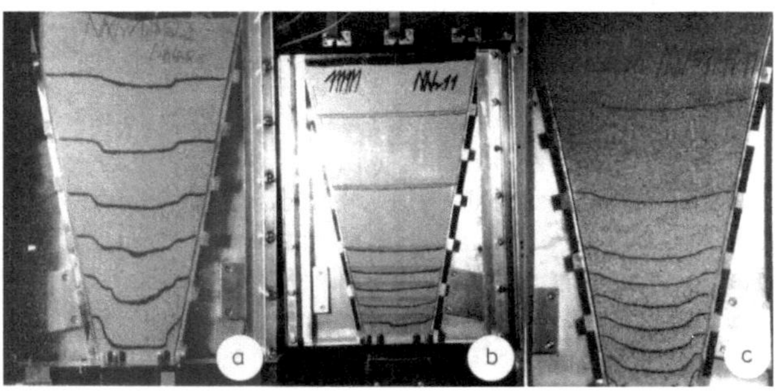

Fig. 2.30. Displacements in a hopper: a) very rough walls and dense sand, b) smooth walls, dense sand, c) very rough walls and loose sand (Tejchman 1989)

For ECT measurements, the model silo contained cohesionless sand with 2 different initial unit weights γ_d: initially loose (γ_d=15.6 kN/m^3) and initially dense (γ_d=17.2 kN/m^3) with a mean grain diameter of d_{50}=0.8 mm. An initially dense sand was obtained by filling the silo using the so-called "rain method" (through a vertically movable sieve). In turn, an initially loose sand was obtained by filling the silo from a pipe located directly above the upper sand surface which was vertically lifted during filling. Smooth and very rough silo walls were used. A wall surface with a high roughness r_w was obtained by gluing a sand paper to the interior wall surfaces ($r_w \approx d_{50}$). The images (so-called tomograms) were gathered using an ECT system. The measurements were taken by applying a two-plane sensor with 12-electrodes in each plane. The two sensors were placed at the silo outer wall at the height of h=0.5 m and h=1.5 m (measured to the center of the sensors) from the bottom. The measurement results were converted into the dielectric permittivity profiles in a silo cross-section by means of the reconstruction method called Linear Back Projection Method (Jaworski and Dyakowski 2001).

The PIV experiments were carried out with two small rectangular plane perspex model silos (mass flow silo and funnel flow silo). The height of the mass flow silo was h=0.34 m and the width was b=0.09 m. The height of the funnel flow silo was h=0.29 m and the width b=0.15 m. The wall thickness was 10 mm. The tests were performed with initially loose (γ_d=14.6 kN/m^3) and dense sand (γ_d=17.5 kN/m^3). Smooth and very rough walls were used. The silos were emptied gravitationally throw the rectangular outlet with the width of b_o=5 mm. The mean outflow velocity was approximately 10 mm/s.

The PIV system interprets differences in light intensity as a gray-scale pattern recorded at each pixel on CCD-camera (Charge Coupled Device). Two functions are of a major importance for PIV: image field intensity and cross-correlation function. The image intensity field assigns to each point in the image plane a scalar value which reflects the light intensity of the corresponding point in the physical space. A so-called Area Of Interest (AOI) is cut out of the digital image and divided into small sub-areas called Interrogation Cells (patches). If the deformation between two consecutive images is sufficiently small, the patterns of the interrogation cells are supposed not to change their characteristics. A deformation pattern is detected by comparing two consecutive images captured by a camera which remains in a fixed position with its axis oriented perpendicular to the plane of deformation. To find a local displacement between images, a search zone is extracted from the second image. A correct local displacement vector for each interrogation cell is accomplished by means of a cross-correlation function, which calculates simply possible displacements by correlating all gray values from the first image with all gray values from the next image. The correlation plane is evaluated at single pixel intervals (the resolution is equal to separate pixel). The peak in the correlation function indicates that the two images are overlaying each other. Only one displacement vector is calculated within one interrogation cell. The correlation operations are conducted in the frequency domain by taking the Fast Fourier Transform (FFT) of each patch. The procedure is continued by substituting the second image with a subsequent image. Thus, the evolution of displacements in the specimen can be captured. A direct PIV evaluation leads to an Eulerian description since the area of interest and the interrogation cell size are fixed. The relative displacements are next converted into a Lagrangian deformation field yielding total deformations with respect to the initial configuration. The strain vector is calculated with a strain-displacement matrix, thus the volume strain ε_v and deviatoric strain ε_p are calculated. In order to eliminate high frequency noise, a smoothing filter is applied to each vector field by taking an average value of the neighborhood of pixels.

The void ratio changes during silo flow are shown in a 3D representation in Figures 2.31 and 2.32 (using the ECT technique). The void ratio changes along two different cross-sections are expressed by different colors. A decrease of void ratio (contractancy) is denoted by the sign (↑) and (+). In turn, an increase of void ratio (dilatancy) is marked by the sign (↓) and (–). The magnitude of the changes is given for the wall and mid-region by numbers (in %). The changes of the void ratio are related to the initial state (reached during filling).

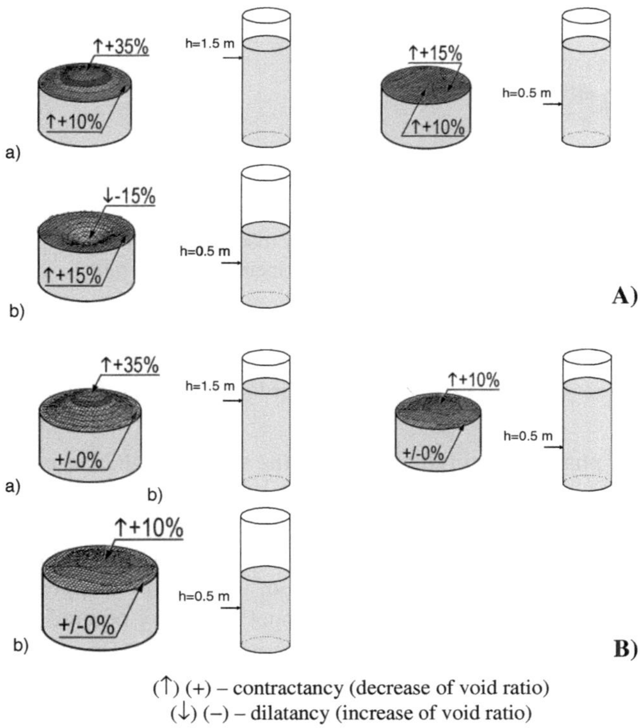

Fig. 2.31. Void ratio distribution in the cylindrical model silo in the cross-section at the height of h=1.5 m and h=0.5 m after: a) 4 s, b) 7 s of emptying (initially loose sand, d_0=0.07 m): A) smooth walls, B) very rough walls (using ECT) (Niedostkiewicz and Tejchman 2007)

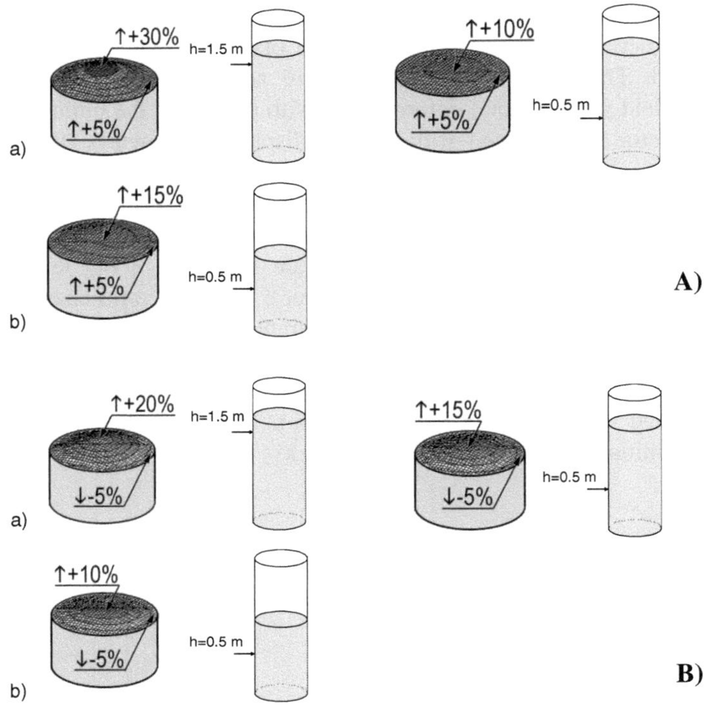

(↑) (+) – contractancy (decrease of void ratio)
(↓) (–) – dilatancy (increase of void ratio)

Fig. 2.32. Void ratio distribution in the cylindrical model silo in the cross-section at the height of h=1.5 m and h=0.5 m after: a) 4 s, b) 7 s of emptying (initially dense sand, d_0=0.07 m): A) smooth walls, B) very rough walls (using ECT) (Niedostkiewicz and Tejchman 2007)

In the initial phase of emptying silo with the smooth wall, loose sand was subject to densification in two cross-sections of the cylinder (h=1.5 m and h=0.5 m from the bottom of the silo) (Figure 2.31A). After 4 s of the emptying process, the void ratio decreased by 10-35% (as compared to the initial value) at the height of 1.5 m, and by 10-15% at the height of 0.5 m. The core of funnel flow appeared after 7 s of emptying at the height of 0.5 m when the upper boundary of sand was at the height of 1.3 m. During funnel flow (after 7 s of emptying), the void ratio in the core was larger by 15% as compared to the initial one (material dilated), whereas close to the silo walls, the void ratio decreased by 15% (material contracted). The width of the core in funnel flow, located in the middle of the silo was approximately equal to the width of the outlet of d_o=0.07 m. During initial and advanced flow in a silo with smooth wall, initially dense sand was subject to contractancy in both cross-sections (Figure 2.32A). The core of funnel flow on appeared at the height of 0.5 m later than for the loose sand (i.e. after 11 s) with the top level of the material being at the height of 0.9 m. After 7 s of flow (during funnel flow), the void ratio was smaller by

15% in the core of funnel flow as compared to the filling state, while near the walls, it was smaller by 15%. The width of the core of funnel flow was smaller than for loose sand by 20%. The wall shear zone could not be recognized at smooth walls.

During silo experiments with very rough walls, the strong interaction effects between the silo fill and silo structure did not occur (Tejchman and Gudehus 1993, Tejchman 1998), (Figure 2.31B and 2.32B). The total emptying time was about 4 s longer than in a silo with smooth walls. Mass flow still took place in the upper part of the silo, but the material moved lightly faster beyond the shear zone than along the wall. On one hand, the shear zone formation contributed to a significant increase of frequencies of pulsations due to the presence of additional horizontal and rotational stress waves. The frequency of these waves was greater than of longitudinal stress waves due to a shorter way of propagation. As a results, the resultant frequency was significantly higher than natural frequencies of the silo structure. On the other hand, the shear zone dampened the dynamic pulsations. The behaviour of initially loose sand in the mid-region of the silo at the beginning of silo emptying (t=1 s) was similar to that in the case of smooth walls. Close to the walls, no volume changes were observed during the whole flow. After 4 s of flow, the void ratio in the mid-region was smaller by 10%-35% as compared to the initial state. After 7 s of flow, the material in the core of funnel flow was denser. The width of the pronounced shear zone along the silo wall was about 20 mm ($25\times d_{50}$). For initially dense sand (Figure 2.32B), at the beginning of silo emptying, the void ratio in the mid-region was smaller by 15-20% compared to the initial value (Figure 2.32Ba). The width of the shear zone at the wall was smaller than in loose sand, i.e. 15 mm ($19\times d_{50}$). In the wall shear zone, the material experienced dilatancy (void ratio was larger by 5% as compared to the initial state).

Figure 2.33 shows the displacements in sand obtained with coloured layers in a mass and funnel flow silo. During mass flow, shear zones along bin walls occurred. The width of the wall shear zone was insignificant at smooth walls in the bin (about t_s=3 mm), and noticeable at very rough walls in the bin: about t_s=16 mm (with initially loose sand) and about t_s=12 mm (with initially dense sand). The initial density influenced the width of the moving material in the core in a funnel flow silo, which changed from 65 mm up to 107 mm (in initially loose sand) and from 56 mm up to 100 mm (in initially dense sand). The shape of the upper surface was similar in both silos with very rough walls. The effect of the wall roughness on the flow pattern in a funnel flow silos was insignificant. On the basis of coloured layers, sand flow seems to be almost symmetric, in particular, in an initially loose sand.

The evolutions of the volumetric strain ε_v and deviatoric strain ε_p in sand in the silo up to h=0.20 m using PIV during first 7 s of mass flow are shown in Figures 2.34 and 2.35. The magnitude of strains (expressed by a color intensity scale) is attached to Figures (for t=7 s). The magnitude of contractancy (volume decrease) is denoted by the numbers with the sign (-) and dilatancy (volume increase) by the numbers with the sign (+). The calculated values are assigned to the colours. The colour division was always similar.

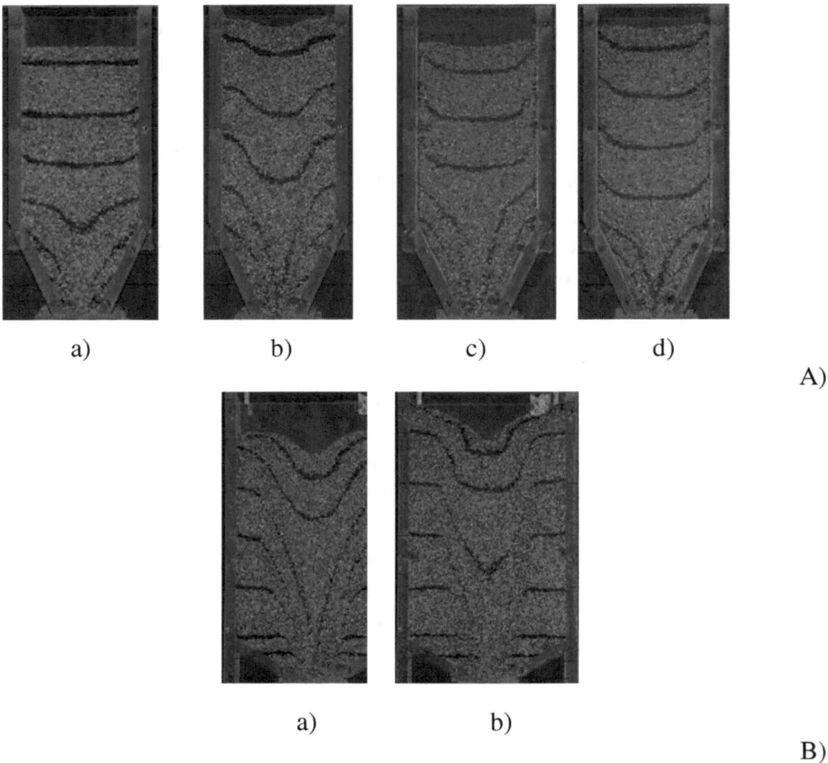

Fig. 2.33. Displacements of sand after 3 s of flow in a mass flow silo (A) and after 7 s of flow in a funnel flow silo (B): a) initially loose sand, smooth walls, b) initially dense sand, smooth walls c) initially loose sand, very rough walls, d) initially dense sand, very rough walls

The results in the silo with smooth walls during mass flow (Figs.2.34a and 2.34b) show that the distribution of the volumetric and deviatoric strain on the sand surface is non-uniform. The strains were significantly larger in initially dense sand (3-5 times). The minimum volumetric strain (contractancy) and maximum volumetric strain (dilatancy) were about -0.06% and +0.09% (initially loose sand) and -0.28% and +0.3% (initially dense sand), respectively (at t=7 s). The deviatoric strain changed between 0%-0.09% (initially loose sand) and 0%-0.5% (initially dense sand). Densification zones were mixed with loosening ones. The location of dilatancy zones was, in particular, very non-uniform in initially loose sand. Their area was about 30% of the entire sand area. In the case of initially dense sand, several curvilinear dilatant zones occurred in the hopper and one pronounced parabolic in the bin. The dilatant zones were created in the neighborhood of the outlet. The material was sheared not only inside of the material (as in the case of initially loose sand) but also along the walls. The distribution of the volume and deviatoric strain in a model silo with very rough walls was less non-uniform (Figure 2.34c and 2.34d) than in a silo with smooth walls. The strains increased with decreasing initial density. The minimum volumetric strain (contractancy) and maximum volumetric strain (dilatancy) were about -0.015% and +0.04% (loose sand), and 0.1% and +0.4% (dense sand), respectively (at t=7 s). The deviatoric strain changed between 0%-0.06% (initially loose sand) and 0%-0.3% (initially dense sand), respectively. The dilatant and contractant zones appeared in the entire silo. The shape of dilatant zones was chaotic in initially loose and medium dense sand. In initially dense sand, their shape was more regular (parabolic) in the bin. The material was subject to shearing at walls and inside of the fill mainly in initially dense sand.

For funnel flow (Fig.2.35), the distribution of the volume and deviatoric strain was also non-uniform in the core. It was similar independently of the wall roughness and initial sand density. In all cases, the dilatans zones were connected to shear zones. The magnitudes of strain were of the same order. The width of the moving material in the core in a funnel flow silo was larger with smooth walls.

The results of PIV were qualitatively compared with the images of volume changes inside of sand obtained with X-rays method. The loosening regions were only detected during flow with an initially dense specimen. The radigraphs of silo flow of initially dense sand in a mass flow silo with smooth walls (Figure 2.36A) revealed that a symmetrical pair of curvilinear dilatant rupture zones was created in the neighborhood of the outlet (loosening is marked by a bright shadow). The zones propagated upward, crossed each other around the symmetry of the silo, reached the walls and subsequently were reflected from them. This process repeated itself until the zones reached the free boundary. Similar volume changes were observed by Michalowski (1990).

In turn, in the case of the initial mass flow in a silo with very rough walls and dense sand (Figure 2.36B), curvilinear almost symmetric dilatant zones occurred in the material core above the outlet. They propagated upwards (some of them or crossed each other). Two different regions could be observed in the material (one core region in motion and two almost motionless regions at walls). The initial flow pattern was similar to that in a funnel flow silo (Michalowski 1990). Later, during advanced flow, a pronounced loosening region above the outlet and dilatant shear zones along the walls appeared.

In the experiments using PIV, the distribution of the volume strain in dense sand with smooth walls (Figure 2.34a and 2.34b) was generally different than that measured using X-rays method (Figure 2.36A). There were more curvilinear dilatant zones appearing above the outlet in the hopper which were also very non-regular. The pronounced volume changes were also observed along these zones. However, the shape of the parabolic dilatant zone in the bin was similar when using both experimental methods. In the case of very rough walls, the parabolic shape of dilatant zones in the bin and the magnitude of dilatancy in the hopper were also approximately similar.

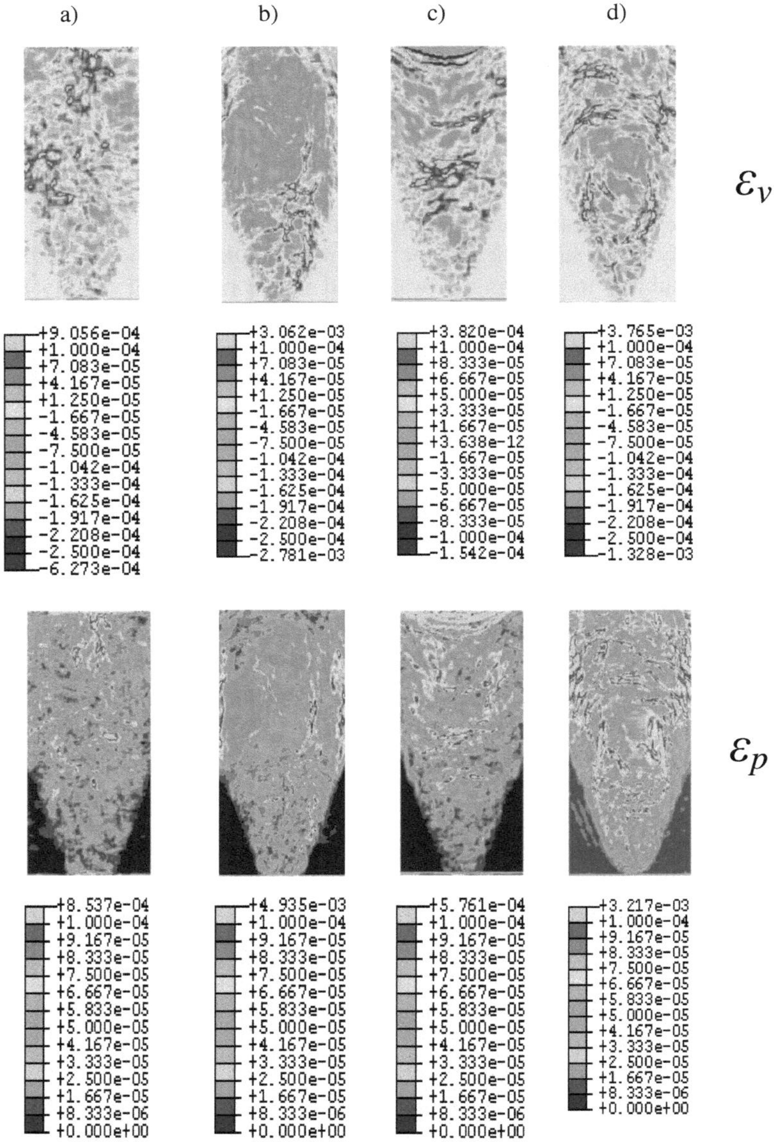

Fig. 2.34. Evolution of the volume strain ε_v and deviatoric strain ε_p after 7 s of emptying of sand (mass flow silo): a) initially loose sand, smooth walls, b) initially dense sand, smooth walls, c) initially loose sand, very rough walls, d) initially dense sand, very rough walls (using PIV) (Slominski et al. 2007)

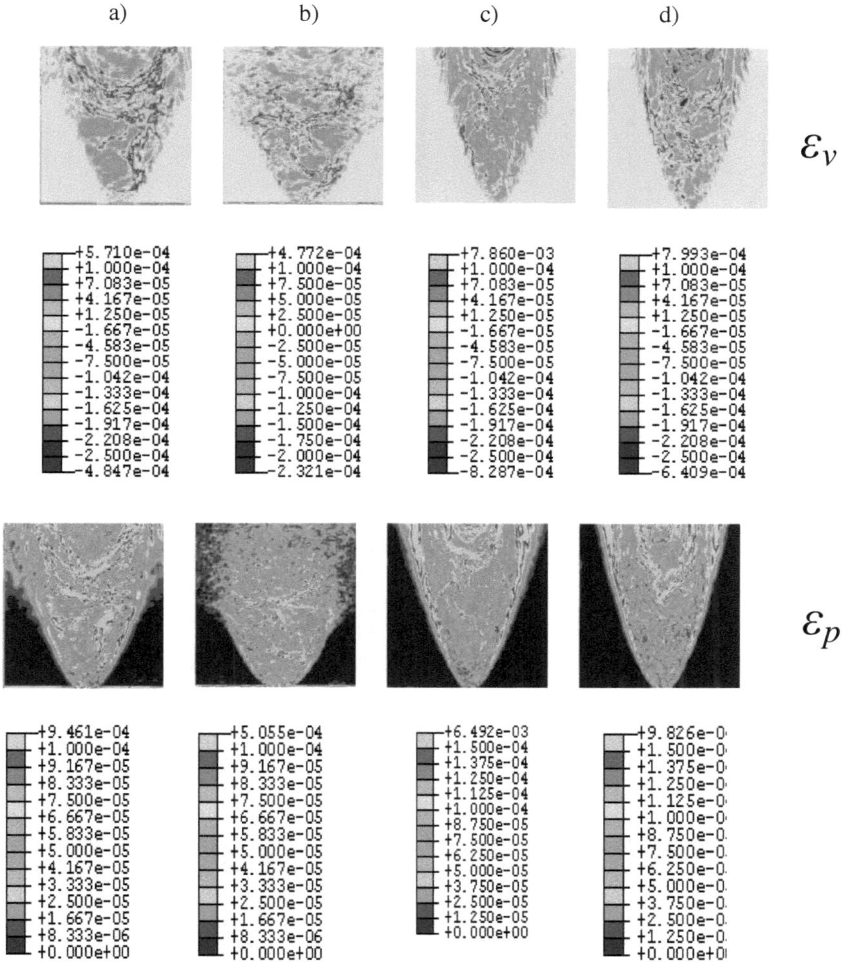

Fig. 2.35. Evolution of the volume strain ε_v and deviatoric strain ε_p after 7 s of emptying of sand (funnel flow silo): a) initially loose sand, smooth walls, b) initially dense sand, smooth walls, c) initially loose sand, very rough walls, d) initially dense sand, very rough walls (using PIV, Slominski et al. 2007)

The reason of differences is caused by the fact that strains in the PIV technique can be traced only on the surface of the granular specimen. In contrast, the volume changes registered with X-rays result from the entire specimen depth. Thus, friction between the wall and rough sand grains influences the results when using PIV.

The experimental results with both non-invasive methods ECT and PIV (which were originally developed in the field of experimental fluid and gas mechanics) show

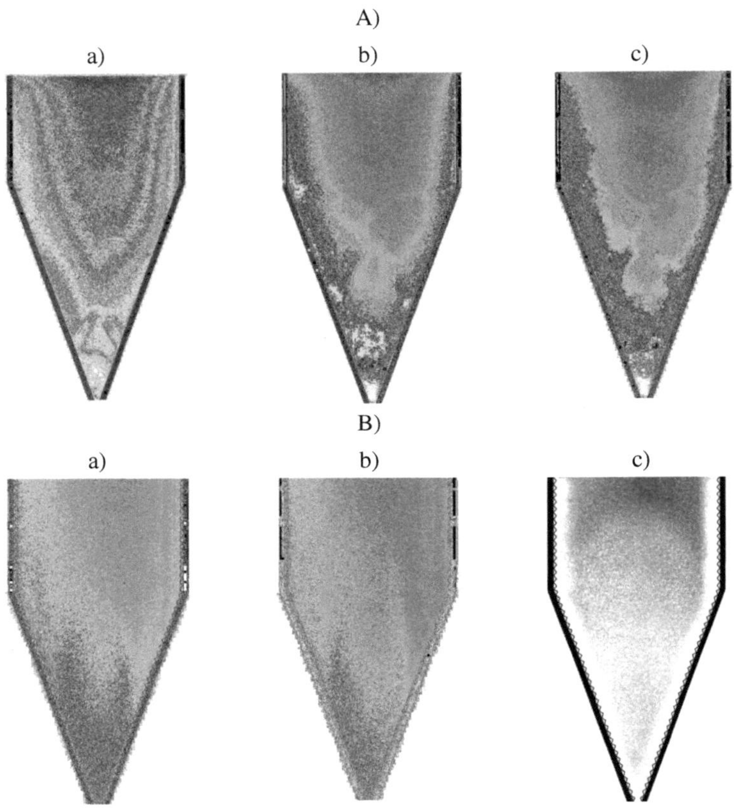

Fig. 2.36. X-radiographs of initially dense sand during mass flow: A) with smooth walls and B) with very rough walls after: a) 2, b) 3, c) 4 and d) 7s of emptying (Slominski et al. 2007)

that the methods can be successfully applied to detect volume changes in cohesionless bulk solids during silo emptying. The ECT method allows us to determine the porosity changes in the entire material volume during flow. It can be used for small and medium size specimens. Thus, it can find application in industrial and semi-industrial flow problems. For such problems, its accuracy is sufficient. The disadvantage of the method is its high cost. In turn, the PIV method is able to determine the porosity changes only on the surface of the granular specimen. It is a very accurate method. However, it can be mainly used for small specimens. The results are influenced by wall friction. Its advantages are simplicity and a low cost.

Figs.2.37-2.39 present radiographs of silo flow in experiments by Michalowski (1984, 1990) and Baxter and Behringer (1990). In the experiment with a mass flow silo, a symmetrical pair of curvilinear dilatant rupture zones was created in the neighbourhood of the outlet. The zones propagated upward, crossed each other around the symmetry of the silo, reached the walls and subsequently were reflected from

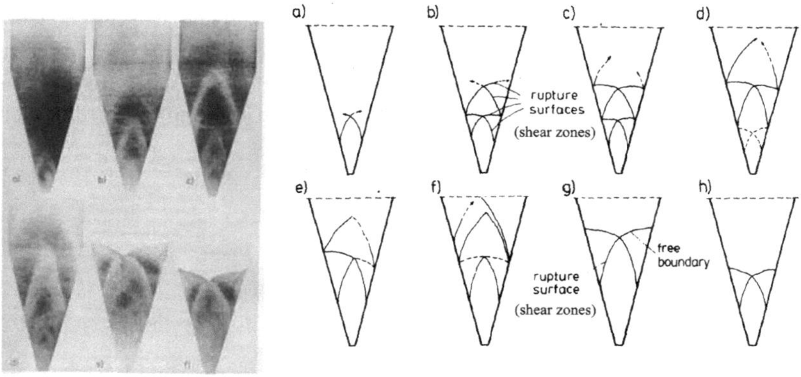

A) B)

Fig. 2.37. Mass flow in silos: A) x-radiographs, B) schematically for different flow time (a-h) (Michalowski 1984)

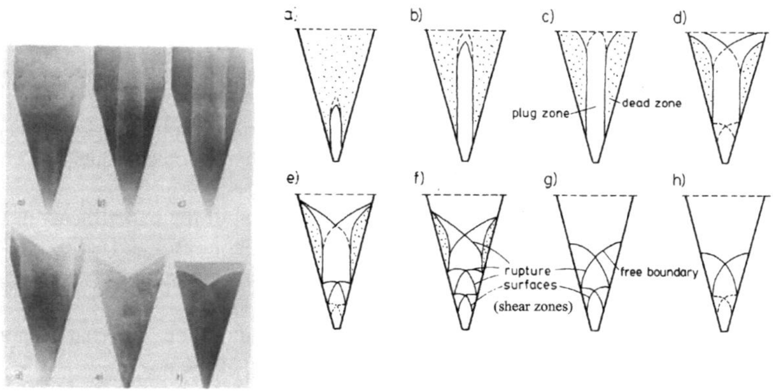

A) B)

Fig. 2.38. Funnel flow in silos: A) x-radiographs, B) schematically for different flow time (a-h) (Michalowski 1984)

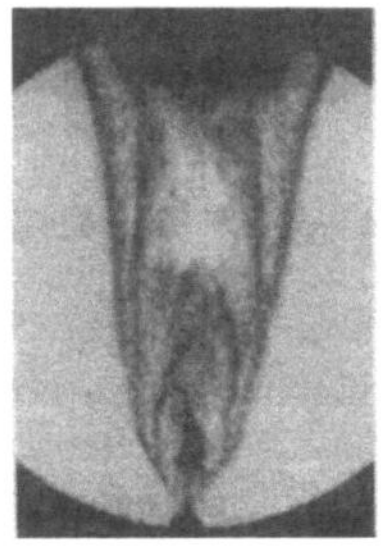

Fig. 2.39. X-radiographs of funnel flow in hoppers (Baxter and Behringer 1990)

them. This process repeated itself until the zones reached the free boundary in the converging hopper or the transition zone in the parallel-converging silo. In the case of a funnel flow silo, the curvilinear dilatant zones in the material core were almost symmetric about a vertical mid-line. Some of them propagated upwards or crossed each other.

The rapture zone occurred mainly in sands with rough grains.

References

Adeosun, A.: Lateral forces and failure patterns in the cutting of sands. Research Project at the University of Cambridge (1968)

Arthur, J.R.F.: Strains and lateral force in sand. PhD Thesis at the University of Cambridge (1962)

Baxter, G.W., Behringer, R.P.: Pattern formation and time-dependence in flowing sand. In: Two Phase Flows and Waves, pp. 1–29. Springer, New York (1990)

Becker, M., Lippmann, H.: Plane plastic flow of granular model material. Arch. Mech. 29, 829–846 (1977)

Bransby, P.L.: Stress and strain in sand caused by rotation of a model wall. PhD Thesis at the University of Cambridge (1968)

Desrues, J.: La localization de la deformation dans les materiaux granulaires. PhD thesis, USMG and INPG, Grenoble, France (1984)

Desrues, J., Chambon, R., Mokni, M., Mazerolle, F.: Void ratio evolution inside shear bands in triaxial sand specimens studied by computed tomography. Géotechnique 46(3), 529–546 (1996)

Desrues, J., Viggiani, G.: Strain localization in sand: overview of the experiments in Grenoble using stereo photogrammetry. Int. J. Numer. Anal. Meth. Geomech. 28(4), 279–321 (2004)

Graf, B.: Theoretische and experimentelle Ermittlung des Vertikaldrucks auf eingebettete Bauwerke. PhD thesis, Karlsruhe University, Heft 96 (1984)

Gudehus, G., Schwing, E.: Standsicherheit historischer Stützwände. Internal Report of the Institute of Soil and Rock Mechanics, University Karlsruhe (1986)

Hammad, W.: Modelisation non lineaire et etude experimentale des bandes de cisaillement dans les sables. PhD thesis, University of Grenoble, France (1991)

Han, C., Vardoulakis, I.: Plane strain compression experiments on water saturated fine-grained sand. Geotechnique 41, 49–78 (1991)

Harris, W.W., Viggiani, G., Mooney, M.A., Finno, R.J.: Use of stereo photogrammetry to analyze the development of shear bands in sand. Geotech. Test. J. 18(4), 405–420 (1995)

James, R.G.: Stress and strain fields in sand. PhD Thesis, University of Cambridge (1965)

Jaworski, A., Dyakowski, T.: Application of electrical capacitance tomography for measurement of gas-solids flow characteristics in a pneumatic conveying system. Measurement Science and Technology 12, 1109–1119 (2001)

Kohse, W.C.: Experimentelle Untersuchung von Scherfugenmustern an Granulaten. Internal Report of the Institute for Rock- and Soil-Mechanics, University of Karlsruhe (2003)

Kozicki, J., Tejchman, J.: Experimental investigations of strain localization in concrete using Digital Image Correlation (DIC) technique. Arch. Hydro-Engng. Environ. Mech. 54(1), 3–25 (2007)

Leśniewska, D.: Analysis of shear band pattern formation in soil. Monograph. Institute of Hydro-Engineering of the Polish Academy of Sciences, Gdansk, Poland (2000)

Lord, J.A.: Stress and strains in an earth pressure problem. PhD Thesis, University of Cambridge (1969)
Lucia, J.B.A.: Passive earth pressure and failure in sand. Research Report of the University of Cambridge (1966)
Marashdeh, Q., Warsito, W., Fan, L.S., Teixeira, F.: Dual imaging modality of granular flow based on ECT sensors. Granular Matter 10, 75–80 (2008)
May, J.: A pilot project on the cutting of soils. Research Report of the University of Cambridge (1967)
Michalowski, R.L.: Flow of granular material through a plane hopper. Powder Technology 39, 29–40 (1984)
Michalowski, R.L.: Strain localization and periodic fluctuations in granular flow processes from hoppers. Geotechnique 40(3), 389–403 (1990)
Michalowski, R.L., Shi, L.: Strain localization and periodic fluctuations in granular flow processes from hoppers. J. Geotech. Geoenviron. Engng. 129(6), 439–449 (2003)
Milligan, G.W.E.: The behaviour of rigid and flexible retaining walls in sand. Geotechique 26(3), 473–494 (1974)
Mokni, M.: Relations entre deformations en masse et deformations localisees dans les materiaux granulaires. PhD thesis, University of Grenoble, France (1992)
Mokni, M., Desrues, J.: Strain localization measurements in undrained plane strain biaxial tests on Hostun RF sand. Mech. Coh.-Frict. Mater. 4, 419–441 (1998)
Nedderman, R.M., Laohakul, C.: The thickness of the shear zone of flowing granular materials. Powder Technology 25, 91–100 (1980)
Niedostatkiewicz, M., Tejchman, J.: Investigations of porosity changes during granular silo flow using Electrical Capacitance Tomography (ECT) and Particle Image Velocimetry (PIV). Particle & Particle Systems Charact. 24(4-5), 304–312 (2007)
Nübel, K.: Experimental and numerical investigation of shear localisation in granular materials. Publication Series of the Institute of Soil and Rock Mechanics, University Karlsruhe, 62 (2002)
Pradhan, T.B.S., Tatsuoka, F.: Experimental stress-dilatancy relations of sand subject to cyclic loading. Soils Found. 29, 45–64 (1989)
Rechenmacher, A.L., Finno, R.J.: Digital image correlation to evaluate shear banding in dilative sands. Geotech. Test. J. 27(1), 13–22 (2004)
Schwing, E.: Standsicherheit historischer Stützwände. PhD thesis, University of Karlsruhe, 121 (1991)
Slominski, C.: Experimental investigation of shear localization using a PIV method. Internal Report of the Institute for Rock- and Soil Mechanics, University of Karlsruhe (2003)
Slominski, C., Niedostatkiewicz, M., Tejchman, J.: Application of particle image velocimetry (PIV) for deformation measurement during granular silo flow. Powder Technology 173(1), 1–18 (2007)
Smith, I.: Stress and strain in a sand mass adjacent to a model wall. PhD thesis, University of Cambridge (1972)
Tan, S., Fwa, T.: Influence of voids on density measurements of granular materials using gamma radiation techniques. Geotech. Test. J. 14(3), 257–265 (1991)
Takahashi, H., Yanai, H.: Flow profile and void fraction of granular solids in a moving bed. Powder Technology 7(4), 205–214 (1973)
Tatsuoka, F., Goto, S., Tanaka, T., Tani, K., Kimura, Y.: Particle size effects on bearing capacity of footing on granular material. In: Asaoka, A., Adachi, T., Oka, F. (eds.) Deformation and Progressive Failure in Geomechanics, pp. 133–138. Pergamon, Oxford (1997)

Tejchman, J.: Scherzonenbildung und Verspannungseffekte in Granulaten unter Berücksichtigung von Korndrehungen. Publication Series of the Institute of Soil and Rock Mechanics, University Karlsruhe, vol. 117, pp. 1–236 (1989)
Tejchman, J., Gudehus, G.: Silo-music and silo-quake, experiments and a numerical Cosserat approach. Powder Technology 76(2), 201–212 (1993)
Tejchman, J.: Modelling of shear localisation and autogeneous dynamic effects in granular bodies. Publication Series of the Institute of Soil and Rock Mechanics, University Karlsruhe, vol. 140, pp. 1–353 (1997)
Tejchman, J.: Silo-quake - measurements, a numerical polar approach and a way for its suppression. Thin-Walled Structures 31(1-3), 137–158 (1998)
Wilde, K., Rucka, M., Tejchman, J.: Silo music – mechanism of dynamic flow and structure interaction. Powder Technology 186, 113–129 (2008)
Vardoulakis, I.: Scherfugenbildung in Sandkörpern als Verzweigungsproblem. PhD thesis, Institute for Soil and Rock Mechanics, University of Karlsruhe, 70 (1977)
Vardoulakis, I., Goldscheider, M., Gudehus, G.: Formation of shear bands in sand bodies as a bifurcation problem. Int. J. Num. Anal. Meth. Geom. 2, 99–128 (1978)
Vardoulakis, I.: Shear band inclination and shear modulus in biaxial tests. Int. J. Num. Anal. Meth. Geomech. 4, 103–119 (1980)
Vardoulakis, I., Graf, B., Gudehus, G.: Trap-door problem with dry sand: a statical approach based upon model test kinematics. Int. J. Numer. Anal. Meth. Geomech. 5, 57–78 (1981)
Yoshida, T., Tatsuoka, F., Siddiquee, M.S.A.: Shear banding in sands observed in plane strain compression. In: Chambon, R., Desrues, J., Vardoulakis, I. (eds.) Localisation and Bifurcation Theory for Soils and Rocks, Balkema, Rotterdam, pp. 165–181 (1994)
Yoshida, N., Arai, T., Onishi, K.: Elasto-plastic Cosserat finite element analysis of ground deformation under footing load. In: Asaoka, A., Adachi, T., Oka, F. (eds.) Deformation and Progressive Failure in Geomechanics, pp. 395–401. Pergamon, Oxford (1997)

3 Theoretical Model

Abstract. This chapter describes shortly the theory of micro-polar hypoplasticity including a characteristic length of micro-structure in the form of a mean grain diameter. First, a non-polar hypoplastic constitutive model formulated at Karlsruhe University by Gudehus (1996) and Bauer (1996) for mainly monotonic deformation paths is summarized. Some results of so-called element tests (oedometric compression, triaxial compression, cyclic simple shearing) by a hypoplastic constitutive model are presented for different initial void ratios and pressure levels. Next, the calibration procedure for hypoplasticity to determine material parameters given by Bauer (1996) and Herle and Gudehus (1999) is outlined. Later, the micro-polar (Cosserat) continuum is presented. The advantages of the micro-polar theory with respect to a conventional (non-polar) continuum to capture shear localization are outlined. The equations of a micro-polar constitutive law proposed by Tejchman (1997) are given which were obtained by enhancement of a non-polar hypoplastic constitutive law of Gudehus (1996) and Bauer (1996) by micro-polar quantities. Finally, the FE implementation of a micro-polar model in a quasi-static regime is depicted.

3.1 Hypoplastic Constitutive Model

Hypoplastic constitutive models (Kolymbas 1977, Gudehus 2007) are an alternative to elasto-plastic formulations (Lade 1977, Vermeer 1982, Pestana and Whittle 1999, Al Hattamleh et al. 2005, Borja and Andrade 2006) for continuum modeling of granular materials. They were formulated by a heuristic process considering the essential mechanical properties of granular materials observed during homogeneous deformations. They describe the evolution of effective stress components with the change of strain components by a differential equation including isotropic linear and non-linear tensorial functions according to the representation theorem by Wang (1970) where the stress changes are a combination of the stress tensor σ_{mn} and rate-of-deformation tensor d_{kl} (ψ_i are scalar functions of invariants and joint invariants of σ_{mn} and d_{kl}):

$$\begin{aligned}\overset{o}{\sigma}_{ij} &= \psi_1\delta_{ij} + \psi_2\sigma_{ij} + \psi_3 d_{ij} + \psi_4\sigma_{ik}\sigma_{kj} + \psi_5 d_{ik}d_{kj} + \psi_6(d_{ik}\sigma_{kj} + \sigma_{ik}d_{kj}) + \\ &\psi_7(d_{ik}d_{km}\sigma_{mj} + \sigma_{ik}d_{km}d_{mj}) + \psi_8(d_{ik}\sigma_{km}\sigma_{mj} + \sigma_{ik}\sigma_{km}d_{mj}) + \\ &\psi_9(d_{ik}d_{km}\sigma_{mn}\sigma_{nj} + \sigma_{ik}\sigma_{km}d_{mn}d_{nj})\end{aligned} \quad . \tag{3.1}$$

J. Tejchman: Shear Localiz. in Gran. Bodies, SSGG, pp. 47–85, 2008.
springerlink.com

The term hypoplasticity was introduced by Dafalias (1986) to describe incremental constitutive relations wherein the plastic strain rate was not defined to a plastic potential surface. The hypoplastic constitutive laws were widely developed at Karlsruhe University (Kolymbas 1977, 2000, Wu 1992, Wu et al. 1996, Gudehus 1996, 2007, Bauer 1996, von Wolffersdorff 1996, Niemunis and Herle 1997, Bauer et al. 2004, Herle and Kolymbas 2004) and at Grenoble University (Darve et al. 1986, 1995, Chambon 1989, 2001, Desrues and Chambon 1989, Lanier et al. 2004).

In this book, a hypoplastic constitutive law formulated at Karlsruhe University by Gudehus (1996) and Bauer (1996) for monotonic deformation paths was used for FE calculations. This hypoplastic model describes the behaviour of so-called simple grain skeletons which are characterized by the following properties (Gudehus 1996):

- The state is fully defined through the skeleton pressure and void ratio (vanishing principal stresses or tensile stresses are not allowed),
- Deformation of the skeleton is due to grain rearrangements (e.g. small strains $<10^{-5}$ due to an elastic behaviour of grain contacts are assumed negligible),
- Grains are permanent (abrasion and crushing are excluded in order to keep the granulometric properties unchanged),
- Maximum, minimum and critical void ratios decrease exponentially with increasing pressure (Fig.3.1),
- The material manifests an asymptotic behaviour for homogeneous and proportional deformation, monotonic and cyclic shearing and proportional compression,
- Rate effects are negligible,
- Physico-chemical effects (capillary and osmotic pressure) and cementation of grain contacts are not taken into account.

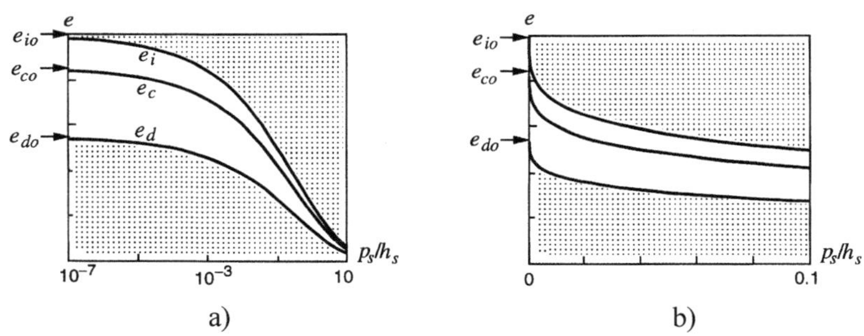

Fig. 3.1. Relationship between void ratios e_i, e_c and e_d and mean pressure p_s in a logarithmic (a) and linear (b) scale (gray zones denote inadmissible states) (Herle and Gudehus 1999)

A striking feature of hypoplasticity is that the stress rate is homogeneous of order 1 in the deformation rate. The hypoplastic constitutive laws are of the rate type. They are capable of describing a number of significant properties of granular materials: non-linear stress-strain relationship, dilatant and contractant behaviour, pressure dependence (barotropy), density dependence (pycnotropy) and material softening. A further feature of hypo-plastic models is the inclusion of critical states, i.e. states in which a grain aggregate can continuously be deformed at constant stress and constant volume. In contrast to elasto-plastic models, a decomposition of deformation components into elastic and plastic parts, the formulation of a yield surface, plastic potential, flow rule and hardening rule are not needed. Moreover, both the coaxiality (understood as a coincidence of the directions of the principal stresses and principal plastic strain increments) and stress-dilatancy rule are not assumed in advance. Although, the hypoplastic models are developed without recourse to concepts of the theory of plasticity, a failure surface, flow rule and plastic potential are obtained as natural outcomes (Chambon 1989, Wu and Niemunis 1996). The hallmark of these models is their simple formulation and procedure for determining material parameters with standard laboratory experiments. The material parameters are related to granulometric properties, viz. size distribution, shape, angularity and hardness of grains (Herle and Gudehus 1999). A further advantage lies in the fact that one single set of material parameters is valid for a wide range of pressures and densities. An exhaustive review of the development of hypoplasticity can be found in Wu and Kolymbas (2000), Tamagnini et al. (2000) and Gudehus (2007).

Stress changes due to the deformation of a granular body can be generally expressed by

$$\overset{o}{\sigma}_{ij} = F(e, \sigma_{kk}, d_{kl}), \tag{3.2}$$

wherein the Jaumann stress rate tensor (objective stress rate tensor) is defined by

$$\overset{o}{\sigma}_{ij} = \dot{\sigma}_{ij} - w_{ik}\,\sigma_{kj} + \sigma_{ik}\,w_{kj}. \tag{3.3}$$

F in Eq.3.2 represents an isotropic tensor-valued function depending on the current void ratio e, the Cauchy skeleton (effective) stress tensor σ_{ij} and the rate of deformations tensor (stretching tensor) d_{kl}. The rate of deformation tensor d_{ij} and the spin tensor w_{ij} are related to the material velocity v as follows:

$$d_{ij} = (v_{i,j} + v_{j,i})/2, \qquad w_{ij} = (v_{i,j} - v_{j,i})/2. \tag{3.4}$$

If the volume of grains remains constant (i.e. grains are incompressible), the rate of the void ratio can be expressed by the evolution equation:

$$\dot{e} = (1+e)d_{kk}. \tag{3.5}$$

For granular materials there does not exist an unique relationship between void ratio and pressure. It means that for the same pressure, a certain range of possible void ratios occurs which is bounded by a maximum e_i and a minimum void ratio e_d. One assumes that the values of the maximum void ratio e_i, minimum void ratio e_d and critical void ratio e_c decrease with increasing pressure σ_{kk} according to the equations (Bauer 1996), Fig.3.1:

$$e_i / e_{i0} = exp[-(-\sigma_{kk} / h_s)^n], \tag{3.6}$$

$$e_d / e_{d0} = exp[-(-\sigma_{kk} / h_s)^n], \tag{3.7}$$

$$e_c / e_{c0} = exp[-(-\sigma_{kk} / h_s)^n], \tag{3.8}$$

wherein e_{i0}, e_{d0} and e_{c0} are the values of e_i, e_d and e_c for $\sigma_{kk}=0$, respectively.

The condition of the incremental non-linearity in hypoplasticity requires that the tensorial function F in Eq.3.2 is not differentiable with respect to d_{ij} only for d_{ij}=0 (Wu 1992). Such requirement results in the following equation where the function F is decomposed into two parts

$$\overset{o}{\sigma}_{ij} = A(e,\sigma_{kl},d_{kl}) + B(e,\sigma_{ij})\,||\,d_{kl}\,||. \tag{3.9}$$

The first term in Eq,3.9 is linear in d_{kl}, while the second term in Eq.3.9 is non-linear in d_{kl} due to $||d_{kl}||$ denoting the Euclidian norm $\sqrt{d_{kl}d_{kl}}$. The stress rates in the first tern of Eq.3.9 show an incrementally linear behaviour, and the responses of the second tern are independent of the stretching. Both terms are positively homogeneous of the first degree in d_{kl} ($\overset{o}{\sigma}_{kl}(\lambda d_{kl}) = \lambda \overset{o}{\sigma}_{kl}\, d_{kl}$) for any scalar λ>0. In this way, Eq.3.9 becomes rate-independent. They are also homogeneous in σ_{kl} ($\overset{o}{\sigma}_{kl}(\lambda \overset{o}{\sigma}_{kl}) = \lambda^{m}\, \overset{o}{\sigma}_{kl}(\overset{o}{\sigma}_{kl})$) for any scalar λ (m denotes the order of homogeneity) to describe a granular property that proportional (i.e straight) strain paths lead to proportional stress paths (Goldscheider 1976, Gudehus 1996), Fig.3.2.

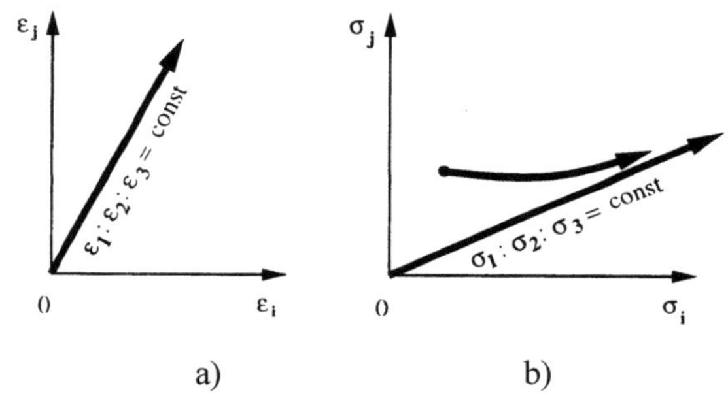

Fig. 3.2. Proportional (i.e. straight) strain paths (a) connected with proportional stress paths (b) (Kolymbas 2000)

The failure surface and flow rule emerge as *by-products* in hypoplasticity (they are not prescribed). They can be calculated from Eq.3.9 (Wu and Niemunis 1996, 1997) (Fig.3.3). There is no need to introduce different functions for loading and unloading.

The failure surface expressed by

$$\overset{o}{\sigma}_{ij} = A(e,\sigma_{kl},d_{kl}) + B(e,\sigma_{ij})\,||\,d_{kl}\,|| = 0 \tag{3.10}$$

can be obtained by rewriting first Eq.3.9 as

$$\overset{o}{\sigma}_{ij} = A'(e,\sigma_{kl})d_{kl} + B(e,\sigma_{ij})\,||\,d_{kl}\,|| = 0\,, \tag{3.11}$$

where A' is the tensor of fourth order and B is the tensor of second order. By defining the flow rule (specifying the direction of stretching) as

$$d_{kl} / \sqrt{d_{kl}d_{kl}} = -A'(e,\sigma_{kl})^{-1} B(e,\sigma_{ij}), \tag{3.12}$$

and by making use of the definition of the norm $\|d_{kl}\| = \sqrt{d_{kl}d_{kl}}$

$$d_{kl}d_{kl} / \| d_{kl} \|^2 = 1, \tag{3.13}$$

the failure surface is described by

$$f(\sigma_{ij}) = B^T(e,\sigma_{ij})(A'^{-1}(e,\sigma_{kl}))^T A'^{-1}(e,\sigma_{kl}) B(e,\sigma_{ij}) - 1 = 0. \tag{3.14}$$

The failure surface is a cone with its apex at the origin of the principle stress component space (σ_1, σ_2, σ_3) (Fig.3.3) (superscript T denotes a transposition). The derived flow rule is non-associated since in general $\partial f(\sigma_{ij})/\partial\sigma_{ij} \neq -A'^{-1}B$.

A response envelope (surface covered by all axisymmetric stress rates corresponding to different stretching rates of unit magnitude $\|d_{kl}\|=1$) introduced by Gudehus (1979) for a linear part of Eq.3.8 forms an ellipsoid with the centre at the origin of the space of stress rates. The cross-section is an ellipse (dashed response envelope $\dot{a}$ in Fig.3.4). The stress rates $\dot{b}$ (Fig.3.4) from a second non-linear part of Eq.3.9 have the same values irrespective of the direction of stretching. The sum of a linear and non-linear part $\dot{t} = \dot{a} + \dot{b}$ causes a shift of the dashed response envelope $\dot{a}$ by the constant value $\dot{b}$ with reference to the initial stress. The resulting response envelope is also elliptic (smooth and convex). The shape and orientation of the response envelope is thus determined by a linear term.

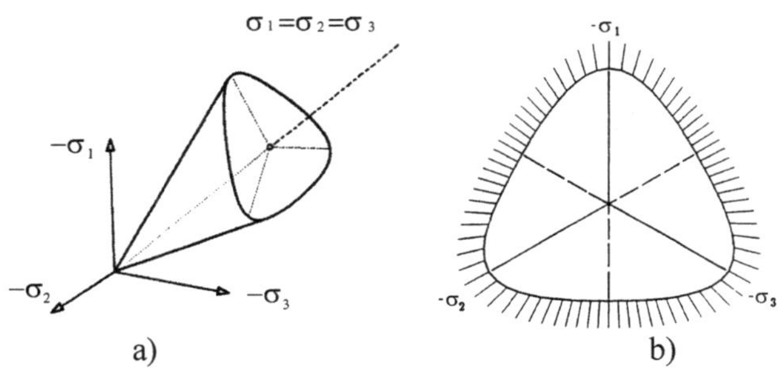

Fig. 3.3. Failure stress surface in the space of principle stresses (a) and failure stress surface and flow rule in the deviatoric plane (Wu and Niemunis 1997)

The possible response envelopes obtained with a hypoplastic constitutive law are depicted together with the failure surface in Fig.3.5. It turns out that the failure surface does not coincide with a bound of the accessible stress states (some stress paths may surpass the failure surface). The bound surface as an intrinsic property of a hypoplastic formulation was theoretically determined by Wu and Niemunis (1997) and is shown in Fig.3.6.

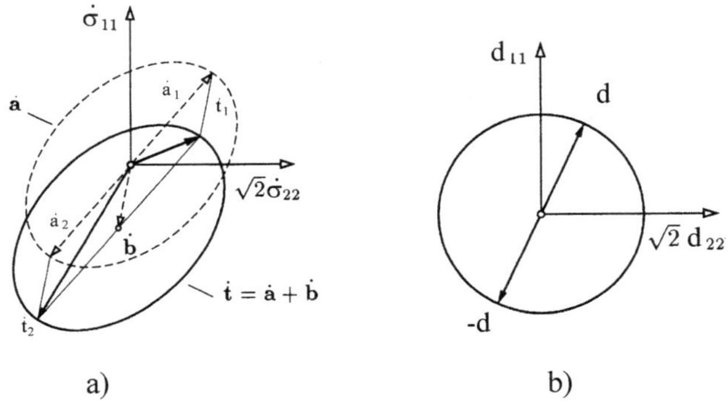

Fig. 3.4. Response envelopes (a) for different axially symmetric stretching units (b) (Bauer 1996)

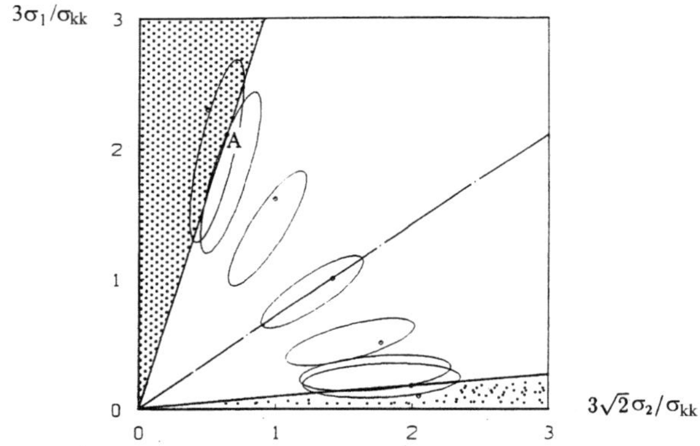

Fig. 3.5. Response envelopes of a hypoplastic constitutive equation (Wu and Niemunis 1997)

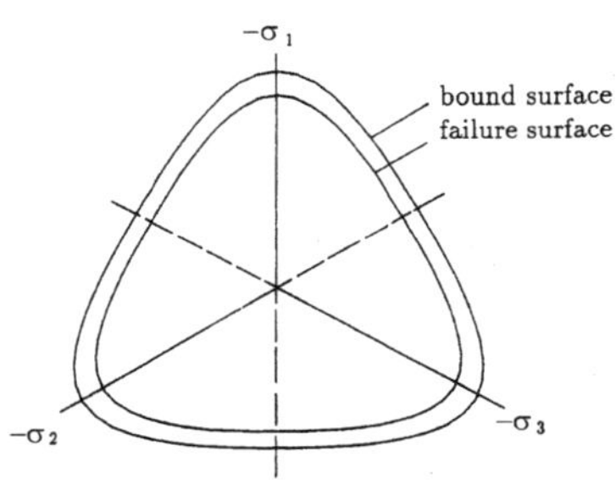

Fig. 3.6. Cross-section of the failure surface and bound surface in a deviatoric plane (Wu and Niemunis 1997)

Triaxial experiments (Fig.3.6) performed by Wu (1992) confirm qualitatively the property that some special stress paths (e.g. path $A \Rightarrow B$ in Fig.3.7a) can lie beyond the failure surface. In these experiments, the specimen was first brought to a hydrostatic stress σ_c=1.0 MPa and then loaded by increasing the axial stress $\sigma_1=\sigma_a$ while keeping the radial stress $\sigma_3=\sigma_r$ constant until failure was reached (point A in Fig.3.7). At the point A, the stress path was changed following the path $A \Rightarrow B$. Next the different stress paths were carried out until an increase of the stress ratio reached its maximum. Thus along the path $A \Rightarrow B$, the stress difference σ_1-σ_3 decreased monotonically (Fig.3.7.b) whereas the stress ratio σ_1/σ_3 increased up to the peak and then decreased (Fig.3.7c).

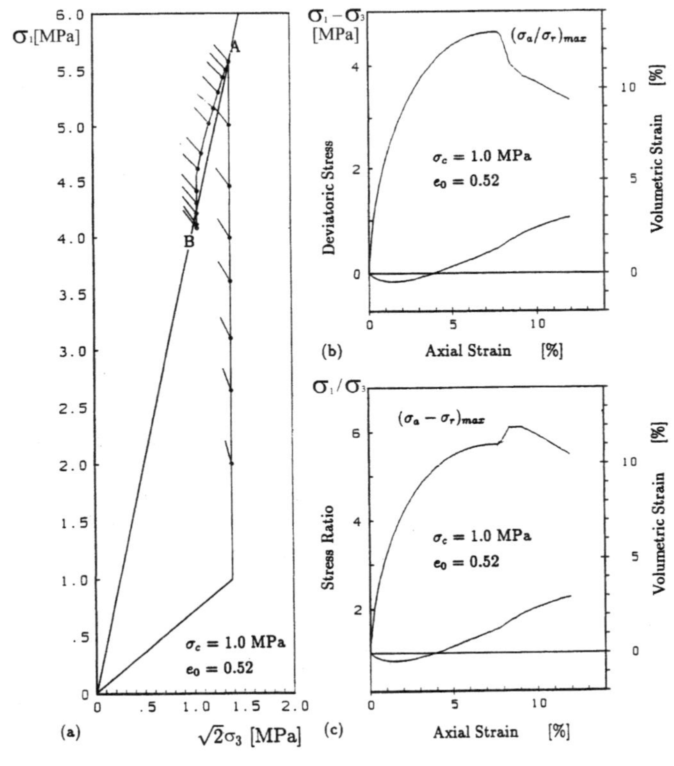

Fig. 3.7. Triaxial compression test: a) stress path, b) stress difference versus axial stress, c) stress ratio versus axial stress (σ_i – principle stresses) (Wu 1992)

The following representation of the general constitutive equation is used (Gudehus 1996, Bauer 1996):

$$\overset{o}{\sigma}_{ij} = f_s \left[L_{ij}(\hat{\sigma}_{kl}, d_{kl}) + f_d N_{ij}(\hat{\sigma}_{ij}) \sqrt{d_{kl} d_{kl}} \right], \tag{3.15}$$

wherein the normalized stress tensor $\hat{\sigma}_{ij}$ is defined by

$$\hat{\sigma}_{ij} = \frac{\sigma_{ij}}{\sigma_{kk}} . \tag{3.16}$$

The tensorial functions L_{ij} and N_{ij} are:

$$L_{ij} = a_1^2 d_{ij} + \hat{\sigma}_{ij} (\hat{\sigma}_{kl} \, d_{kl}) , \qquad N_{ij} = a_1 (\hat{\sigma}_{ij} + \hat{\sigma}_{ij}^*), \tag{3.17}$$

wherein

$$\hat{\sigma}_{ij}^* = \hat{\sigma}_{ij} - \frac{1}{3}\delta_{ij} . \tag{3.18}$$

is the deviator of $\hat{\sigma}_{ij}$ and the dimensionless and positive scalar a_1 is related to stationary states which fulfill the condition for a simultaneous vanishing of the stress rate and volume strain rate independent of the direction of the deviatoric stress.

The scalar factors $f_s = f_s\ (e,\ \sigma_{kk})$ and $f_d = f_d\ (e,\ \sigma_{kk})$ in Eq.3.15 take into account the influence of the density and pressure level on the stress. The density factor f_d resembles a pressure-dependent relative density index and is represented by

$$f_d = (\frac{e - e_c}{e_c - e_c})^{\alpha} , \tag{3.19}$$

where $\propto$ is the material parameter. In a critical state, the density factor is independent of the initial void ratio and pressure (i.e. it is reached independent of the initial void ratio and pressure) (Fig.3.8).

The stiffness factor f_s can be decomposed in three factors:

$$f_s = f_e f_s^* f_b . \tag{3.20}$$

The factor f_e takes into account the fact that the stiffness increases with decreasing initial void ratio

$$f_e = (\frac{e_i}{e})^{\beta} , \tag{3.21}$$

where β is a material parameter.

The factor f_s^* takes into account a decrease of the stiffness with an increase of the stress and an effect of the deviatoric stress direction

$$f_s^* = (1 + \sqrt{\hat{\sigma}_{kl}^* \, \hat{\sigma}_{lk}^*} \, [c_1 + c_2 \cos(3\theta)])^{-1} , \tag{3.22}$$

where θ is the Lode angle (the angle on the deviatoric plane between the stress vector and the axis of the principle stress σ_3).

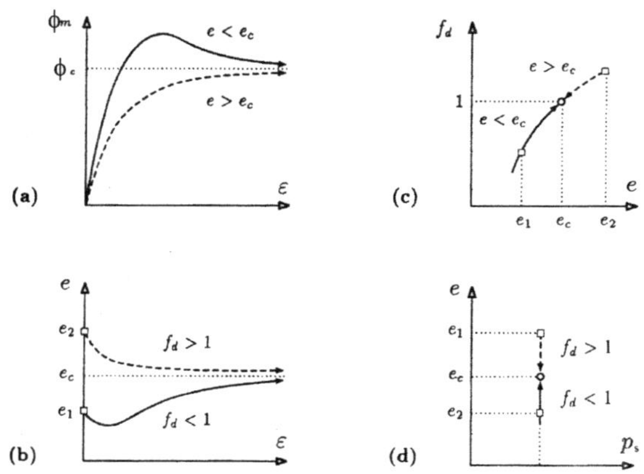

Fig. 3.8. Effect of two different initial void ratios (e_1 and e_2) during monotonic shearing on the mobilized friction angle ϕ_m versus shear strain ε (a), void ratio e versus shear strain ε (b), density factor f_d versus void ratio e (c) and void ratio e versus pressure p_s (d) (Bauer 2000)

$$\cos(3\theta) = -\sqrt{6}\,\frac{(\hat{\sigma}^*_{kl}\,\hat{\sigma}^*_{lm}\,\hat{\sigma}^*_{mk})}{[\hat{\sigma}^*_{kl}\,\hat{\sigma}^*_{kl}]^{1.5}}, \tag{3.23}$$

and the factors c_1 and c_2 are constants.

The pressure dependent factor f_b is

$$f_b = \frac{h_s}{nh_i}\left(\frac{1+e_i}{e_i}\right)\left(-\frac{\sigma_{kk}}{h_s}\right)^{1-n}, \tag{3.24}$$

with

$$h_i = \frac{1}{c_1^2} + \frac{1}{3} - \left(\frac{e_{i0}-e_{d0}}{e_{c0}-e_{d0}}\right)^{\alpha}\frac{1}{c_1\sqrt{3}}. \tag{3.25}$$

The factor f_b is obtained from Eq.3.6 and Eq.3.15 which must coincide for isotropic compression. Thus, the stiffness factor f_s is:

$$f_s = \frac{h_s}{nh_i}\left(\frac{1+e_i}{e_i}\right)\left(\frac{e_i}{e}\right)^{\beta}\left(-\frac{\sigma_{kk}}{h_s}\right)^{1-n}. \tag{3.26}$$

The granulate hardness, h_s, represents a density-independent reference pressure (intrinsic scale of stress) and is related to the entire skeleton (not to single grains).

In the residual (critical) state (at large deformation), the stress rate and volume strain rate vanish ($\overset{o}{\sigma}_{ij} = 0$, $e=e_c$, $f_d=1$) and Eqs.3.15-3.17 reduce to

$$a_1^2 d_{ij} + \hat{\sigma}_{ij}(\hat{\sigma}_{kl}\, d_{kl}) + a_1(\hat{\sigma}_{ij} + \hat{\sigma}^*_{ij})\sqrt{d_{kl}d_{kl}} = 0, \tag{3.27}$$

leading next to an explicit expression for the stationary stress surface

$$a_1 = \sqrt{\hat{\sigma}^*_{kl}\, \hat{\sigma}^*_{lk}}\ , \tag{3.28}$$

which is a cone with its apex at the origin of the principle stress component space (Bauer 1996), Fig.3.9. The parameter a_1 is, thus, the radius of the stress points at the critical state (Fig.3.9). Due to the fact, that the behaviour of granular materials depends upon the direction of the deviatoric stress, the parameter a_1 is described more realistically (with respect to Eq.3.28) by the function proposed by Bauer (1996)

$$a_1^{-1} = c_1 + c_2 \sqrt{\hat{\sigma}^*_{kl}\, \hat{\sigma}^*_{lk}}\ [1 + cos(3\theta)]\ . \tag{3.29}$$

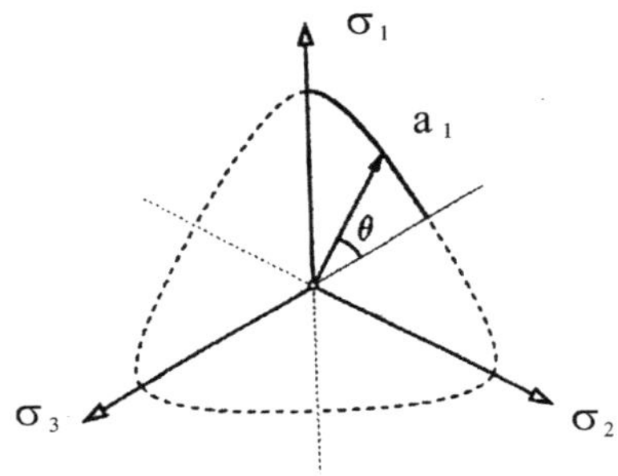

Fig. 3.9. Stationary stress surface in the deviatoric plane (π-plane) (Bauer 2000a)

The constants c_1 and c_2 are calculated under the assumption that the critical friction angle ϕ_c is the same during triaxial compression and extension

$$c_1 = \sqrt{\frac{3}{8}}\frac{(3 - sin\,\phi_c)}{sin\,\phi_c}, \qquad c_2 = \frac{3}{8}\frac{(3 + sin\,\phi_c)}{sin\,\phi_c}. \tag{3.30}$$

For ϕ_c=30°, the parameters c_i are: c_1=3.1 and c_2=2.6. For an isotropic stress state with $\sigma^*_{ij}=0$ and $cos(3\theta)=0$, the parameter $a_1^{-1}=c_1$ in Eq.3.29. If the parameter a_1 is constant and equal to

$$a_1 = \sqrt{\frac{8}{3}}\left(\frac{sin\,\phi_c}{3 - sin\,\phi_c}\right), \tag{3.31}$$

the limit condition in the deviatoric plane is a circle similar to the Drucker-Prager condition (Fig.3.10).

Figs.3.11-3.16 show the results of homogeneous element deformation tests under drained and undrained conditions (oedometric compression and extension, triaxial compression, simple shearing) (Bauer 2000b).

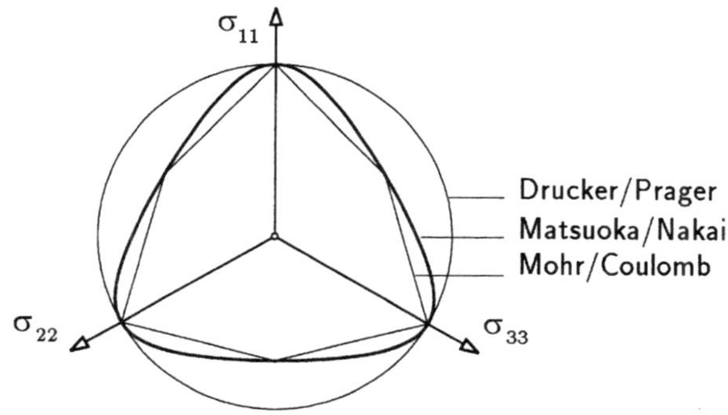

Fig. 3.10. Limit conditions in the deviatoric plane (π- plane) (Bauer 2000a)

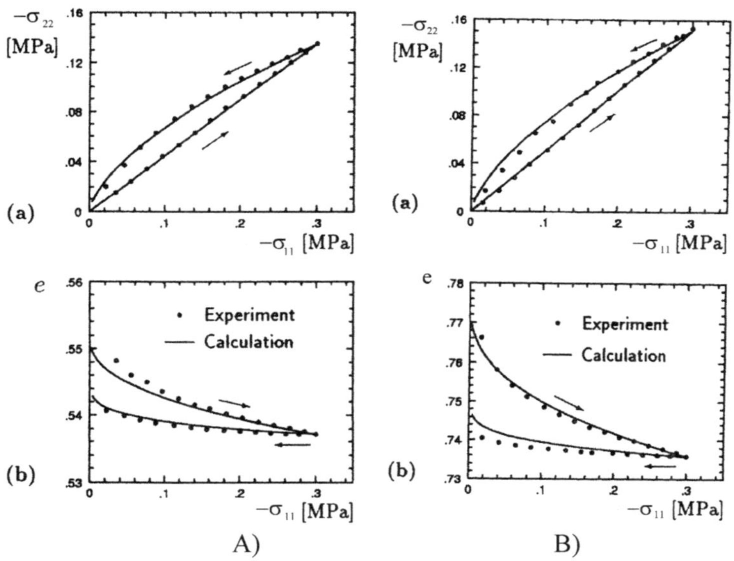

Fig. 3.11. Element tests: oedometric compression and extension of dense sand: A) e_0=0.55, B) e_0=0.77, a) lateral stress versus axial stress, b) void ratio versus axial stress (Bauer 2000b)

Another version of a hypoplastic constitutive law was proposed by von Wolffersdorff (1996) where the critical stress state fulfils the yield condition following Matsuoka and Nakai (1977) (Fig.3.10). The law has the following form:

$$\overset{o}{\sigma}_{ij} = f_s \frac{1}{\sqrt{\hat{\sigma}^*_{kl}\,\hat{\sigma}^*_{lk}}}[\,F^2 d_{ij} + a^2\,\hat{\sigma}_{ij}\,\hat{\sigma}_{kl}\,d_{kl} + f_d\,Fa(\,\hat{\sigma}_{ij} + \hat{\sigma}^*_{ij}\,)\sqrt{d_{kl}d_{kl}}\,]\,, \tag{3.32}$$

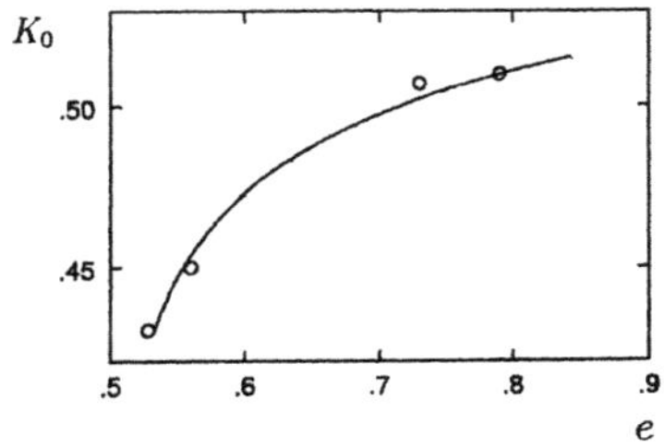

Fig. 3.12. Relationship between stress ratio $K_0=\sigma_{11}/\sigma_{22}$ and initial void ratio (element tests, oedometric compression) (Bauer 2000b)

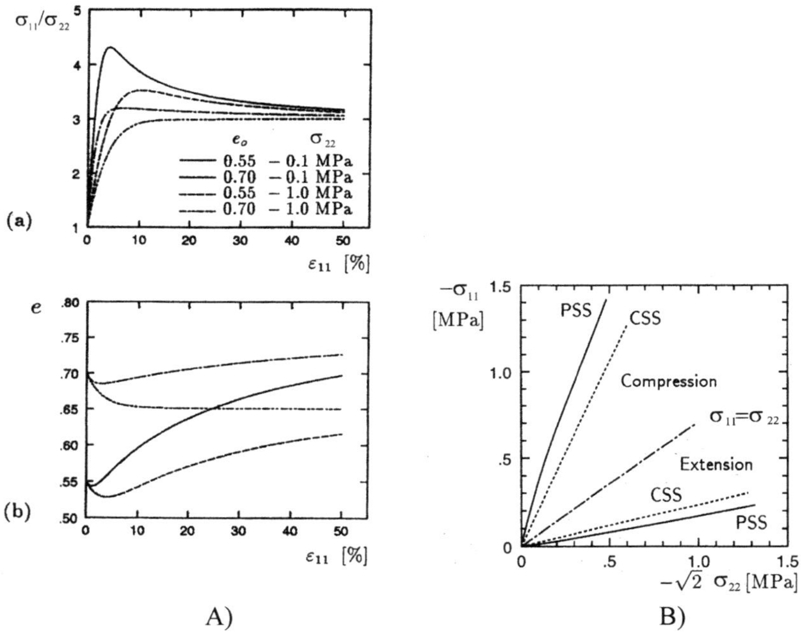

Fig. 3.13. Element tests: triaxial compression under drained conditions for various initial void ratios and lateral pressures: A) stress ratio versus axial stress (a) and void ratio versus axial stress (b), B) peak stress states (PSS) and critical stress states (CSS) in the Rendulic-plane (Bauer 2000b)

$$a = \frac{\sqrt{3}(3-\sin\phi_c)}{2\sqrt{2}\sin\phi_c}, \tag{3.33}$$

$$F = \sqrt{\frac{1}{8}\tan^2\psi + \frac{2-\tan^2\psi}{2+\sqrt{2}\tan\psi\cos 3\theta}} - \frac{1}{2\sqrt{2}}\tan\psi, \tag{3.34}$$

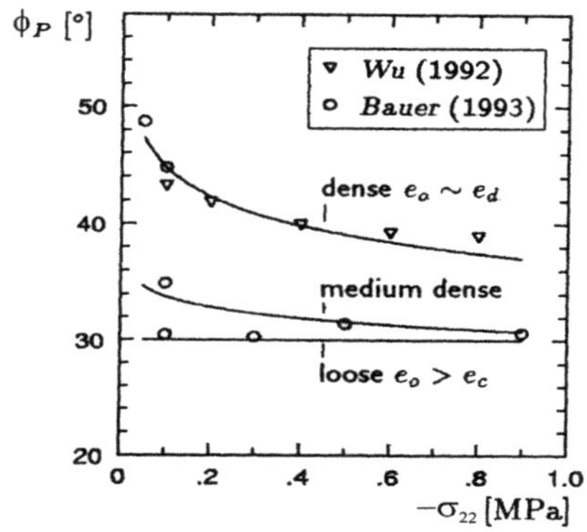

Fig. 3.14. Peak friction angle versus lateral pressure for various initial void ratios during triaxial compression: element test simulations (full lines), experiments (o, Δ) (Bauer 2000b)

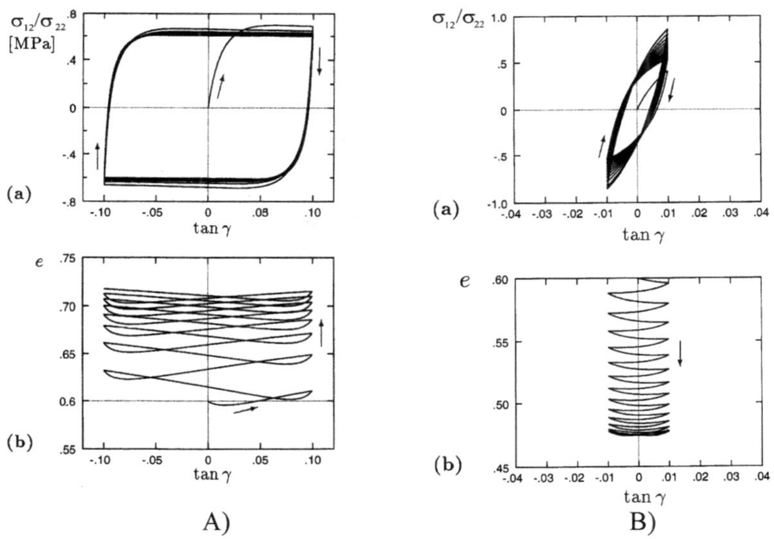

Fig. 3.15. Element tests: cyclic simple shearing (e_o=0.60, σ_{22}=-0.1 MPa) with a large ($\tan\gamma$=±0.1) (A) and small ($\tan\gamma$=±0.01) (B) shear amplitude: a) shear stress ratio versus shear angle, b) void ratio versus shear angle (Bauer 2000b)

$$\tan\psi = \sqrt{3}b_1, \qquad b_1 = \sqrt{\hat{\sigma}^*_{kl}\,\hat{\sigma}^*_{lk}}\ , \tag{3.35}$$

$$f_s = \frac{h_s}{nh_i}\left(\frac{1+e_i}{e_i}\right)\left(\frac{e_i}{e}\right)^{\beta}\left(-\frac{\sigma_{kk}}{h_s}\right)^{1-n}\left[3+a^2-a\,\sqrt{3}\left(\frac{e_{i0}-e_{d0}}{e_{c0}-e_{d0}}\right)^{\alpha}\right]^{-1}, \tag{3.36}$$

$$Fa = \frac{\sin\phi_c}{3-\sin\phi_c}[\, b_1 - \sqrt{\frac{(8/3)-3b_1^2+(\sqrt{6}/2)b_1^3\cos(3\theta)}{1+\sqrt{1.5}b_1\cos(3\theta)}}\,, \tag{3.37}$$

The remaining formulas to calculate the void ratios and density factor are the same as in the law by Gudehus (1996) and Bauer (1996). The parameter F is equal to 1 for triaxial compression.

To increase the application range, a hypoplastic constitutive law has been extended for an elastic strain range (Niemunis and Herle 1997, Niemunis et al. 2005), anisotropy (Wu 1998, Bauer et al. 2004) and for viscosity (Niemunis 2003, Gudehus 2006, Wu 2006). It can be also used for soils with low friction angles (Herle and Kolymbas 2004) and clays (Masin 2005, Huang et al. 2006, Weifner and Kolymbas 2007, 2008, Masin and Herle 2007).

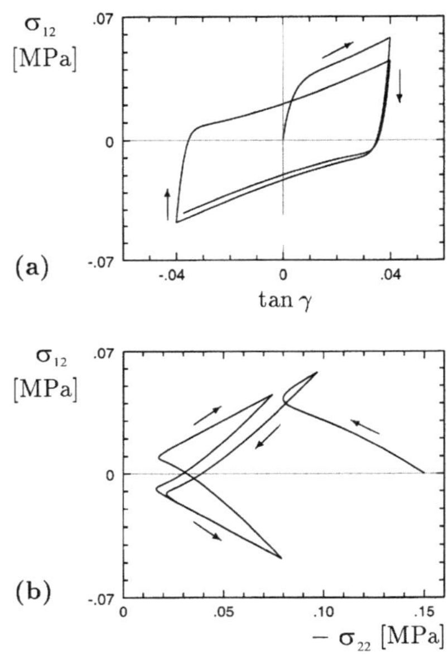

Fig. 3.16. Element tests: cyclic simple shearing (e_o=0.65, $\sigma_{11}=\sigma_{22}=\sigma_{33}$=-0.15 MPa) for a water-saturated specimen without drainage: a) shear stress versus shear angle, b) shear stress versus normal stress (Bauer 2000b)

3.2 Calibration of Hypoplastic Material Parameters

The calibration procedure for sands and gravels (0.1 mm≤d_{50}≤2.0 mm, 1.4≤C_u≤7.2, $C_u=d_{60}/d_{10}$ - non-uniformity coefficient) was given by Herle (1997, 1998, 2000) and Herle and Gudehus (1999). The material parameters depend on granulometric properties involving grain shape and angularity, distribution of grain size represented by a mean grain diameter and the non-uniformity coefficient and grain hardness. There are

8 material parameters in the hypoplastic model: ϕ_c, h_s, n, e_{i0}, e_{d0}, e_{c0}, β and α. They are valid in a pressure range 1 kPa$<p_s=-\sigma_{kk}/3<$1000 kPa. Below it, additional capillary forces due to the air humidity and van der Waals forces may become important, and above it, grain crushing.

Critical (residual) friction angle ϕ_c

The critical state is reached during large monotonic shearing if both the stress rate state and volumetric deformation rate vanish. The critical friction angle ϕ_c can be obtained from the angle of repose of a dry granular material from cone pluviation tests (Fig.3.17) (segregation effects can be neglected). In a heap, a critical state is reached due to large deformation. The values of ϕ_c obtained in this way correspond well to results of different shear tests. The critical friction angle is more influenced by a mean grain diameter d_{50} than by a non-uniformity coefficient C_u (Fig.3.18). It increases with increasing grain size due to an increase of the rolling resistance (a small increase for $d_{50}<0.2$ mm may be due to capillarity resulting from air humidity). If the portion of grains with a diameter $d<0.1$ mm is too large, conventional shear tests are recommended for the determination of ϕ_c.

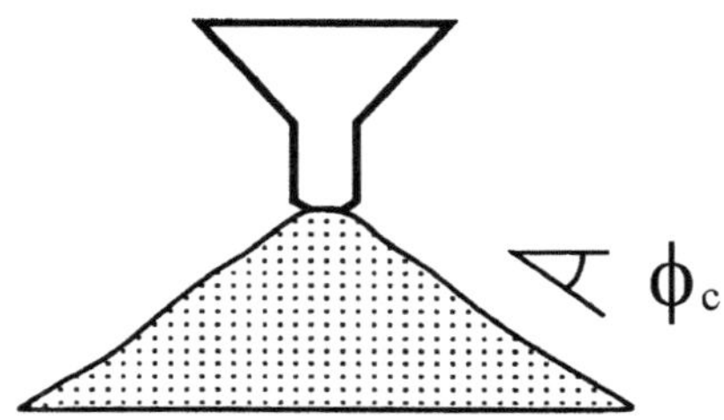

Fig. 3.17. Angle of repose (Herle and Gudehus 1999)

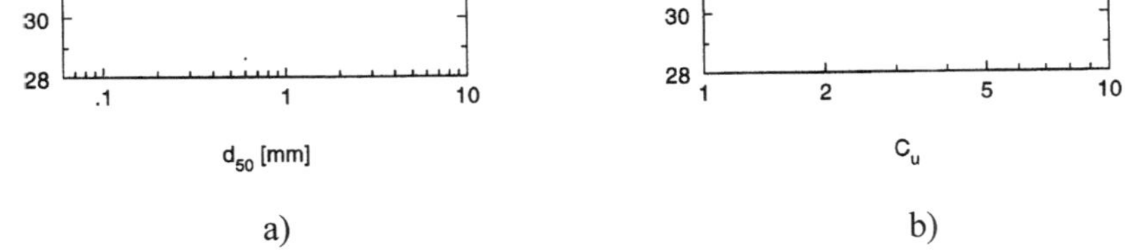

Fig. 3.18. Relationship between the angle of repose ϕ_c and mean grain diameter d_{50} (a) and between the angle of repose ϕ_c and non-uniformity coefficient C_u (b) (Herle and Gudehus 1999)

Granulate hardness h_s and exponent n

The parameters h_s and n (Eq.3.26) are estimated from a single oedometric compression test (or from an isotropic compression test) with an initially loose specimen. The parameter h_s (scaling factor for pressure p_s) reflects the slope of the curve in a semi-logarithmic representation, and the parameter n its curvature (Fig.3.19).

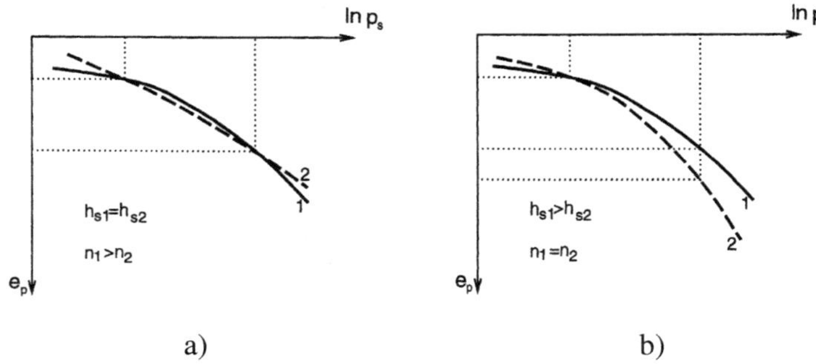

Fig. 3.19. Influence of material constants n and h_s on compression curves for two different granulates: a) $n_1>n_2$, b) $h_{s1}>h_{s2}$

The exponent n can be calculated as:

$$n = \frac{ln(\frac{e_1 C_{c2}}{e_2 C_{c1}})}{ln\frac{p_{s1}}{p_{s2}}}, \tag{3.38}$$

where $p_s=-\sigma_{kk}/3$ and the coefficient C_c is the compression index defined as (Fig.3.20)

$$C_c = \frac{\Delta e}{\Delta\, ln\frac{p_s}{p_{so}}}. \tag{3.39}$$

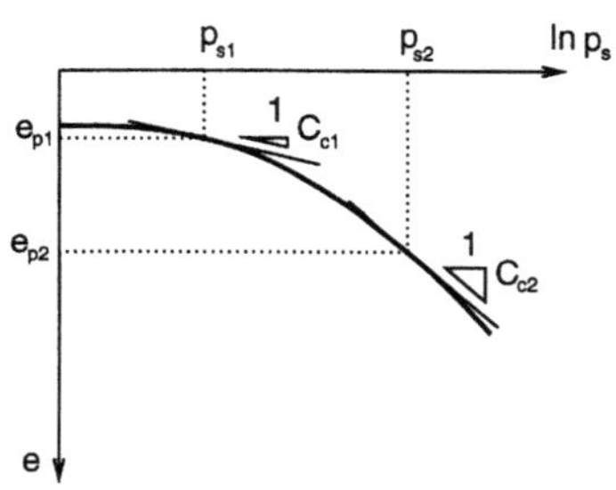

Fig. 3.20. Determination of the parameter n from the measured values on boundaries of the investigated pressure range ($p_s=-\sigma_{kk}/3$)

The parameter h_s can be obtained with the help of Eq.3.38 from:

$$h_s = 3 p_s \left(\frac{ne}{C_c} \right)^{1/n} . \tag{3.40}$$

The parameter n increases with decreasing C_u, increasing angularity and increasing d_{50}. The value of the parameter n=0.666 can be considered as an upper bound corresponding to a regular packing of elastic spheres. The lower bound of $n \to 0$ ($C_c \approx 0$) (n=0.10) represents a straight compression line in a semi-logarithmic plot. If grain crushing takes place, the values of h_s and n can significantly change.

Minimum void ratio e_{d0} at zero pressure

The minimum void ratio e_{d0} at zero pressure can be calculated from Eq.3.7

$$e_{d0} = e_c \exp\left[\left(\frac{3 p_s}{h_s} \right)^n \right] , \tag{3.41}$$

if the values of e_d, h_s and n are known. The best densification of a granular material can be reached by means of cyclic shearing with a small amplitude under constant pressure. In a practice, a minimum void ratio e_{min} is determined from index tests (ASTM D 1253, DIN18126). The densification methods proposed in standards are performed at a certain pressure level but they are not so effective as cyclic shearing. Therefore the measured value of e_{min} lies slightly above e_{d0}. However, one can assume that $e_{d0} \cong e_{min}$. The values of e_{d0} depend on non-uniformity and grain shape (Fig.3.21).

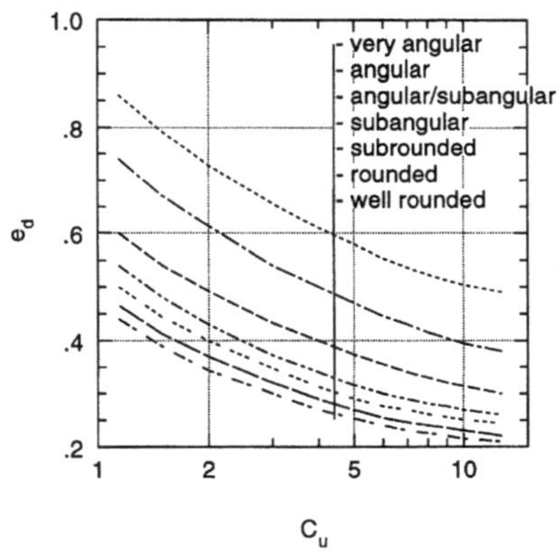

Fig. 3.21. Relationship between the minimum void ratio e_d at pressure p_s=55 kPa and non-uniformity coefficient C_u and grain angularity (Youd 1973)

Maximum void ratio e_{i0} at zero pressure

The maximum void ratio at zero pressure e_{i0} can be imagined as the maximum void ratio which is reached during an isotropic consolidation of a grain suspension in a gravity-free space. Thus, it is impossible to determine experimentally the value of e_{i0}. There are several proposals how to determine a maximum void ratio e_{max} in the

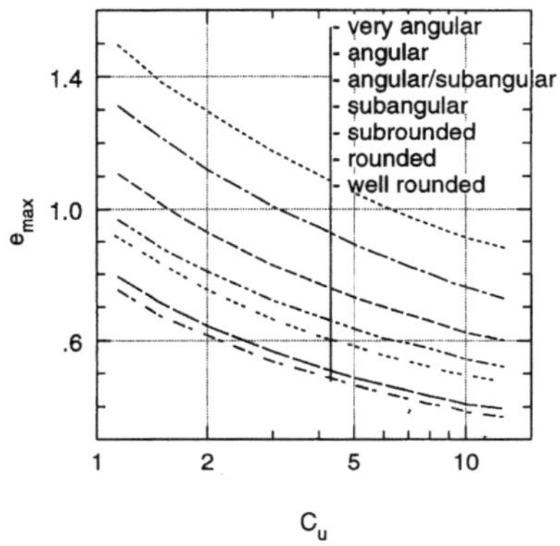

Fig. 3.22. Relationship between the minimum void ratio e_{max} and non-uniformity coefficient C_u and grain angularity (Youd 1973)

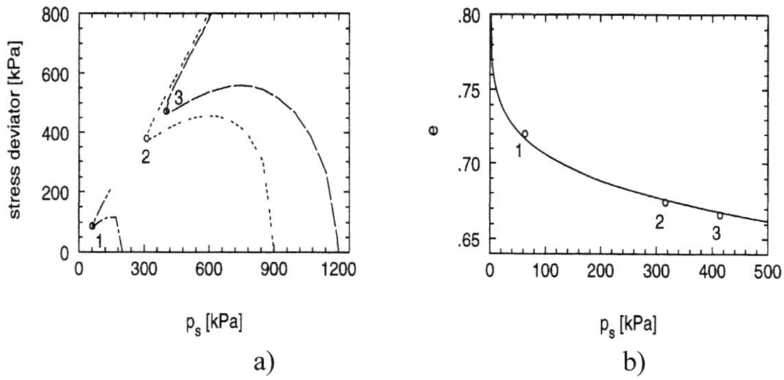

Fig. 3.23. Determination of e_c from a quasi-steady state in triaxial undrained tests: a) stress deviator versus mean pressure, b) void ratio versus mean pressure (Herle 1997)

laboratory (ASTM D 1251, DIN18126). A comparison of experimental values of e_{max} and theoretical values of e_{io} (obtained with spheres and cubes) yields the ratio $e_{i0}/e_{max}\approx 1.2$ in the case of identical spheres and $e_{i0}/e_{max}\approx 1.3$ in the case of identical cubes, respectively. The values of e_{max} depend both on non-uniformity and grain shape (Fig.3.22).

Critical void ratio e_{c0} at zero pressure

The critical void ratio e_{c0} at zero pressure can be calculated analogously to e_{d0} from Eq.3.8

$$e_{c0} = e_c \, exp[(\frac{3p_s}{h_s})^n] \,. \tag{3.42}$$

A critical state in undrained triaxial tests is considered to be sufficient for the determination of e_{c0} (Fig.3.23). The value of e_{c0} corresponds well to the value e_{max} from index tests due to the fact that during an e_{max}-test, a state close to the critical one is reached (large deformations during a steady stress state close to zero pressure).

Parameters α and β

The parameters α and β are found from a triaxial test with a dense specimen (they reflect the height and position of the peak value of the stress-strain curve). During shearing of an initially dense grain skeleton at constant mean pressure, a peak value of the friction angle ϕ_p and later a residual (critical) value ϕ_c can be observed. The peak value decreases with increasing C_u, grain angularity and with decreasing d_{50}. The difference between the peak and residual value increases with a decrease of the pressure-dependent relative void ratio

$$r_e = (e - e_d)/(e_c - e_d)\,, \tag{3.43}$$

and is controlled by the exponent α which is nearly proportional to both friction angles. Fig.3.24 shows the relationship between α, ϕ_p and ϕ_c for two different r_e-values.

The parameter β is important if the void ratio e is substantially lower than e_i (dense granulate). It takes into account the fact that the incremental stiffness increases with a decrease of void ratio for the same stress state. It can be calculated from the ratio of the incremental stiffness moduli $E = \dot{\sigma}_{11}/d_{11}$ at the same mean pressures but at two different void ratios $e_1 = e_{loose}$ and $e_2 = e_{dense}$ during isotropic compression:

$$\beta = \frac{ln(\beta_0 \frac{E_2}{E_1})}{ln(\frac{e_1}{e_2})} \tag{3.44}$$

with

$$\beta_o = \frac{3 + a_1^2 - a_1\sqrt{3} f_{d1}}{3 + a_1^2 - a_1\sqrt{3} f_{d2}}\,. \tag{3.45}$$

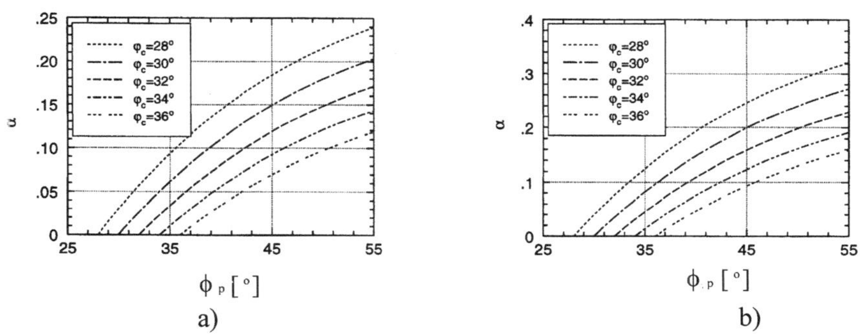

Fig. 3.24. Relationship between the parameter α and friction angles ϕ_p and ϕ_c for two different r_e-values of Eq.3.43: a) r_e=0.2, b) r_e=0.3 (Herle and Gudehus 1999)

Usually, the stiffness modulus E increases proportionally with decreasing e (thus, $\beta \cong 1$ for many sands). The effect of the parameter β on the factor f_s (Eq.3.26) is rather low.

In this book, the FE analyses were carried out mainly with so-called Karlsruhe sand using a set of material parameters proposed by Bauer (1996): e_{i0}=1.3, e_{d0}=0.51, e_{c0}=0.82, ϕ_c=30°, h_s=190 MPa, n=0.5, β=1.0 and α=0.25-0.30. These parameters (in particular h_s and n) differ from the hypoplastic ones given by Herle and Gudehus (1999) (Tab.3.1) due to an assumption of a larger pressure range during oedometric compression (Fig.3.25). The material parameters for well-graded granular materials were given by Rondon et al. (2007).

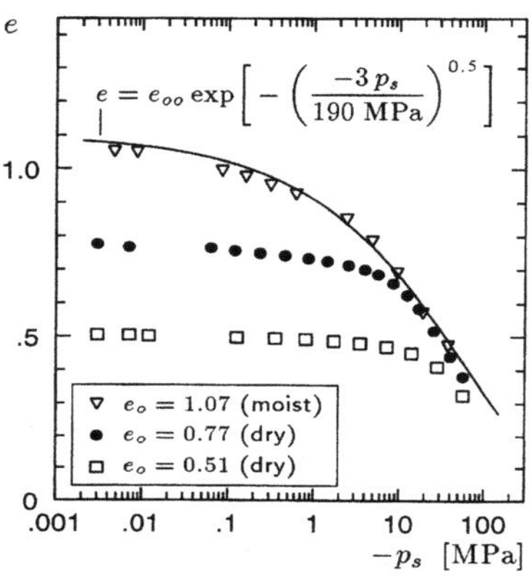

Fig. 3.25. Oedometric compression behaviour of medium quartz sand (experiments for different initial void ratios and fitting curve) (e – void ratio, p_s – mean pressure) (Bauer 1996, 2000a)

Table 3.1. Hypoplastic material parameters for different sands (Herle and Gudehus 1999)

Material	ϕ_c [°]	h_s [MPa]	n	e_{d0}	e_{c0}	e_{c0}	α	β
Hochstetten gravel	36	32000	0.18	0.26	0.45	0.50	0.10	1.9
Hochstetten sand	33	1500	0.28	0.55	0.95	1.05	0.25	1.0
Hostun sand	32	1000	0.29	0.61	0.96	1.09	0.13	2.0
Karlsruhe sand	30	5800	0.28	0.53	0.84	1.0	0.13	1.0
Lausitz sand	33	1600	0.19	0.44	0.85	1.0	0.25	1.0
Toyura sand	30	2600	0.27	0.61	0.98	1.1	0.18	1.1

3.3 Micro-polar Continuum

A micro-polar continuum which is a continuous collection of particles behaving like rigid bodies combines two kinds of deformations at two different levels, viz: micro-rotation at the particle level and macro-deformation at the structural level. Thus, a micro-polar (Cosserat) continuum differs from a non-polar one by the presence of additional independent rotations (Cosserat and Cosserat 1909, Günther 1958, Schäfer 1962, 1967 Kessel 1964, Mindlin 1964, Neuber 1966, Bogdanova-Bontscheva and Lippmann 1975, Becker and Lippmann 1977, Kanatani 1979, Mühlhaus 1987, 1990). For plane strain or axial-symmetry, each material point has three degrees of freedom: two translations u_1 and u_2, and one rotation ω^c (Fig.3.26). The rotation ω^c is related with the micro-rotation of the micro-elements and is not determined from displacements as in a non-polar continuum

$$\omega_{ij} = 0.5(u_{i,j} - u_{j,i}). \tag{3.46}$$

Deformation is described by the following six quantities (which are considered here as small increments):

$$\varepsilon_{11} = u_{1,1}, \qquad \varepsilon_{22} = u_{2,2}, \tag{3.47}$$

$$\varepsilon_{12} = u_{1,2} + \omega^c, \qquad \varepsilon_{21} = u_{2,1} - \omega^c, \tag{3.48}$$

$$\kappa_1 = \omega^c_{,1}, \qquad \kappa_2 = \omega^c_{,2}. \tag{3.49}$$

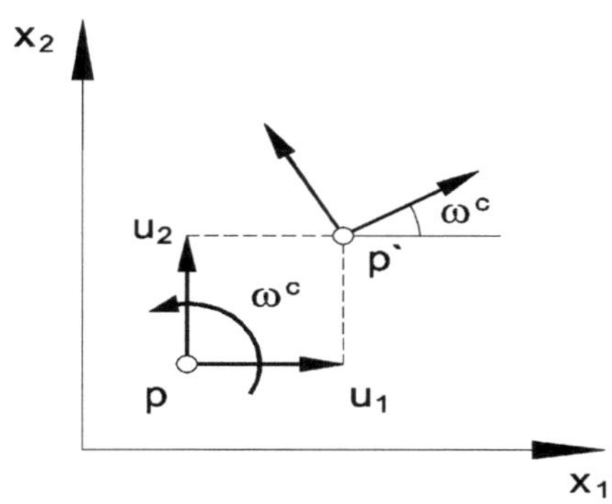

Fig. 3.26. Degrees of freedom in a plane Cosserat continuum: u_1 - horizontal displacement, u_2 - vertical displacement, ω^c - Cosserat rotation, P –material point

ε_{ij} are components of the deformation tensor and κ_i are components of the curvature vector. The extensions ε_{11} and ε_{22} are defined similarly as in a non-polar continuum. In turn, the deformations ε_{12} and ε_{21} can be viewed as a relative deformation relating the displacement gradient and the micro-rotation (Fig.3.27); in contrast to a non-polar continuum ε_{12} is not generally equal to ε_{21}. The curvatures κ_1

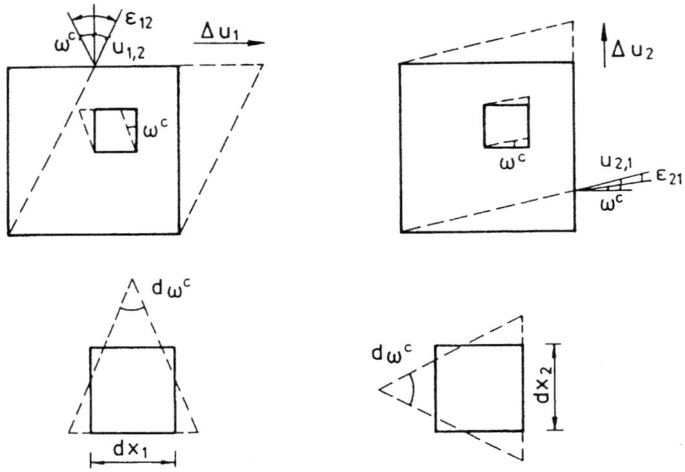

Fig. 3.27. Shear deformations ε_{12} and ε_{21} (a) and curvatures κ_1 and κ_2 (b) in plane Cosserat continuum

and κ_2 describe the gradients of the micro-rotation. ε_{ij} and κ_i are invariant with respect to rigid body motions (Günther 1958, Mühlhaus 1989a, 1989b).

The deformation tensor ε_{ij} can be decomposed into a symmetric part E_{ij} and a skew symmetric part W_{ij}-W_{ij}^c:

$$\varepsilon_{ij} = E_{ij} + W_{ij} - W_{ij}^c , \tag{3.50}$$

where

$$E_{ij} = 0.5(u_{i,j} + u_{j,i}), \tag{3.51}$$

$$W_{ij} = \omega_{ij} = 0.5(u_{i,j} - u_{j,i}), \tag{3.52}$$

$$W_{12}^c = -\omega^c , \qquad W_{21}^c = \omega^c . \tag{3.53}$$

The quantities E_{ij} and W_{ij} denote the symmetric and skew symmetric part of the displacement gradient, respectively, and W_{ij}^c denotes the skew symmetric tensor corresponding to the Cosserat rotation ω^c. E_{ij} is the deformation tensor and W_{ij} is the rotation tensor which are characteristic of a non-polar continuum. The skew symmetric part W_{ij}-W_{ij}^c describes the difference between the macro- and micro-rotation. If $W_{ij}=W_{ij}^c$, ε_{ij} reduces to E_{ij} and the kinematics of a non-polar continuum is retrieved. If $W_{ij}<W_{ij}^c$, an overall negative (clockwise) Cosserat rotation emerges.

The conditions for compatibility of deformations and curvatures are (Günther 1958):

$$\varepsilon_{11,2} - \varepsilon_{12,1} + \kappa_1 = 0 , \tag{3.54}$$

$$\varepsilon_{22,1} - \varepsilon_{21,2} - \kappa_2 = 0\,, \tag{3.55}$$

$$\kappa_{1,2} - \kappa_{2,1} = 0\,. \tag{3.56}$$

The six deformation quantities are energy-conjugate with the six stress quantities. Four components of ε_{ij} are associated with four components of the stress tensor σ_{ij} (plane strain) or five components of the stress tensor σ_{ij} (axisymmetric case) which is now in general non-symmetric. The curvatures κ_i are associated with couple stresses m_i. Figs.3.28 and 3.29 show stresses σ_{ij} and couple stresses m_i at an infinitesimal plane strain and axisymmetric element.

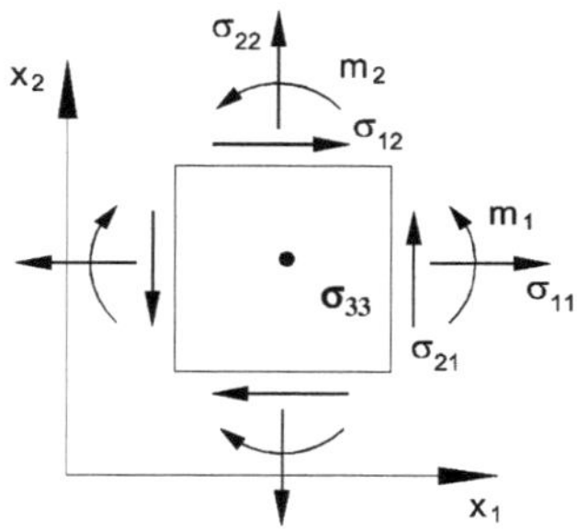

Fig. 3.28. Plane Cosserat continuum: stresses σ_{ij} and couple stresses m_i at an element (without volume body forces and body moment)

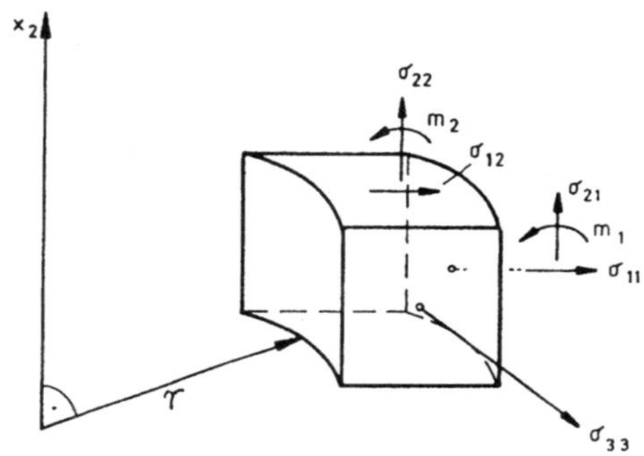

Fig. 3.29. Axisymmetric Cosserat continuum: stresses σ_{ij} and couple stresses m_i at an element (without volume body forces and body moment), r – distance from the symmetry axis

Force and moment equilibrium require for plane strain

$$\sigma_{11,1} + \sigma_{12,2} - f_1^B = 0, \tag{3.57}$$

$$\sigma_{21,1} + \sigma_{22,2} - f_2^B = 0, \tag{3.58}$$

$$m_{1,1} + m_{2,2} + \sigma_{21} - \sigma_{12} - m^B = 0, \tag{3.59}$$

and for an axisymmetric case

$$\sigma_{11,1} + \sigma_{12,2} + \frac{\sigma_{11} - \sigma_{33}}{r} - f_1^B = 0, \tag{3.60}$$

$$\sigma_{21,1} + \sigma_{22,2} + \frac{\sigma_{21}}{r} - f_2^B = 0, \tag{3.61}$$

$$m_{1,1} + m_{2,2} + \sigma_{21} - \sigma_{12} - m^B = 0, \tag{3.62}$$

where f_i^B and m^B are the volume body forces and volume body moment, respectively.

The representation of stresses in the Mohr's plane leads to a circle whose center is displaced along the shear ordinate (the shift of the center is a measure of the loss of symmetry), Fig.3.30. The coordinates of the center σ_c and τ_c are

$$\sigma_c = \frac{\sigma_{11} + \sigma_{22}}{2}, \qquad \tau_c = \frac{\sigma_{12} - \sigma_{21}}{2} \tag{3.63}$$

and the circle's radius is

$$r = \sqrt{\left(\frac{\sigma_{11} - \sigma_{22}}{2}\right)^2 + \left(\frac{\sigma_{12} + \sigma_{21}}{2}\right)^2}\,. \tag{3.64}$$

The principal stresses σ_1 and σ_2 are expressed by the relationship

$$\sigma_{1,2} = \frac{\sigma_{11} + \sigma_{22}}{2} \pm \sqrt{\left(\frac{\sigma_{11} - \sigma_{22}}{2}\right)^2 + \left(\frac{\sigma_{12} + \sigma_{21}}{2}\right)^2}\,. \tag{3.65}$$

The normal stress σ, shear stress τ and couple stress m at an arbitrary surface with the normal vector n_i ($\cos\theta$, $\sin\theta$) and tangent vector t_i ($-\sin\theta$, $\cos\theta$) are

$$\sigma = \sigma_c + r\cos 2\theta, \quad \tau = \tau_c - r\sin 2\theta \quad \text{and} \quad m = m_1 \cos\theta + m_2 \sin\theta, \tag{3.66}$$

where θ is the counterclockwise rotation of the normal vector of the surface from the maximum stress direction $\sigma_{max}=\sigma_c+r$.

Using the generalized form of the virtual work principle, the equilibrium equations (Eqs.3.57-3.62) can be expressed as

$$\int_B (\sigma_{ij}\delta\varepsilon_{ij} + m_i\delta\kappa_i)dV = \int_B (f_i^B \delta u_i + m^B \delta\omega^c)dV + \int_{\partial_1 B} t_i \delta u_i dA + \int_{\partial_2 B} m\delta\omega^c dA, \tag{3.67}$$

where $t_i=\sigma_{ij}\, n_j$ and $m=m_i\, n_i$. t_i and m are prescribed boundary tractions and moment on the boundary $\partial_1 B$ and $\partial_2 B$ with the normal vector n_i, $\delta\varepsilon_{ij}$ and $\delta\kappa_i$ denote virtual

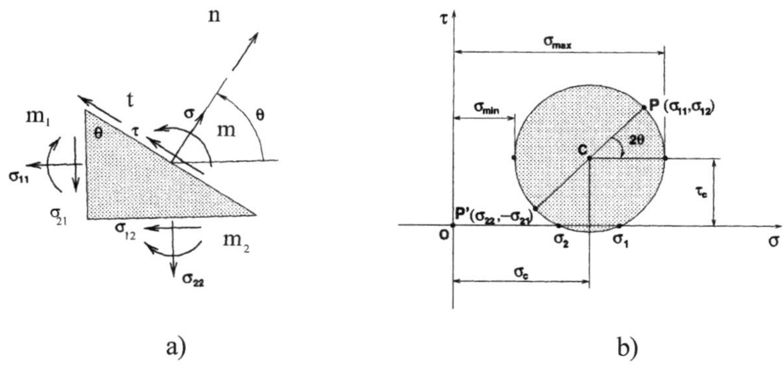

Fig. 3.30. Stresses and couple stresses at an arbitrary plane (a) and Mohr's circle of a non-symmetric state of stress (b) (Iordache and Willam 1998)

deformations and curvatures, respectively, δu_i are virtual displacements, $\delta\omega^c$ is a virtual Cosserat rotation, A stands for the surface, and V denotes the volume. Virtual displacements and Cosserat rotations vanish on those parts of the boundary where kinematic boundary conditions are prescribed. The virtual work principle is used to formulate the FE equations of motion in a micro-polar continuum (Mühlhaus 1989a, 1990, Tejchman 1989, 1997, de Borst 1991, Sluys 1992, Dietsche 1993, Dietsche et al. 1993, Steinmann 1995, Murakami and Yoshida 1997, Groen 1997, Iordache and Willam 1998). As a consequence of micro-rotations and couple stresses, the constitutive equation is endowed with a characteristic length corresponding to the mean grain diameter.

A micro-polar model has good physical grounds to describe the behaviour of granulates since it takes into account rotations and couple stresses which are observed during shearing but remain negligible during homogeneous deformation without shear zones (Uesugi 1987, Uesugi et al. 1988, Tejchman 1989, Oda et al. 1982, Oda 1993). Its other advantages are: the characteristic length is directly related to the mean grain diameter and realistic wall boundary conditions at the interface of granulate with a structure can be derived (Section 4.12). Pasternak and Mühlhaus (2001) have demonstrated that the additional rotational degree of freedom of a Cosserat continuum arises naturally by mathematical homogenization of an originally discrete system of spherical grains with contact forces and contact moments. Ehlers et al. (2003) have shown that a particle ensemble has the character of a micro-polar Cosserat continuum and the couple stresses naturally result only from the eccentricities of normal contact forces. The Cosserat model is only suitable for shear dominated problems but not for tension (decohesion) dominated applications. Thus, the propagation of cracks in cohesive soils cannot be modeled. The implementation of a polar model into standard FE codes is not quite straightforward. A micro-polar continuum can be conceived as a continuum approximation of elastic lattices whose members possess a finite bending stiffness (Bazant 1971). A Timoshenko beam can be also considered as a specific one-dimensional version of a Cosserat continuum (Tejchman 1989).

3.4 Micro-polar Hypoplastic Constitutive Model

The version used in the FE calculations can be summarised for plane strain as follows (Tejchman 1997, Tejchman and Gudehus 2001, Tejchman 2002, Tejchman and Górski 2007, 2008):

$$\overset{o}{\sigma}_{ij} = f_s [L_{ij}(\hat{\sigma}_{kl}, \hat{m}_k, d^c_{kl}, k_k d_{50}) + f_d N_{ij}(\hat{\sigma}_{ij})\sqrt{d^c_{kl} d^c_{kl} + k_k k_k d^2_{50}}] , \tag{3.68}$$

$$\overset{o}{m}_i / d_{50} = f_s [L^c_i(\hat{\sigma}_{kl}, \hat{m}_k, d^c_{kl}, k_k d_{50}) + f_d N^c_i(\hat{m}_i)\sqrt{d^c_{kl} d^c_{kl} + k_k k_k d^2_{50}}] , \tag{3.69}$$

$$\overset{o}{\sigma}_{ij} = \dot{\sigma}_{ij} - w_{ik}\sigma_{kj} + \sigma_{ik} w_{kj} , \tag{3.70}$$

$$\overset{o}{m}_i = \dot{m}_i - 0.5 w_{ik} m_k + 0.5 m_k w_{ki} , \tag{3.71}$$

$$\dot{e} = (1 + e) d_{kk} , \tag{3.72}$$

$$d^c_{ij} = d_{ij} + w_{ij} - w^c_{ij} , \qquad k_i = w^c_{,i} , \qquad w^c_{kk} = 0, \qquad w^c_{21} = -w^c_{12} = w^c , \tag{3.73}$$

$$d_{ij} = (v_{i,j} + v_{j,i}) / 2, \qquad w_{ij} = (v_{i,j} - v_{j,i}) / 2 , \tag{3.74}$$

$$\hat{\sigma}_{ij} = \frac{\sigma_{ij}}{\sigma_{kk}} , \tag{3.75}$$

$$\hat{m}_k = \frac{m_k}{\sigma_{kk} d_{50}} , \tag{3.76}$$

$$f_s = \frac{h_s}{n h_i} (\frac{1 + e_i}{e_i})(\frac{e_i}{e})^{\beta} (- \frac{\sigma_{kk}}{h_s})^{1-n} , \tag{3.77}$$

$$h_i = \frac{1}{c_1^2} + \frac{1}{3} - (\frac{e_{i0} - e_{d0}}{e_{c0} - e_{d0}})^{\alpha} \frac{1}{c_1 \sqrt{3}} , \tag{3.78}$$

$$f_d = (\frac{e - e_d}{e_c - e_d})^{\alpha} \tag{3.79}$$

$$e_i = e_{i0} \exp[-(-\sigma_{kk} / h_s)^n], \tag{3.80}$$

$$e_d = e_{d0} \exp[-(-\sigma_{kk} / h_s)^n], \tag{3.81}$$

$$e_c = e_{c0} \exp[-(-\sigma_{kk} / h_s)^n], \tag{3.82}$$

$$L_{ij} = a_1^2 d_{ij}^c + \hat{\sigma}_{ij}(\hat{\sigma}_{kl}\, d_{kl}^c + \hat{m}_k\, k_k d_{50}), \tag{3.83}$$

$$L_i^c = a_1^2 [k_i d_{50} + \hat{m}_i(\hat{\sigma}_{kl}\, d_{kl}^c + \hat{m}_k\, k_k d_{50})], \tag{3.84}$$

$$N_{ij} = a_1(\hat{\sigma}_{ij} + \hat{\sigma}_{ij}^*), \tag{3.85}$$

$$N_i^c = a_1^2 a_c\, \hat{m}_i, \tag{3.86}$$

$$\hat{\sigma}_{ij}^* = \hat{\sigma}_{ij} - \frac{1}{3}\delta_{ij}, \tag{3.87}$$

$$a_1^{-1} = c_1 + c_2\sqrt{\hat{\sigma}_{kl}^*\, \hat{\sigma}_{lk}^*}\,[1 + \cos(3\theta)], \tag{3.88}$$

$$\cos(3\theta) = -\frac{\sqrt{6}}{[\hat{\sigma}_{kl}^*\, \hat{\sigma}_{kl}^*]^{1.5}}(\hat{\sigma}_{kl}^*\, \hat{\sigma}_{lm}^*\, \hat{\sigma}_{mk}^*), \tag{3.89}$$

$$c_1 = \sqrt{\frac{3}{8}}\frac{(3 - \sin\phi_c)}{\sin\phi_c}, \qquad c_2 = \frac{3}{8}\frac{(3 + \sin\phi_c)}{\sin\phi_c}. \tag{3.90}$$

wherein m_i - Cauchy couple stress vector, $\overset{o}{m}_i$ - Jaumann couple stress rate vector, d_{ij}^c - polar rate of deformation tensor, k_i - rate of curvature vector, w^c - rate of Cosserat rotation, a_c - micro-polar constant and d_{50} - mean grain diameter of sand.

A micro-polar constant can be determined from an inverse identification process of experimental data. It can be correlated with the grain roughness and estimated with a numerical analysis for shearing of a narrow granular strip between two very rough boundaries (Section 4.2). It can be assumed as constant (e.g. a_c=2) or can be connected to the parameter a_1^{-1} (e.g. $a_c=a_1^{-1}$) which lies empirically in the range 3.0-4.5 for the usual friction angles of granulates (25°-35°). In both cases, the stresses and couple stresses are similar and manifest an asymptotic behaviour along very rough boundaries (Tejchman 1997). In the FE calculations, the expression $a_c=a_1^{-1}$ was mainly assumed. Thus, the micro-polar hypoplastic model has 9 material constants (1 additional as compared to a non-polar one): e_{i0}, e_{d0}, e_{c0}, ϕ_c, h_s, α, β, n and d_{50}.

An alternative for L_i^c and N_i^c (Eq.3.84 and 3.86) are the functions (Tejchman 1997, Maier 2002)

$$L_i^c = a_1^2 k_i d_{50} + \hat{m}_i(\hat{\sigma}_{kl}\, d_{kl}^c + \hat{m}_k\, k_k d_{50}\,), \tag{3.91}$$

$$N_i^c = a_1 a_c\, \hat{m}_i, \tag{3.92}$$

or (Tejchman et al. 1999)

$$L_i^c = a_1^2 k_i d_{50} + \hat{m}_i(\hat{\sigma}_{kl}\, d_{kl}^c + \hat{m}_k\, k_k d_{50}\,), \tag{3.93}$$

$$N_i^c = a_1^2 a_c\, \hat{m}_i, \tag{3.94}$$

or (Nübel 2002, Huang et al. 2002, Huang and Bauer 2003)

$$L_i^c = a_c^2 k_i d_{50} + \hat{m}_i(\hat{\sigma}_{kl}\, d_{kl}^c + \hat{m}_k\, k_k d_{50}\,), \tag{3.95}$$

$$N_i^c = 2a_c\, \hat{m}_i, \tag{3.96}$$

where the Lode angle of Eq.3.89 is calculated with symmetric shear stresses (obtained as the mean value of two non-symmetric stresses). In Eqs.3.92, 3.94 and 3.96, the micro-polar constant a_c was assumed to have a constant value.

The micro-polar extension of the hypoplastic law was achieved analogously to Mühlhaus's formulation (1990). First, the term in the non-polar function L_{ij} with the power of the stress ratio tensors and the non-polar modulus of the deformation rate $||d_{kl}||$ were extended by the micro-quantities $m_k k_k d_{50}$ and $k_k k_k d_{50}^2$, respectively. Therein, the polar deformation rates d_{ij}^c were used. The non-polar function N_{ij} was left unchanged. The polar function L_i^c was similarly defined as L_{ij}. However, the function N_i^c had to be assumed in another way than N_{ij} since the evolution of couple stresses during shearing is different from that of stresses due to their skew symmetry and lack of sign restriction (Tejchman 1989). Assuming that the material has an asymptotic behaviour both for stresses and couple stresses during monotonic shearing, the function N_i^c was found by fitting the numerical results for shearing of an infinite layer between two very rough walls (Tejchman 1989, 1994) with a theoretical solution within a micro-polar elastic continuum (Schäfer 1962, Tejchman 1994) and with a numerical solution within a micro-polar elasto-plastic continuum (Tejchman 1994, 1997). The linear term $f_s L_i^c$ in Eq.3.69 causes an increase of couple stresses, and the non-linear term $f_s f_d N_i^c \sqrt{d_{kl}^c d_{kl}^c + k_k k_k d_{50}^2}$ in Eq.3.69 reduces them to reach a stationary value during stationary shearing. Other linear and non-linear representations of N_i^c were tested in FE calculations. However, the most realistic results were obtained with the given linear representation of N_i^c. In addition, the function N_i^c in the micro-polar hypoplastic law was verified by results of other boundary value problems involving shear localization. If the characteristic length d_{50} becomes infinitely small, the micro-polar hypoplastic model reduces to the non-polar one since the effect of polar quantities disappears. In this case, the same results are obtained in calculations with a micro-polar and a non-polar hypoplastic law when using the same material parameters.

The capability of a micro-polar hypoplastic model has also been demonstrated in solving boundary value problems in quasi-static and dynamic regimes involving localization by other researchers (Zaimi 1998, Wehr 1999, Huang et al. 2002, Nübel 2002, Maier 2002, Huang and Bauer 2003, Gudehus and Nübel 2004). A close agreement between calculations and experiments was achieved.

The FE calculations showed also that the thickness of shear zones did not depend upon the mesh discretisation if the size of finite elements in the shear zone was not greater than five times the mean grain diameter when using triangular finite elements with linear shape functions for displacements and a Cosserat rotation (Tejchman and Bauer 1996, Tejchman 1997). Numerical calculations by Sluys (1992), Groen (1997) and Maier (2002) within a micro-polar continuum also indicate that convergence to a unique solution can only be obtained when the element size is small enough compared to the width of the localized zone.

3.5 FE Implementation

FE equation of motion in quasi-static regime

The formulation of FEM within a micro-polar continuum was performed in an analogous manner as within non-polar continuum. Applying the virtual work principle (Eq.3.67), a conventional equation of motion governing the response of a system of finite elements in a quasi-static regime is obtained (Mühlhaus 1990, Tejchman 1997):

$$KU = R, \tag{3.97}$$

where

$$R = R_B + R_S - R_I + R_C. \tag{3.98}$$

The matrix K

$$K = \sum_m \int_{V^{(m)}} B^{(m)^T} E B^{(m)} dV^{(m)} = \sum_m K^{(m)}, \tag{3.99}$$

is the stiffness of the element assemblage (m denotes the number of the finite element). The load vector R

$$R = R_B + R_S - R_I + R_C \tag{3.100}$$

includes the effect of volume body forces and a volume body moment

$$R_B = \sum_m \int_{V^{(m)}} H^{(m)^T} f^{B^{(m)}} dV^{(m)} = \sum_m R_B^{(m)}, \tag{3.101}$$

the effect of surface forces and surface moments

$$R_S = \sum_m \int_{V^{(m)}} H^{S(m)^T} f^{S^{(m)}} dS^{(m)} = \sum_m R_S{}^{(m)}, \tag{3.102}$$

the effect of initial stresses and moments

$$R_I = \sum_m \int_{V^{(m)}} B^{(m)^T} \sigma^{I^{(m)}} dV^{(m)} = \sum_m R_I{}^{(m)}, \tag{3.103}$$

and the effect of concentrated forces and moments

$$R_c = \sum_m F^{(m)}. \tag{3.104}$$

The vector U is the nodal point displacement vector

$$U = \sum_m U^{(m)}. \tag{3.105}$$

$V^{(m)}$ is the volume of the element m, $S^{(m)}$ denotes the element surface, $H^{(m)}$ is the displacement interpolation matrix and $H^{S(m)}$ stands for the surface displacement interpolation matrix.

The matrix $B^{(m)}$ is the deformation-displacement matrix which associates the strain vector $\varepsilon^{(m)}$ with the nodal displacement vector $U^{(m)}$

$$\varepsilon^{(m)} = B^{(m)} U^{(m)}, \tag{3.106}$$

where

$$\varepsilon^{(m)} = [\,\varepsilon_{11}, \varepsilon_{22}, \varepsilon_{12}, \varepsilon_{21}, \kappa_1, \kappa_2\,]^T, \tag{3.107}$$

and

$$U^{(m)} = [\,u_1^1, u_2^1, \omega^{c^1}, u_1^2, u_2^2, \omega^{c^2}, \ldots, u_1^N, u_2^N, \omega^{c^N}\,]^T. \tag{3.108}$$

N denotes the node number.

The stress vector $\sigma^{(m)}$ is related to the strain vector $\varepsilon^{(m)}$ by

$$\sigma^{(m)} = E^{(m)} \varepsilon^{(m)}, \tag{3.109}$$

where $E^{(m)}$ is the material stress-strain matrix and

$$\sigma^{(m)} = [\,\sigma_{11}, \sigma_{22}, \sigma_{12}, \sigma_{21}, m_1, m_2\,]^T. \tag{3.110}$$

The displacement vector $u^{(m)}$ is connected to the nodal point displacement vector $U^{(m)}$ through the displacement interpolation matrix $H^{(m)}$

$$u^{(m)} = H^{(m)} U^{(m)}, \tag{3.111}$$

wherein

$$u^{(m)} = [u_1, u_2, \omega^c]^T . \qquad (3.112)$$

The vectors $f^{B(m)}$, $f^{S(m)}$, $\sigma^{I(m)}$ and $F^{(m)}$ include:

the volume body forces f_i^B and the volume body moment m^B

$$f^{B(m)} = [f_1^B, f_2^B, m^B]^T , \qquad (3.113)$$

the surface forces f_i^S and the surface moment m^S

$$f^{S(m)} = [f_1^S, f_2^S, m^S]^T , \qquad (3.114)$$

the initial stresses σ_{ij}^I and initial couple stresses m_i^I

$$\sigma^{I(m)} = [\sigma_{11}^I, \sigma_{22}^I, \sigma_{12}^I, \sigma_{21}^I, m_1^I, m_2^I]^T , \qquad (3.115)$$

and the concentrated forces F_i and concentrated moment M

$$F^{(m)} = [F_1, F_2, M]^T . \qquad (3.116)$$

The displacement vector u, the rate of deformation vector d, the time derivative stress vector $\dot{\sigma}$ and the Jaumann stress rate vector $\overset{o}{\sigma}$ can be introduced as the pseudo-vectors:

$$u = [u_1, u_2, \omega^c]^T , \qquad (3.117)$$

$$d = [d_{11}, d_{22}, d_{12}, d_{21}, k_1, k_2]^T , \qquad (3.118)$$

$$\dot{\sigma} = [\dot{\sigma}_{11}, \dot{\sigma}_{22}, \dot{\sigma}_{12}, \dot{\sigma}_{21}, \dot{m}_1, \dot{m}_2]^T , \qquad (3.119)$$

$$\overset{o}{\sigma} = [\overset{o}{\sigma}_{11}, \overset{o}{\sigma}_{22}, \overset{o}{\sigma}_{12}, \overset{o}{\sigma}_{21}, \overset{o}{m}_1, \overset{o}{m}_2]^T . \qquad (3.120)$$

The volume body moment (Eq.3.67) was neglected. The initial calculations have shown that the expression

$$m^B = \pm \gamma_d \frac{d_{50}}{2} \qquad (3.121)$$

has an insignificant effect on results (γ_d - volume weight, d_{50} - mean grain diameter) (Tejchman 1989).

Element displacement interpolation matrix

The calculations were performed with quadrilateral elements composed of four diagonally crossed triangles using linear-shaped functions (Bathe 1982) for the interpolation of displacements and a Cosserat rotation. Using these elements, the effect of volumetric locking can be avoided (Nagtegaal et al. 1974, Groen 1997). The displacement interpolation matrix H is

$$\begin{bmatrix} h_i & 0 & 0 & h_j & 0 & 0 & h_k & 0 & 0 \\ 0 & h_i & 0 & 0 & h_j & 0 & 0 & h_k & 0 \\ 0 & 0 & h_i & 0 & 0 & h_j & 0 & 0 & h_k \end{bmatrix}, \tag{3.122}$$

where

$$h_i = \frac{a_i + b_i x_1 + c_i x_2}{2\Delta}, \tag{3.123}$$

$$h_j = \frac{a_j + b_j x_1 + c_j x_2}{2\Delta}, \tag{3.124}$$

$$h_k = \frac{a_k + b_k x_1 + c_k x_2}{2\Delta}, \tag{3.125}$$

$$a_i = x_1^j x_2^k - x_1^k x_2^j, \tag{3.126}$$

$$a_j = x_1^k x_2^i - x_1^i x_2^k, \tag{3.127}$$

$$a_k = x_1^i x_2^j - x_1^j x_2^i, \tag{3.128}$$

$$b_i = x_2^j - x_2^k, \tag{3.129}$$

$$b_j = x_2^k - x_2^i, \tag{3.130}$$

$$b_k = x_2^i - x_2^j, \tag{3.131}$$

$$c_i = x_1^k - x_1^j, \tag{3.132}$$

$$c_j = x_1^i - x_1^k, \tag{3.133}$$

$$c_k = x_1^j - x_1^i. \tag{3.134}$$

x_1 and x_2 are the coordinates inside of the element, x_i^i, x_i^j and x_i^k are the coordinates of the element nodes i, j and k, respectively, and Δ is the element area. Dietsche (1993) investigated the effect of different shape functions within a Cosserat continuum and showed that the application of triangular elements with linear shape functions yielded mesh-independent FE results.

Strain-displacement matrix

The strain-displacement matrix B is obtained by an appropriate differentiation and combining rows of the matrix H (Eq.3.122):

$$\begin{bmatrix} b_i/2\Delta & 0 & 0 & b_j/2\Delta & 0 & 0 & b_k/2\Delta & 0 & 0 \\ 0 & c_i/2\Delta & 0 & 0 & c_j/2\Delta & 0 & 0 & c_k/2\Delta & 0 \\ c_i/2\Delta & 0 & h_i & c_j/2\Delta & 0 & h_j & c_k/2\Delta & 0 & h_k \\ 0 & b_i/2\Delta & -h_i & 0 & b_j/2\Delta & -h_j & 0 & b_k/2\Delta & -h_k \\ 0 & 0 & b_i/2\Delta & 0 & 0 & b_j/2\Delta & 0 & 0 & b_k/2\Delta \\ 0 & 0 & c_i/2\Delta & 0 & 0 & c_j/2\Delta & 0 & 0 & c_k/2\Delta \end{bmatrix}. \tag{3.135}$$

In contrast to a non-polar continuum, the matrix B is not constant since it includes the linear shape functions h_i, h_j and h_k which depend on the coordinates x_1 and x_2. To obtain the element stiffness matrix K (Eq.3.99), the polynomials h_i, h_j and h_k must be integrated. The linear polynomials can be integrated exactly by the Gauss quadrature (Bathe 1982). In the FE analyses, the integration was carried out either at one Gauss point located in the middle of each finite element or at three Gauss points located at the centre of each side of each triangular finite element.

A so-called "Updated Lagrangian" formulation was used to take into account large deformations and curvatures (Bathe 1982). As objective measures for stress and couple stress, the Jaumann stress rate tensor and the Jaumann couple stress rate vector were used (Mühlhaus 1989a, 1989b), Eqs.3.70 and 3.71. During calculations, the varying configuration and varying volume of the body were taken into account.

For time integration, a simple one-step Euler forward scheme was used. The element stress tensor σ in the step $t+\Delta t$ can be calculated as follows

$$\sigma(t+\Delta t)=\sigma(t)+\int_t^{t+\Delta t}\dot{\sigma}dt=\sigma(t)+\dot{\sigma}(t)\Delta t\ , \tag{3.136}$$

where Δt is the time increment.

Stress-strain matrix

The elements of the stiffness matrix $\partial\overset{o}{\sigma}/\partial d$ are:

$$\frac{\partial\overset{o}{\sigma}_{ij}}{\partial d_{mn}}=f_s[\,a_1^2\delta_{ij}\delta_{mn}+\hat{\sigma}_{ij}\hat{\sigma}_{mn}+f_d a_1(\hat{\sigma}_{ij}+\hat{\sigma}^*_{ij})\frac{d_{mn}}{\sqrt{d_{kl}d_{kl}+k_k k_k d_{50}^2}}\,]\ , \tag{3.137}$$

$$\frac{\partial\overset{o}{\sigma}_{ij}}{\partial(k_m d_{50})}=f_s[\,\hat{\sigma}_{ij}\hat{m}_m+f_d a_1(\hat{\sigma}_{ij}+\hat{\sigma}^*_{ij})\frac{k_m d_{50}}{\sqrt{d_{kl}d_{kl}+k_k k_k d_{50}^2}}\,]\ , \tag{3.138}$$

$$\frac{\partial(\overset{o}{m}_i/d_{50})}{\partial d_{mn}}=f_s a_1^2[\,\hat{m}_i\hat{\sigma}_{mn}+f_d a_c\hat{m}_i\frac{d_{mn}}{\sqrt{d_{kl}d_{kl}+k_k k_k d_{50}^2}}\,]\ , \tag{3.139}$$

$$\frac{\partial(\overset{o}{m_i}/d_{50})}{\partial(k_m d_{50})} = f_s a_1^2 [\,\delta_{im} + \hat{m}_i \hat{m}_m + f_d a_c \hat{m}_i \frac{k_m d_{50}}{\sqrt{d_{kl} d_{kl} + k_k k_k d_{50}^2}}\,] \,. \tag{3.140}$$

For the solution of the system of non-linear quasi-static equations (Eq.3.92), a modified Newton-Raphson scheme with line search (Bathe 1982) was used with an initial global stiffness matrix calculated with only two first terms of the constitutive equations which are linear in d_{ij}^c and $\kappa_k d_{50}$ (Eqs.3.83-3.86). The procedure was found to yield a sufficiently accurate and fast convergence. The magnitude of the maximum out-of-balance force at the end of each calculation step was smaller than 1-2% of the calculated total force along the boundary. Due to the presence of non-linear terms (taking into account the direction of the deformation rate) in the constitutive model (Eqs.3.83-3.86) and material softening, this procedure turned out to be more efficient than the full Newton-Raphson method (when taking into account all terms in Eqs.3.136-3.140). To accelerate the calculations in the softening regime, the initial (first) increments of displacements and the Cosserat rotation $\Delta U^{(1)}$ in each calculation step 'n+1' were assumed to be equal to the converged increments from the previous step (Vermeer and van Langen 1989, Tejchman 1989):

$$^{n+1}\Delta U^{(1)} = {}^{n}U - {}^{n-1}U \,, \tag{3.141}$$

$$^{n+1}U^{(1)} = {}^{n}U + {}^{n+1}\Delta U^{(1)} \,, \tag{3.142}$$

$$K\,^{n+1}\Delta U^{i+1} = {}^{n+1}R - {}^{n+1}F^{i} \,, \qquad i=1,2,3,\ldots \tag{3.143}$$

$$^{n+1}U^{(i+1)} = {}^{n+1}U^{(i)} + \beta\,^{n+1}\Delta U^{(i+1)} \,, \tag{3.144}$$

$$K = \int_V B^T E B dV \,, \qquad F = \int_V B^T \sigma dV \,, \tag{3.145}$$

wherein F is the vector of nodal point forces equivalent to the element stresses and couple stresses. The scalar multiplier β in the line search procedure (Eq.3.144) (Bathe 1982) was assumed to be 0.5 (on the basis of preliminary calculations, Tejchman 1997). The iteration steps were performed using translational and rotational convergence criteria. For the translational degrees of freedom:

$$\left\|\Delta u^k\right\| \le b_1 \tag{3.146}$$

and for the rotational degrees of freedom

$$\left\|\Delta \omega^{ck}\right\| \le b_2 \,. \tag{3.147}$$

Δu^k and $\Delta(\omega^c)^k$ are the incremental displacement and rotation vectors in the k'th iteration, respectively and $\|\cdot\|$ is the Euclidean norm of vector. The constants b_1 and b_2 were assumed be means of preliminary calculations (Tejchman 1997).

To prevent both eventual inadmissible stresses in a compressive regime (caused by the fact that a flow surface is not used in hypoplasticity) and tensile stresses (which are excluded in cohesionless granulates), small displacement increments were assumed in FE calculations. In addition, a sub-stepping algorithm was used (Tejchman 1989, 1997, de Borst and Heeres 2002, Huang and Bauer 2003). The increments of deformation d_{ij} and curvature k_k were divided into smaller parts (usually divided by k=100-1000) within each iteration step. At singular regions, the parameter k was increased even up to e.g. 20000 to avoid tensile normal stresses. However, if tensile normal stresses were obtained, the normal stresses were replaced either by very small compressive values (e.g. equal to 10^{-6} times the initial values) or zero values.

The wall boundary conditions within a micro-polar hypoplasticity are described in Section 4.12.

References

Al Hattamleh, O., Muhunthan, B., Zbib, H.M.: Stress distribution in granular heaps using multi-slip formulation. Int. J. Numer. Analyt. Meth. Geomech. 29, 713–727 (2005)

Bathe, K.J.: Finite Element Procedures in Engineering Analysis. Prentice-Hall, Inc., Englewood Cliffs (1982)

Bauer, E.: Calibration of a comprehensive hypoplastic model for granular materials. Soils Found. 36(1), 13–26 (1996)

Bauer, E.: Conditions of embedding Casagrande's critical states into hypoplasticity. Mech. Cohes. Frict. Mater. 5, 125–148 (2000a)

Bauer, E.: Modelling of the pressure and density sensitive behaviour of sand within the framework of hypoplasticity. Task Quarterly 4(3), 367–389 (2000b)

Bazant, Z.: Micropolar medium as a model for buckling of grid frameworks. Developments in mechanics. In: Proc. 12th Midwestern Mechanics Conference, University of Notre Dam, pp. 587–593 (1971)

Becker, M., Lippmann, H.: Plane plastic flow of granular model material. Arch. Mech. 29, 829–846 (1977)

Bogdanova-Bontscheva, N., Lippmann, H.: Rotations symmetrisches ebenes Fliessen eines granularen Modellmaterials. Acta Mech. 21, 93–113 (1975)

Borja, R.I., Andrade, J.E.: Critical state plasticity. Part VI: Meso-scale finite element simulation of strain localization in discrete granular materials. Comp. Meths. Appl. Mech. Engng. 195, 5115–5140 (2006)

de Borst, R.: Simulation of strain localization: a reappraisal of the Cosserat continuum. Engng. Comput. 8, 317–332 (1991)

de Borst, R., Heeres, O.M.: A unified approach to the implicit integration of standard, non-standard and viscous plasticity models. Int. J. Num. Anal. Meth. Geomech. 26(11), 1059–1070 (2002)

Chambon, R.: Une classe de lois de compartement incrementelement non lineaire pour les sols non visqueux, resolution de quelques problemes de coherences. C.R. Acad. Sci. 308, 1571–1576 (1989)

Chambon, R.: Incremental behaviour of a simple deviatoric constitutive CLoE model. In: Mühlhaus, H.B., et al. (eds.) Bifurcation and Localisation Theory in Geomechanics, Swets, Zeitlinger, Lisse, pp. 21–28 (2001)

Cosserat, E., Cosserat, F.: Theorie des Corps Deformables. A. Hermann and Fils, Paris (1909)

Dafalias, Y.F.: Bounding surface plasticity: I. Mathematical foundation and hypoplasticity. J. Engrg. Mech. ASCE 112, 966–980 (1986)

Darve, F., Flavigny, E., Rojas, E.: A class of incrementally non-linear constitutive relations and applications to clays. Comp. Geotech. 2, 43–66 (1986)

Darve, F., Flavigny, E., Megachou, M.: Yield surfaces and principle of superposition revisited by incrementally non-linear constitutive relations. Int. J. Plasticity 11(8), 927–948 (1995)

Desrues, J., Chambon, R.: Shear band analysis for granular materials – the question of incremental linearity. Ing. Arch. 59, 187–196 (1989)

Dietsche, A.: Lokale Effekte in linear-elastischen und elasto-plastischen Cosserat-Continua. Dissertation at the Karlsruhe University, pp.1-141 (1993)

Dietsche, A., Steinmann, P., Willam, K.: Micropolar elastoplasticity and its role in localization. Int. J. Plasticity 9(8), 813–831 (1993)

Ehlers, W., Ramm, E., Diebels, S., D'Addetta, G.A.: From particle ensembles to Cosserat continua: homogenisation of contact forces towards stresses and couple stresses. Int. J. Solids Structures 40, 6681–6702 (2003)

Goldscheider, M.: Grenzbedingung und Fliessregel von Sand. Mech. Res. Comm. 3, 463–468 (1976)

Groen, A.E., Three-dimensional elasto-plastic analysis of soils. PhD Thesis, Delft University, pp. 1-114 (1997)

Gudehus, G.: A comparison of some constitutive laws for soils under radially symmetric loading and unloading. In: Proc. 3th Int. Conf. Num. Meths. in Geomechanics, pp. 1309–1323 (1979)

Gudehus, G.: A comprehensive constitutive equation for granular materials. Soils Found. 36(1), 1–12 (1996)

Gudehus, G.: Seismo-hypoplasticity with a granular temperature. Granular Matter 8, 93–102 (2006)

Gudehus, G.: Physical Soil Mechanics. Springer, Heidelberg (2008)

Günther, W.: Zur Statik und Kinematik des Cosserat-Kontinuums. Abh. Braunschweigische Wiss. Gessellsch. 10, 195–213 (1958)

Herle, I.: Hypoplastizität und Granulometrie einfacher Korngerüste. PhD thesis, University of Karlsruhe, 142 (1997)

Herle, I.: A relation between parameters of a hypoplastic constitutive model and grain properties. In: Adachi, T., Oka, F., Yashima, A. (eds.) Localisation and Bifurcation Theory for Soils and Rocks in Gifu, Balkema, pp. 91–99 (1998)

Herle, I., Gudehus, G.: Determination of parameters of a hypoplastic constitutive model from properties of grain assemblies. Mech. Cohes.-Frict. Mater. 4(5), 461–486 (1999)

Herle, I.: Granulometric limits of hypoplastic models. Task Quarterly 4(3), 389–408 (2000)

Herle, I., Kolymbas, D.: Hypoplasticity for soils with low friction angles. Comp. Geotech. 31, 365–373 (2004)

Huang, W., Nübel, K., Bauer, E.: A polar extension of hypoplastic model for granular material with shear localization. Mech. Mater. 34, 563–576 (2002)

Huang, W., Bauer, E.: Numerical investigations of shear localization in a micro-polar hypoplastic material. J. Num. Anal. Meth. Geomech. 27, 325–352 (2003)

Huang, W., Bauer, E., Sloan, S.W.: Behaviour of interfacial layer along granular soil-structure interfaces. Struct. Eng. Mech. 15(3), 315–329 (2003)

Huang, W.X., Wu, W., Sun, D.A., Sloan, S.: A simple hypoplastic model for normally consolidated clay. Acta Geotech. 1, 15–27 (2006)

Iordache, M.M., Willam, K.J.: Localized failure analysis in elastoplastic Cosserat continua. Comp. Meth. Appl. Mech. Engrg. 151, 559–586 (1998)

Kanatani, K.: A micropolar continuum theory for granular materials. Int. J. Engng. Sci. 17, 419–432 (1979)

Kessel, S.: Lineare Elastizitätstheorie des anisotropen Cosserat-Kontinuums. Abh. Braunschweigische Wiss. Gess. 16, 1–22 (1964)

Kolymbas, D.: A rate-dependent constitutive equation for soils. Mech. Res. Comm. 6, 367–372 (1977)

Kolymbas, D.: Introduction to hypoplasticity. Advances in Geotechnical Engineering and Tunneling, Balkema, Rotterdam, Brookfield (2000)

Lade, P.V.: Elasto-plastic stress-strain theory for cohesionless soil with curved yield surfaces. Int. J. Solid Structures 13, 1019–1035 (1977)

Lanier, J., Caillerie, D., Chambon, R.: A general formulation of hypoplasticity. Int. J. Numer. Anal. Meth. Geomech. 28(15), 1461–1478 (2004)

Maier, T.: Numerische Modellierung der Entfestigung im Rahmen der Hypoplastizität. PhD Thesis, Dortmund University, Germany (2002)

Masin, D.: A hypoplastic constitutive model for clays. Int. J. Numer. Anal. Meths. Geomech. 29, 311–336 (2005)

Masin, D., Herle, I.: Improvement of a hypoplastic model to predict clay behaviour under undrained conditions. Acta Geotech (in press, 2007)

Matsuoka, H., Nakai, T.: Stress-strain relationship of soil based on the SMP. In: Proc. IX Int. Conf. Soil Mech. Found. Eng., Tokyo, pp. 153–162 (1977)

Mindlin, R.D.: Microstructure in linear elasticity. Arch. Ration. Mech. Anal. 16, 51–78 (1964)

Murakami, A., Yoshida, N.: Cosserat continuum and finite element analysis. In: Asaoka, A., Adachi, T., Oka, F. (eds.) Deformation and Progressive Failure in Geomechanics, pp. 871–876. Pergamon, Oxford (1997)

Mühlhaus, H.B.: Berücksichtigung von Inhomogenitäten im Gebirge im Rahmen einer Kontinuumstheorie. Veröffentlichung des Instituts für Boden- und Felsmechanik, Universität Karlsruhe, vol. 106, pp. 1-65 (1987)

Mühlhaus, H.B.: Application of Cosserat theory in numerical solutions of limit load problems. Ing. Arch. 59, 124–137 (1989a)

Mühlhaus, H.B.: Stress and couple stress in a layered half plane with surface loading. Int. J. Numer. Anal. Met. Geomech. 13, 545–563 (1989b)

Mühlhaus, H.B.: Continuum models for layered and blocky rock. In: Hudson, J.A., Fairhurst, C. (eds.) Comprehensive Rock Engineering, vol. 2, pp. 209–231. Pergamon Press, Oxford (1990)

Nagtegaal, J.C., Parks, D.M., Rice, J.R.: On numerically accurate finite element solutions in fully plastic range. Comp. Meth. Appl. Mech. Engng. 4, 153–177 (1974)

Neuber, H.: Über Probleme der Spannungskonzentration im Cosseratkörper. Acta Mech. 2, 48–69 (1966)

Niemunis, A., Herle, I.: Hypoplastic model for cohesionless soils with elastic strain range. Mech. Cohes. -Frict. Mater. 2, 279–299 (1997)

Niemunis, A.: Extended hypoplastic models for soils. Monograph, Gdansk University of Technology (2003)

Niemunis, A., Wichtmann, T., Triantafyllidis, T.: A high-cycle accumulation model for sand. Comp. Geotech. 32(4), 245–263 (2005)

Nübel, K., Gudehus, G.: Evolution of localised shearing: dilation and polarization in grain skeleton. In: Kishino (ed.) Powder and Grains, Swets, Zeitlinger, Lisse, pp. 289–292 (2001)

Nübel, K.: Experimental and numerical investigation of shear localisation in granular materials. Publication Series of the Institute of Soil and Rock Mechanics, University Karlsruhe, 62 (2002)

Oda, M., Konishi, J., Nemat-Nasser, S.: Experimental micromechanical evaluation of strength of granular materials, effects of particle rolling. Mech. Mater. 1, 269–283 (1982)

Oda, M.: Micro-fabric and couple stress in shear bands of granular materials. In: Thornton, C. (ed.) Powders and Grains, Rotterdam, Balkema, pp. 161–167 (1993)

Pasternak, E., Mühlhaus, H.-B.: Cosserat continuum modelling of granulate materials. In: Valliappan, S., Khalili, N. (eds.) Computational Mechanics – New Frontiers for New Millennium, pp. 1189–1194. Elsevier Science, Amsterdam (2001)

Pestana, J.M., Whittle, A.J.: Formulation of a unified constitutive model for clays and sands. Int. J. Num. Anal. Meth. Geomech. 23, 1215–1243 (1999)

Rondon, H.A., Wichtmann, T., Triantafyllidis, T., Lizcano, A.: Hypoplastic material constants for a well-graded granular material for base and subbase layers of flexible pavements. Acta Geotechnica 2, 113–126 (2007)

Schäfer, H.: Versuch einer Elastizitätstheorie des zweidimensionalen ebenen Cosserat-Kontinuums. In: Festschrift Tolmien, W. (ed.) Miszellaneen der Angewandten Mechanik, pp. 277–292. Akademie-Verlag, Berlin (1962)

Schäfer, H.: Das Cosserat-Kontinuum. Zeitschrift für Angewandte Mathematik und Mechanik 8, 485–498 (1967)

Sluys, L.J.: Wave propagation, localisation and dispersion in softening solids. PhD Thesis, Delft University of Technology (1992)

Steinmann, P.: Theory and numerics of ductile micropolar elastoplastic damage. Int. J. Num. Meth. in Engng. 38, 583–606 (1995)

Tamagnini, C., Viggiani, C., Chambon, R.: A review of two different approaches to hypoplasticity. In: Kolymbas, D. (ed.) Constitutive Modeling of Granular Materials, pp. 107–145. Springer, Heidelberg (2000)

Tejchman, J.: Scherzonenbildung und Verspannungseffekte in Granulaten unter Berücksichtigung von Korndrehungen. Publication Series of the Institute of Soil and Rock Mechanics, University Karlsruhe, vol. 117, pp. 1–236 (1989)

Tejchman, J.: Numerical study on localised deformation in a Cosserat continuum. In: Chambon, R., Desrues, J., Vardoulakis, I. (eds.) Localisation and Bifurcation Theory for Soils and Rocks, Balkema, Rotterdam, pp. 257–275 (1994)

Tejchman, J., Bauer, E.: Numerical simulation of shear band formation with a polar hypoplastic model. Comp. Geotech. 19(3), 221–244 (1996)

Tejchman, J.: Modelling of shear localisation and autogeneous dynamic effects in granular bodies. Publication Series of the Institute of Soil and Rock Mechanics, University Karlsruhe, vol. 140, pp. 1–353 (1997)

Tejchman, J., Herle, I., Wehr, J.: FE-studies on the influence of initial void ratio, pressure level and mean grain diameter on shear localisation. Int. J. Num. Anal. Meth. Geomech. 23, 2045–2074 (1999)

Tejchman, J.: Shearing of an infinite narrow granular layer between two boundaries. In: Mühlhaus, H.B. (ed.) Bifurcation and Localisation Theory in Geomechanics, Swets, Zeitlinger, Lisse, pp. 95–103 (2001)

Tejchman, J., Gudehus, G.: Shearing of a narrow granular strip with polar quantities. Int. J. Num. Anal. Meth. Geomech. 25, 1–28 (2001)

Tejchman, J.: Patterns of shear zones in granular materials within a polar hypoplastic continuum. Acta Mech. 155(1-2), 71–95 (2002a)

Tejchman, J., Bauer, E., Wu, W.: Effect of texturial anisotropy on shear localization in sand during plane strain compression. Acta Mech. 1-4, 23–51 (2007)

Tejchman, J., Górski, J.: Computations of size effects in granular bodies within micro-polar hypoplasticity during plane strain compression. Int. J. Solids Structures 45(6), 1546–1569 (2008)

Uesugi, M.: Friction between dry sand and construction. PhD thesis, Tokyo Institute of Technology (1987)

Uesugi, M., Kishida, H., Tsubakihara, Y.: Behaviour of sand particles in sand-steel friction. Soils Found. 28(1), 107–118 (1988)

Wang, C.C.: A new representation theorem for isotropic functions. J. Rat. Mech. Anal. 36, 166–223 (1970)

Wehr, J.: Granulatumhüllte Anker und Nägel. Publication Series of the Institute of Soil and Rock Mechanics, 144. Karlsruhe University, Germany (1999)

Weifner, T., Kolymbas, D.: A hypoplastic model for clay and sand. Acta Geotech. 2, 103–112 (2007)

Weifner, T., Kolymbas, D.: Review of two hypoplastic equations for clay considering axisymmetric element deformations. Comp. Geotech. (2008), doi:10.1016/j.compgeo.2007.12.001

von Wolffersdorff, P.A.: A hypoplastic relation for granular materials with a predefined limit state surface. Mech. Cohes. -Frict. Mater. 1, 251–271 (1996)

Wu, W.: Hypoplastizität als mathematisches Modell zum mechanischen Verhalten granularer Stoffe. Publication Series of the Institute of Soil and Rock Mechanics, University Karlsruhe, 129 (1992)

Wu, W., Niemunis, A.: Failure criterion, flow rule and dissipation function derived from hypoplasticity. Mech. Cohes. -Frict. Mater. 1, 145–163 (1996)

Wu, W., Bauer, E., Kolymbas, D.: Hypoplastic constitutive model with critical state for granular materials. Mech. Mater. 23, 45–69 (1996)

Wu, W., Niemunis, A.: Beyond failure in granular materials. Int. J. Num. and Anal. Meths. in Geomech. 21(2), 153–174 (1997)

Wu, W.: Hypoplastizität als mathematisches Modell zum mechanischen Verhalten granularer Stoffe. Publication Series of the Institute of Soil and Rock Mechanics, University Karlsruhe, 129 (1992)

Wu, W., Kolymbas, D.: Hypoplasticity then and now. In: Kolymbas, D. (ed.) Constitutive Modeling of Granular Materials, pp. 57–105. Springer, Heidelberg (2000)

Wu, W.: On high-order hypoplastic models for granular materials. J. Engng. Math. 56, 23–34 (2006)

Vermeer, P.A.: A five-constant model unifying well-established concepts. In: Gudehus, G., Darve, F., Vardoulakis, I. (eds.) Proc. Int. Workshop on Constitutive Relations for Soils, Balkema, pp. 175–197 (1982)

Vermeer, P.A., van Langen, H.: Soil collapse computations with finite elements. Ing. Arch. 59, 221–236 (1989)

Zaimi, S.A.: Modelisation de l'coulement des charges dans le haut fourneau. PhD Thesis, Ecole Centrale Paris, France (1998)

Youd, T.L.: Factors controlling maximum and minimum densities of sands. Evaluation of relative density and its role in geotechnical projects involving cohesionless soils. STP 523, 98–112 (1973)

4 Finite Element Calculations: Preliminary Results

Abstract. This chapter presents preliminary numerical results of different quasi-static boundary value problems in granular bodies including shear localization. Calculations were carried out with the finite element method on the basis of a micro-polar hypoplastic model. The following problems were considered: plane strain compression test, monotonic and cyclic shearing of an infinite layer, biaxial compression test, strip foundation, earth pressure, direct and simple shear test, wall direct shear test and contractant shear zones. Attention was paid to the thickness and spacing of shear zones. Numerical solutions were compared with corresponding laboratory tests.

4.1 Plane Strain Compression Test

The FE calculations of plane strain compression tests were performed with a sand specimen which was h_o=14 cm high and b=2 cm wide (Tejchman et al. 1999, Tejchman 2004c, 2006). In total, 320 quadrilateral elements (0.25×0.35 cm) divided into 1440 triangular elements were used. The dimensions of finite elements were approximately $5\times d_{50}$. to properly capture shear localization (Tejchman and Bauer 1996). The inclination of the mesh lines to the bottom was equal to 54.5°. In this case, the calculations were exceptionally carried out with small deformations and curvatures.

As the initial stress state, the K_0-state with $\sigma_{22}=\sigma_c+\gamma_d x_2$, $\sigma_{11}=\sigma_c+K_0\gamma_d x_2$, $\sigma_{33}=\sigma_{11}$, $\sigma_{12}=\sigma_{21}=0$ and $m_i=0$ were assumed in the sand specimen, where σ_c denotes the confining pressure, x_2 is the vertical coordinate measured from the top of the specimen, γ_d denotes the initial volume weight and K_0=0.45 is the pressure coefficient at rest (σ_{11} - horizontal normal stress, σ_{22} - vertical normal stress, σ_{33} – horizontal normal stress perpendicular to the plane of deformation, σ_{12} - horizontal shear stress, σ_{21} - vertical shear stress, Fig.3.28).

The quasi-static deformation in sand was initiated through a constant vertical displacement increment prescribed to the nodes along the upper edge of the specimen Δu ($\Delta u/h_0$=0.0001). The boundary conditions of the sand specimen were: along the vertical sides traction and moment free while the top and the bottom smooth, i.e. $u_2=n\Delta u$, σ_{12}=0, m_2=0 (top) and u_2=0, σ_{12}=0, m_2=0 (bottom), n denotes the number of the time step. To preserve the stability of the specimen against the sliding along the top boundary,

J. Tejchman: Shear Localiz. in Gran. Bodies, SSGG, pp. 87–211, 2008.
springerlink.com

the node in the middle of the upper edge was kept fixed. To numerically obtain a shear zone inside the specimen, a weak element with a large initial void ratio, e_0=1.0, was inserted in the middle of the left side of the specimen. The global stiffness matrix (calculated with two first terms of the constitutive equations, Eqs.3.137-3.140) was updated every 20 steps. The following material constants for so-called Karlsruhe sand were used: e_{i0}=1.3, e_{d0}=0.51, e_{c0}=0.82, ϕ_c=30°, h_s=190 MPa, α=0.3, n=0.5, d_{50}=0.5 mm and a_c=5 (Eqs.3.68-3.90).

Fig.4.1 presents the effect of the initial void ratio e_0, confining pressure σ_c and mean grain diameter d_{50} on the calculated normalised load-displacement curves. All curves for e_0<0.8 increase first, show a pronounced peak, drop later and reach almost the same residual state. In turn, the curves for $e_0 \geq$0.80 (loose sand) continuously increase.

The mean mobilized angle of internal friction for the entire specimen was calculated from the formula

$$\phi = \arcsin\frac{\sigma_1 - \sigma_2}{\sigma_1 + \sigma_2}. \tag{4.1}$$

It is in the residual state for all curves about ϕ_{cr}=33°-34°, and is different then the assumed critical angle of internal friction, ϕ_c=30° (Eq.3.90). In Eq.4.1, $\sigma_1 = P/(bl)$ denotes the vertical principle stress (P - vertical force on the top, $\sigma_2=\sigma_c$ - horizontal principal stress, b – initial specimen width, l=1.0 m. The normalised load-displacement curves for large confining pressure and initially dense sand correspond qualitatively to the curves for initially loose sand and low confining pressure, respectively.

The peak angle of internal friction for initially dense sand (e_0=0.60) at low confining pressure (σ_c=0.2 MPa) is ϕ^p=45°, and is reached for the vertical deformation of u_2/h_0=3.5% (Fig.4.1a). The residual angle of internal friction is then ϕ_{cr}=34° (u_2/h_0=10%). In turn, the internal friction angle at peak for dense sand (e_0=0.60) for large confining pressure (σ_3=2.0 MPa) is ϕ^p=38°, and is reached for the normalised vertical deformation of u_2/h_0=8% (Fig.4.1b). In this case, the residual angle of internal friction is also ϕ_{cr}=34° (u_2/h_0=17%). The numerical results demonstrate evidently that the lower the initial void ratio and confining pressure, and the larger the mean grain diameter, the higher the maximum vertical force. The lower the initial void ratio, confining pressure, mean grain diameter, the larger the rate of material softening. With an increase of the initial void ratio, the confining pressure and mean grain diameter, the vertical displacement related to the peak becomes larger. The peak friction angle and residual friction angle are higher for a larger grain diameter (Fig.4.1c). It is due to that the polar granular body is stiffer than a non-polar one using the same constants, and its stiffness increases with increasing mean grain diameter. The work of a micro-polar continuum is augmented, namely, by couple stresses, curvatures and Cosserat rotations which depend upon the mean grain diameter (Eq.3.67). Thus, the additional degree of freedom of a micro-polar continuum in the form of a Cosserat rotation releases the additional resistance against itself due to the presence of couple stresses (this corresponds for instance to the over-stiffening of a hinge joint of a frame by prescribing an additional moment).

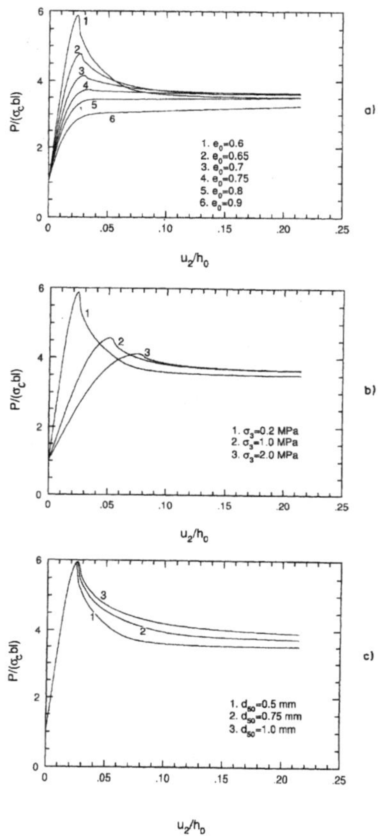

Fig. 4.1. Calculated normalised load-displacement curves: a) for different initial void ratios e_0 (σ_c=0.2 MPa, d_{50}=0.5 mm), b) for different confining pressures σ_c (e_0=0.60, d_{50}=0.5 mm), c) for different mean grain diameters d_{50} (e_0=0.6, σ_c=0.2 MPa): *P* - vertical force on the top, *b*=2.0 cm – initial specimen width, h_0=14.0 cm - initial specimen height, *l*=1.0 m - specimen length, u_2 - vertical displacement of the top

The obtained results of internal friction angles at peak and in the residual state, and the corresponding displacements compare well with experimental results presented by Vardoulakis (1980) and Yoshida et al. (1994). However, the shape of the calculated load-displacement curves differs slightly. Contrary to the experiments (Fig.2.3), the calculated stiffness is too small at the beginning of loading and too high close to the peak of the load-displacement curve.

Figs.4.2-4.11 demonstrate the results for initially medium dense sand (e_0=0.60, d_{50}=0.50 mm) under low confining pressure (σ_c=0.2 MPa). The calculated displacements, Cosserat rotations and void ratios are depicted in Figs.4.2-4.7. The magnitude of the Cosserat rotation (Fig.4.3) is marked by circles with a maximum diameter corresponding to the maximum Cosserat rotation in the given step. In turn, the grade of an increase of the void ratio (Fig.4.4) is marked by a dark region. The darker is the region, the greater sand loosening. At the beginning of the compression, two shear zones are created expanding outward from the weakest element. They can be recognized from the distribution of the Cosserat rotation (Fig.4.3a). Afterwards, and up to the end, only one shear zone dominates (Figs.4.2-4.4). The complete shear zone is already noticeable shortly after the peak. It is characterised both by a concentration of shear deformations (Fig.4.2) and a Cosserat rotation (Fig.4.3), and an increase of the void ratio (Fig.4.4). The thickness of the shear zone on the basis of displacements and Cosserat rotations is about $t_{sh}=12\times d_{50}$. This result is in accordance with the observed thickness from the experiments, $t_{sh}=13\times d_{50}$ (Vardoulakis 1980) and $9\times d_{50}$ (Yoshida et al. 1994). The calculated inclination of the shear zone from the bottom ($\theta\cong 57^{\circ}$) is higher than the mesh alignment ($\theta=54.5^{\circ}$). It is similar as in the experiments by Yoshida et al. (1994) (Tab.2.3) but smaller than the observed one in the experiments by Vardoulakis (1980) (Tab.2.1) for the same conditions. The thickness of the shear zone on the basis of an increase of the void ratio is larger (Fig.4.5) since dense granular material dilates already before the shear zone is created (Herle 1997). The Cosserat rotations are only noticeable in the shear zone (Fig.4.5) and they appear only when a shear zone is created (Fig.4.6). Outside the shear zone, they are negligible (Figs.4.3 and 4.5). Thus, the Cosserat rotations are a good criterion to detect shear zones.

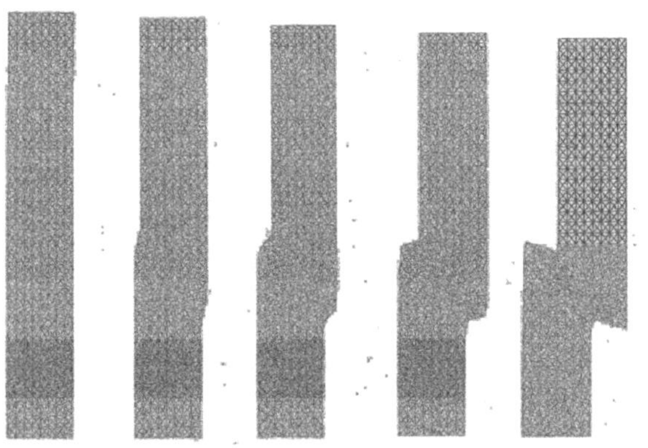

a) b) c) d) e)

Fig. 4.2. Deformed mesh during the vertical displacement of the top u_2 (e_0=0.60, σ_c=0.2 MPa, d_{50}=0.5 mm): *a)* u_2/h_0=0.018, b) u_2/h_0=0.036, c) u_2/h_0=0.055, d) u_2/h_0=0.072, e) u_2/h_0=0.108

The void ratio changes across the shear zone from 0.64 up to 0.82 (Figs.4.5 and 4.7). Outside the shear zone, the void ratio is e=0.597, and is slightly lower than its initial value of e_0=0.60, because the granular material undergoes contractancy at the beginning of shearing (Fig.4.7). The largest void ratio in the shear zone corresponds to the critical value e_c calculated by Eq.3.82. This result confirms the experimental findings by Desrues et al. (1996) that the void ratio in the shear zone reaches its critical value.

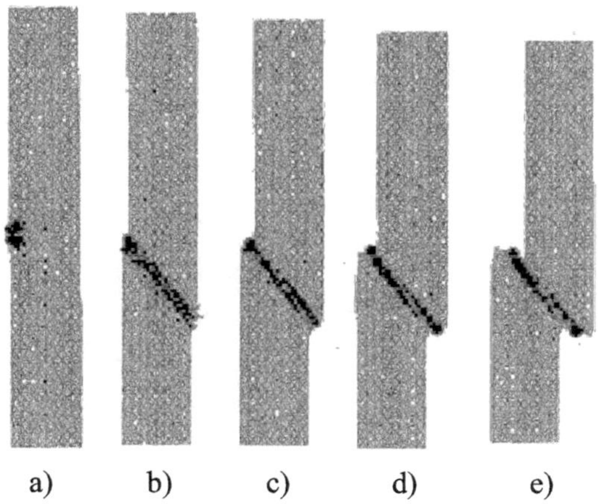

Fig. 4.3. Distribution of the Cosserat rotation during vertical displacement of the top u_2 (e_0=0.60, σ_3=0.2 MPa, d_{50}=0.5 mm): *a)* u_2/h_0=0.018, b) u_2/h_0=0.036, c) u_2/h_0=0.055, d) u_2/h_0=0.072, e) u_2/h_0=0.108

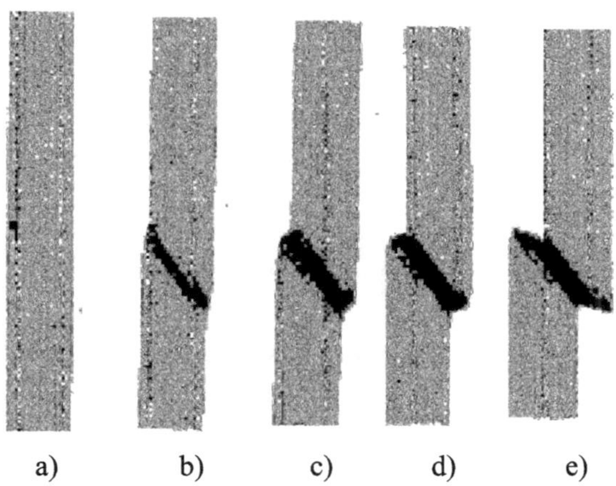

Fig. 4.4. Distribution of the void ratio e during vertical displacement of the top u_2 (e_0=0.60, σ_3=0.2 MPa, d_{50}=0.5 mm): *a)* u_2/h_0=0.018, b) u_2/h_0=0.036, c) u_2/h_0=0.055, d) u_2/h_0=0.072, e) u_2/h_0=0.108

The behaviour of stresses, couple stresses and Cosserat rotations versus the vertical normalised displacement of the top of the specimen both in the shear zone and outside the shear zone is presented in Fig.4.8. All stresses and couple stresses reach a residual state. Their shape is similar to the shape of the load-displacement curve. In the shear zone, the stress tensor is non-symmetric ($\sigma_{12}\neq\sigma_{21}$) and the couple stresses are noticeable. It is in agreement with numerical calculations using a descrete element method (Nakase and Motegi 1998, Oda and Kazama 1998, Iwashita and Oda 1998, Oda and Iwashita 2000). Outside the shear zone, the stress tensor is symmetric ($\sigma_{12}=\sigma_{21}$) and

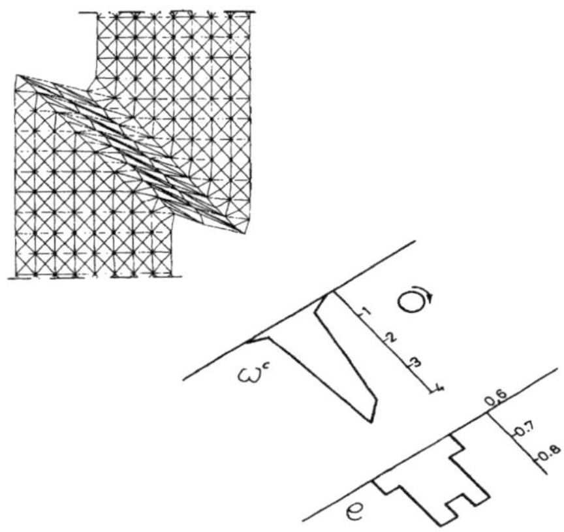

Fig. 4.5. Distribution of the Cosserat rotation ω^c and void ratio e across the shear zone in the middle of the specimen (u_2/h_0=0.2)

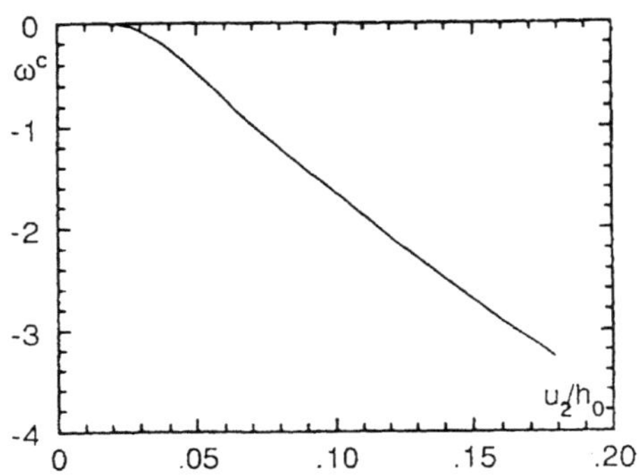

Fig. 4.6. Evolution of the Cosserat rotation ω^c in the shear zone (e_0=0.60, σ_c=0.2 MPa, d_{50}=0.5 mm)

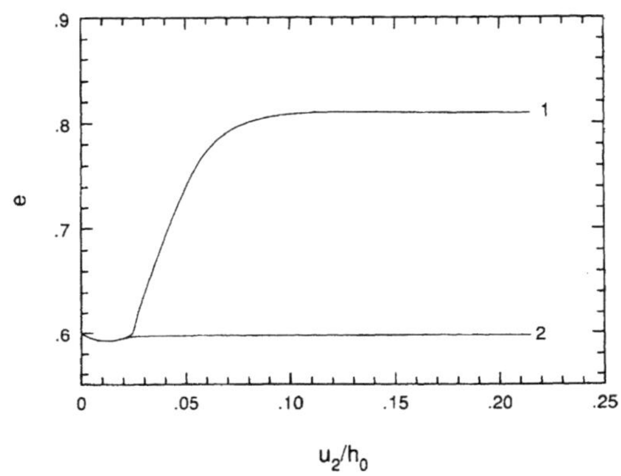

Fig. 4.7. Evolution of the void ratio e in the shear zone (1) and outside the shear zone (2) (u_2 - vertical displacement of the top, h_0 - initial specimen height) (e_0=0.60, σ_3=0.2 MPa, d_{50}=0.5 mm)

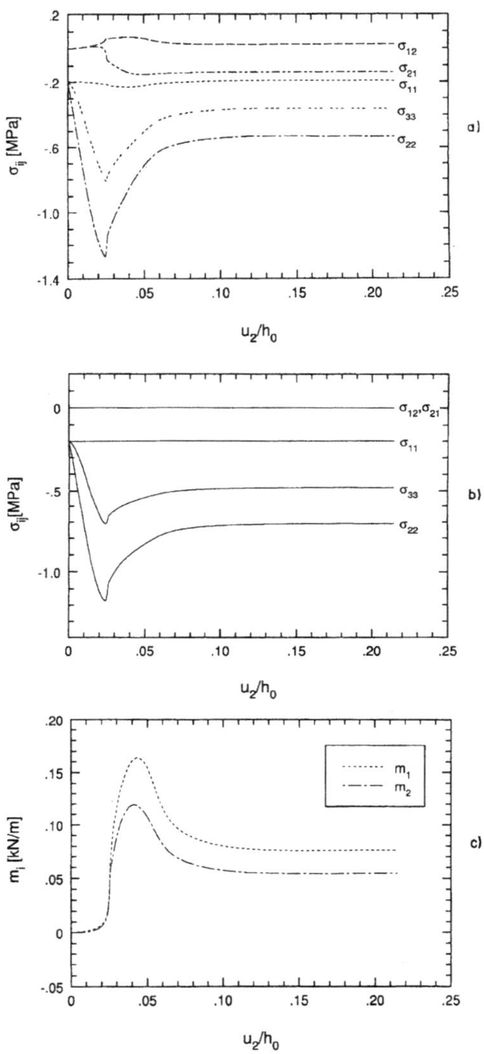

Fig. 4.8. Evolution of the stresses σ_{ij} and couple stresses m_i in the shear zone (a,c) and development of stresses σ_{ij} beyond the shear zone (b), (u_2 - vertical displacement of the top, h_0 - initial specimen height) (e_0=0.60, σ_c=0.2 MPa, d_{50}=0.5 mm)

the couple stresses disappear (Fig.4.8b). The occurrence of the non-symmetry of the stress tensor ($\sigma_{12} \neq \sigma_{21}$) and the appearance of the couple stresses take place immediately with the onset of the shear zone formation.

It can be concluded by comparing Fig.4.1a, Fig.4.8a and Fig.4.8c that the shear zone is created slightly before the peak of the load-displacement curve occurs. Thus, this result coincides with the solution of a bifurcation analysis (Vardoulakis 1980).

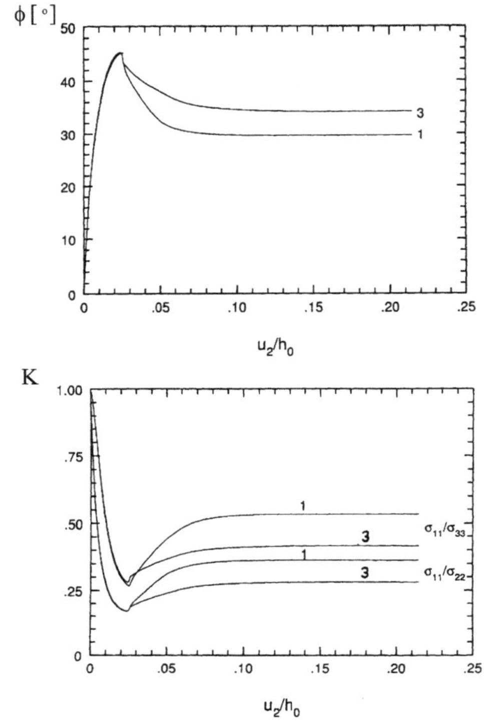

Fig. 4.9. Evolution of the internal friction angle ϕ and the stress ratio K: 1) in the shear zone, 3) outside the shear zone (u_2 - vertical displacement of the top, h_0 - initial specimen height) (e_0=0.60, σ_c=0.2 MPa, d_{50}=0.5 mm)

The residual angle of internal friction ϕ_{cr} is 30° in the shear zone and 34° outside the shear zone (Fig.4.1a). It was calculated by Eq.4.1. The stress ratios σ_{11}/σ_{22} and σ_{11}/σ_{33} are in the global coordinate system in a residual state (Fig.4.9b): 0.35 and 0.52 (shear zone), and 0.28 and 0.42 (beyond the shear zone), respectively. In turn, the stress ratios and the normalised couple stress along the shear zone (local coordinate system) are in the residual state: $\sigma'_{11}/\sigma'_{22}$=0.83, $\sigma'_{11}/\sigma'_{33}$=0.90, and $m'_2=m'_2/(\sigma'_{kk}d_{50})$=0.20, respectively. The values of σ'_{ii} and m'_2 in the shear zone were obtained after the rotation of σ_{ij} and m_i by the angle α=90°-θ (θ - the inclination of the shear zone from the bottom):

$$\sigma'_{11} = \sigma_{11}\cos^2\alpha + \sigma_{22}\sin^2\alpha + (\sigma_{12} + \sigma_{21})\sin\alpha\cos\alpha\,, \tag{4.2}$$

$$\sigma'_{22} = \sigma_{11}\sin^2\alpha + \sigma_{22}\cos^2\alpha - (\sigma_{12} + \sigma_{21})\sin\alpha\cos\alpha\,, \tag{4.3}$$

$$m'_2 = m_2\cos\alpha + m_1\sin\alpha\,, \tag{4.4}$$

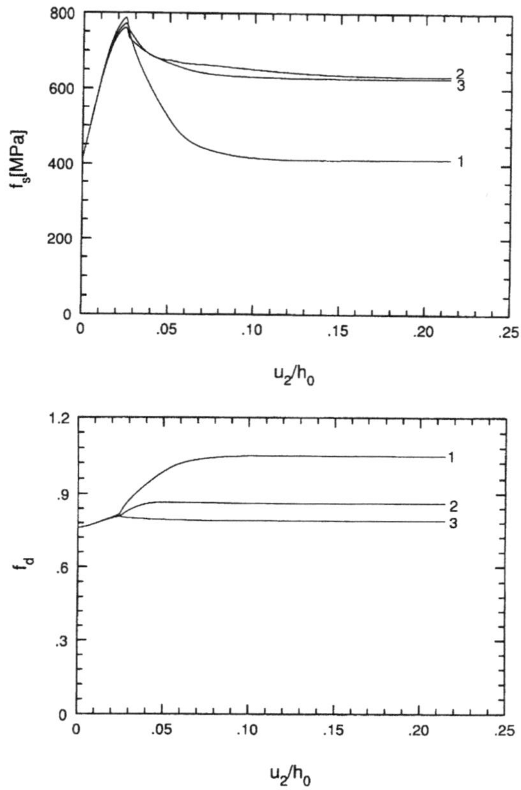

Fig. 4.10. Evolution of the stiffness factor f_s and density factor f_d: 1) in the shear zone, 2) at the boundary of the shear zone, 3) outside the shear zone (u_2 - vertical displacement of the top, h_0 - initial specimen height) (e_0=0.60, σ_c=0.2 MPa, d_{50}=0.5 mm)

The evolution of the stiffness factor f_s (Eq.3.77), density factor f_d, (Eq.3.79), parameter a_1 (Eq.3.85) and Lode angle θ (Eq.3.89) during a plane strain compression is demonstrated in Figs.4.10 and 4.11. The stiffness factor f_s behaves as the load-displacement curve. It is much smaller in the shear zone than in the remaining region at the residual state. The density factor is f_d=1 in the shear zone because $e=e_c$. The coefficient a_1 is approximately 0.25 in the entire specimen in the residual state. The Lode angle θ changes its sign during shearing both within and beyond the shear zone.

The influence of the initial void ratio, confining pressure and mean grain diameter on the Cosserat rotations inside the specimen during the compression is presented in Figs.4.12-4.14. In all cases for sand with e_0<0.80, the granular material experiences softening connected with dilatancy, and a single shear zone is created. The thickness of the shear zone increases with increasing initial void ratio, confining pressure and mean grain diameter.

The thickness (determined from the Cosserat rotations) is about $t_{sz}\cong 12\times d_{50}$ (e_0=0.60), $t_{sz}\cong 16\times d_{50}$ (e_0=0.65), $t_{sz}\cong 20\times d_{50}$ (e_0=0.70) and $t_{sz}\cong 30\times d_{50}$ (e_0=0.75) for

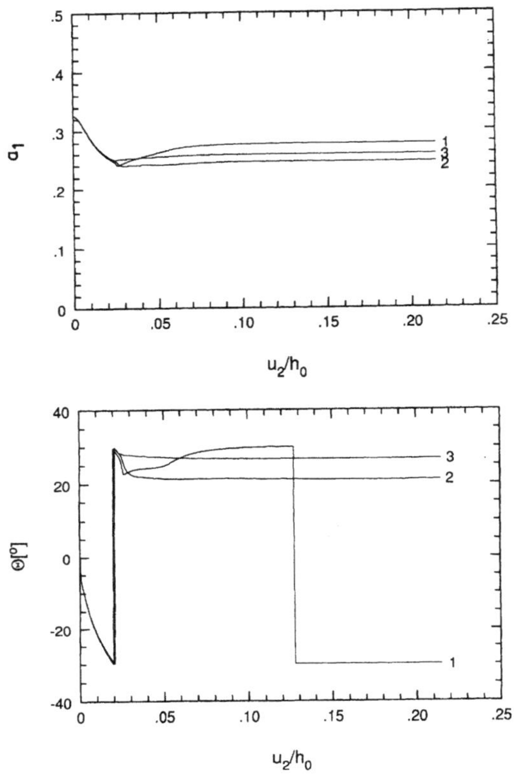

Fig. 4.11. Evolution of the coefficient a_1 and Lode angle Θ: 1) in the shear zone, 2) at the boundary of the shear zone, 3) outside the shear zone (u_2 - vertical displacement of the top, h_0 - initial specimen height) (e_0=0.60, σ_c=0.2 MPa, d_{50}=0.5 mm)

σ_c=0.2 MPa and d_{50}=0.5 mm. In the case of initially loose sand ($e_0 \geq e_c \geq 0.80$), the specimen does not undergo material softening and dilatancy (Fig.4.1 and 4.15), and the shear zone is not created. For a larger confining pressure: $t_{sz} \cong 17 \times d_{50}$ (σ_c=1.0 MPa) and $t_{sz} \cong 23 \times d_{50}$ (σ_c=2.0 MPa). For a larger mean grain diameter: $t_{sz} \cong 8 \times d_{50}$ (d_{50}=0.5 mm) and $t_{sz} \cong 11 \times d_{50}$ (d_{50}=1.0 mm). An increase of the thickness of the shear zone with increasing e_0, σ_c and d_{50} corresponds to a decrease of the rate of softening. The material becomes softer and thus, a larger deformation can develop. An increase of the thickness of the shear zone with an increase of the mean grain diameter was observed in the plane strain compression tests (Vardoulakis 1980, Yoshida et al. 1994). In turn, an increase of the thickness of a shear zone with increasing void ratio was observed by Tejchman (1989) during silo mass flow with controlled outflow velocity in a model silo with parallel very rough walls.

In contrast to the plane strain compression tests by Yoshida et al. (1994), which show a decrease of the thickness of a shear zone with increasing stress for large confining pressures, the FE results show the opposite. It may be due to grain crushing

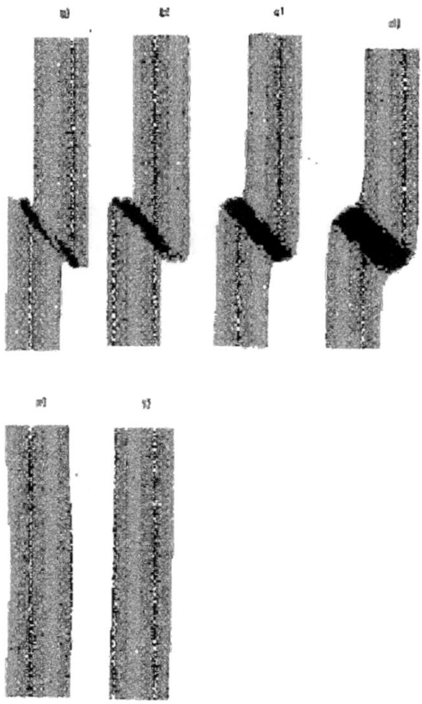

Fig. 4.12. Deformed meshes and Cosserat rotations in the residual state (u_2/h_0=0.1) for different initial void ratios (σ_c=0.2 MPa, d_{50}=0.5 mm): a) e_0=0.60, b) e_0=0.65, c) e_0=0.70, d) e_0=0.75, e) e_0=0.80, f) e_0=0.90

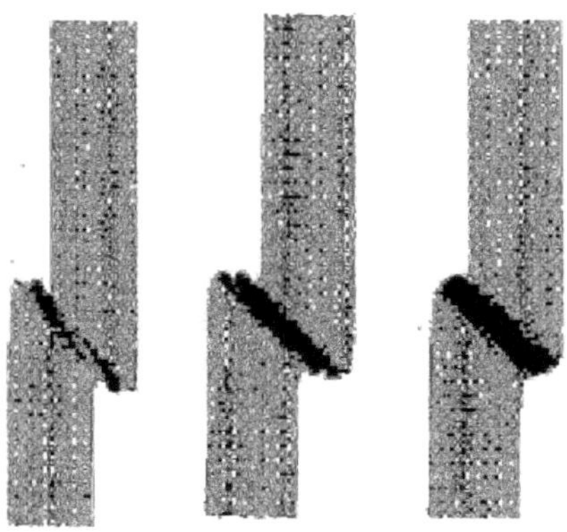

a) b) c)

Fig. 4.13. Deformed meshes and Cosserat rotations in the residual state (u_2/h_0=0.1) for different confining pressures (e_0=0.60, d_{50}=0.5 mm): a) σ_c=0.2 MPa, b) σ_c=1.0 MPa, c) σ_c=2.0 MPa

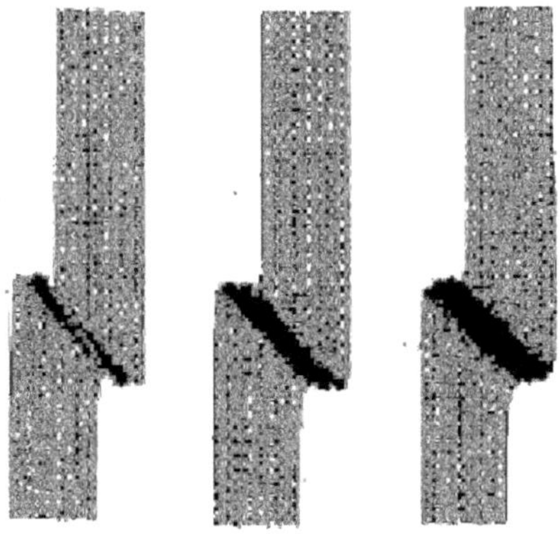

a) b) c)

Fig. 4.14. Deformed meshes and Cosserat rotations in the residual state (u_2/h_0=0.1) for different mean grain diameters (e_0=0.60, σ_c=0.2 MPa): a) d_{50}=0.5 mm, b) d_{50}=0.75 mm, c) d_{50}=1.0 mm

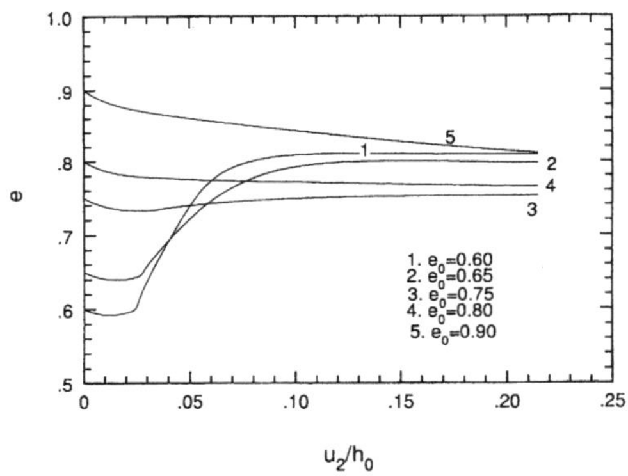

Fig. 4.15. Evolution of the void ratio *e* at the place of the shear zone formation (u_2 - vertical displacement of the top, h_0 - initial specimen height) (σ_c=0.2 MPa, d_{50}=0.5 mm)

which is not taken into account in the constitutive relation. The grain crushing causes the formation of a narrower shear zone since its thickness becomes smaller with decreasing grain size. On the other hand, the friction tests along a very rough wall for a low stress level (Hassan 1995, Löffelmann 1989) confirm the results of the FE analysis in this Section.

The behaviour of the void ratio at the place of the shear zone is shown in Fig.4.15. From the distribution of the void ratio in the shear zone in the course of shearing, it can be seen that sand for e_0<0.80 first experiences slight densification, later loosening and finally does not experience any deformation in the residual state. For e_0≥0.80 the material continuously densifies.

Moreover, the FE calculations show that the inclination of the shear zone measured from the bottom decreases with increasing lateral pressure, mean grain diameter and

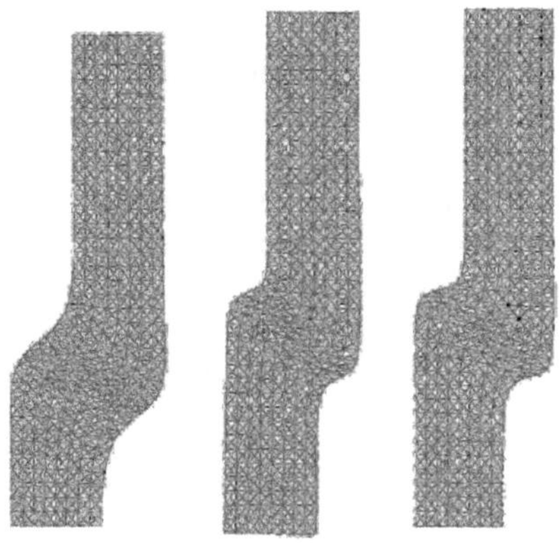

Fig. 4.16. Effect of the initial void ratio, pressure level, mean grain diameter on the shear zone inclination: a) e_0=0.75, σ_c=0.2 MPa, d_{50}=0.5 mm, b) e_0=0.60, σ_c=2.0 MPa, d_{50}=0.5 mm, c) e_0=0.60, σ_c=0.2 MPa, d_{50}=1.0 mm

initial void ratio (Fig.4.16). This finding is in agreement with experiments (Yoshida et al. 1994, Desrues and Hammad 1989, Pradhan 1997). For dense sand (e_0=0.60), the inclination is about $\theta \cong 57^{\circ}$. For initially loose sand (e_0=0.75), the inclination is about $\theta \cong 48^{\circ}$. For a larger lateral pressure (σ_3=2.0 MPa) and a larger mean grain diameter (d_{50}=1.0 mm), the inclination is about $\theta \cong 54^{\circ}$.

The results show that the location of the imperfection in sand determines the position of the shear zone. If a dense element (e_0=0.65) is inserted in medium dense sand (e_0=0.70), no shear localization appears. If the initial void ratio only slightly differs from the initial void ratio of the loose specimen, a shear zone is also not created. In both cases, the imperfection is too weak to induce a shear zone. The micro-polar model is, namely, less sensitive to imperfections than the non-polar one due to the presence of polar quantities.

The calculations were also carried out with a different amplitude and frequency of fluctuations of the change of the current void ratio during plane strain compression (Tejchman 2006). The pulsation of the current void ratio did not affect the shear localization (its thickness and inclination). However, it influenced the void ratio changes beyond the shear zone.

The following observations follow from the FE calculations:

- The micro-polar approach is effective as a regularisation method when shear localization is dominant.
- The Cosserat rotations and couple stresses are only noticeable in shear zones. The Cosserat rotation and the increasing void ratio are suitable indicators for shear zones. The polar granular body is stronger and stiffer than a non-polar one. It is also less sensitive to imperfections in a granular body.
- The calculated thickness of shear zones increases with increasing initial void ratio, pressure level and mean grain diameter.

- The inclination of shear zones from the minor principle stress decreases with increasing initial void ratio, pressure level and mean grain diameter.
- The rate of material softening decreases with increasing initial void ratio, pressure level and mean grain diameter.
- The void ratio in the shear zone approaches its critical value.
- The location of the shear zone is very sensitive to the position of the imperfection in the specimen and its initial void ratio.
- The smaller is the micro-polar parameter, the larger shear zone thickness.
- The spacing of single imperfections does not influence the thickness of shear zones.

Similar FE analyses were carried out by Maier (2002) using a hypoplastic constitutive law by von Wolffersdorff (1996) enhanced by polar terms (Eqs.4.46 and 4.47). The dimensions of the granular specimen were: 140 mm (height) and 40 mm (width). The top and bottom were smooth. The results showed that the effect of the mesh discretization (with a fixed mesh alignment of 54°) on the shear zone width and load-displacement curve was insignificant. The mesh alignment did not influence the shear zone thickness and slightly influenced the shear zone inclination. The shear zone inclination decreased with increasing d_{50}. The vertical force on the top increased with increasing d_{50}. The shear zone thickness increased linearly with increasing d_{50}. The larger is the micro-polar constant a_c, the smaller the polar effect on the load-displacement curve and the narrower the shear zone.

4.2 Monotonic Shearing of an Infinite Layer

The FE calculations of simple shearing (Tejchman 2000, Tejchman and Gudehus 2001) were mainly performed for a sand strip of h=50 mm height. The study was performed with only one element column with a width of b=10 cm, consisting of 20 quadrilateral horizontal elements composed of four diagonally crossed triangles. Thus, the height of the elements was $5\times d_{50}$ for the smallest mean grain diameter considered in the analysis (d_{50}=0.5 mm). For the largest mean grain diameter (d_{50}=10 mm), the height of the elements was $d_{50}/4$. Initial FE calculations showed that the number of element columns does not influence the results if displacements and rotations along both sides of the column are the same. The calculations were carried out with large deformations and curvatures (Section 3.5).

As the initial stress state in the granular strip, a K_o-state without polar quantities (σ_{22}=-10 kPa, $\sigma_{11}=\sigma_{33}$=-3 kPa, $\sigma_{12}=\sigma_{21}=m_1=m_2=0$) was assumed. Gravity was neglected.

A quasi-static shear deformation was initiated through constant horizontal displacement increments prescribed at the nodes along the top of the strip. Both bottom and top were very rough. The boundary conditions were along the bottom: $u_1=0$, $u_2=0$ and $\omega^c=0$, and along the top: $u_1=n\Delta u$, ω^c=0, and $\sigma_{22}=p$. n denotes the number of time steps, Δu is the constant displacement increment in one step and p is the vertical uniform pressure prescribed to the top of the strip. Thus, full shearing of sand along both boundaries was assumed. The displacement increments were chosen as $\Delta u/h$=0.0005. About 2000 steps were performed. The following material constants for so-called

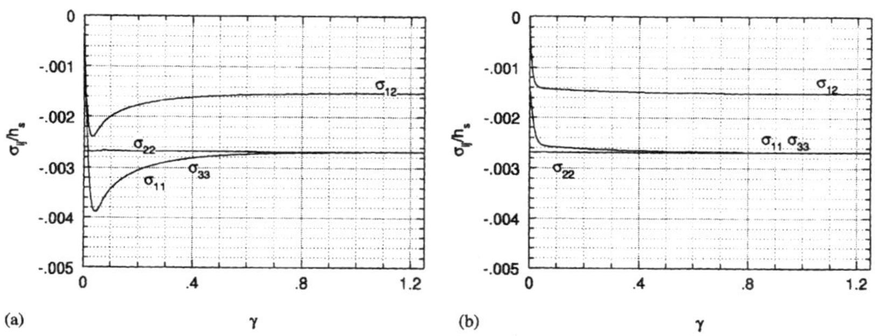

Fig. 4.17. Numerical results of simple shearing within a non-polar continuum: evolution of normalised stresses σ_{ij}/h_s versus shear deformation γ with dense (a) and loose sand (b) under constant pressure p=500 kPa

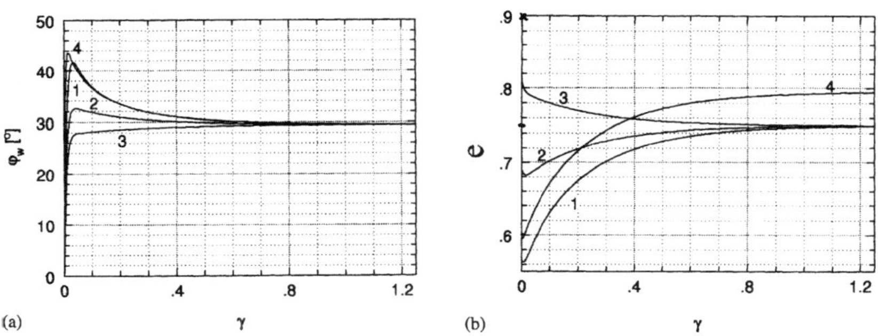

Fig. 4.18. Numerical results of simple shearing within a non-polar continuum: evolution of wall friction angle φ_w versus γ (a), and evolution of void ratio e versus γ (b): 1. e_0=0.60, p=500 kPa, 2. e_0=0.75, p=500 kPa, 3. e_0=0.90, p=500 kPa, 4. e_0=0.60, p=50 kPa

Karlsruhe sand were used: e_{i0}=1.3, e_{d0}=0.51, e_{c0}=0.82, ϕ_c=30°, h_s=190 MPa, α=0.3, n=0.5, d_{50}=0.5 mm and $a_c=a_1^{-1}$.

The calculations were performed with different initial void ratios of sand (e_0=0.60, e_0=0.75, e_0=0.80 and e_0=0.90), vertical pressures (p=50 kP, p=500 kPa and p=5000 kPa) and mean grain diameters (d_{50}=0.5 mm (h/d_{50}=100), d_{50}=5 mm (h/d_{50}=10), d_{50}=10 mm (h/d_{50}=5)). In addition, the effect of the grain roughness, expressed by the micro-polar constant a_c was studied by assuming $a_c=0.5a^{-1}$ (Eq.3.86) (other material constants were left unchanged).

The global stiffness matrix (calculated with only two first terms of the constitutive equations which are linear in d_{kl}^c and kd_{50}, Eqs.3.137-3.140) was updated every 100 steps.

Numerical results of simple shearing (so-called element tests) are summarized in Figs.4.17 and 4.18. The sand element was subject first to compression by pressure p and then to shearing with free dilatancy. Presented are the evolution of normalised stresses σ_{ij}/h_s during shearing for dense sand (e_0=0.60) and very loose sand (e_0=0.90) under constant pressure p=500 kPa (Fig.4.17), and the changes both of the mobilised friction angle $\varphi_w = arctan(\sigma_{12}/\sigma_{22})$ and void ratio e for different values of e_0 and p versus shear deformation γ (Fig.4.18). The angle φ_w (called wall friction angle as it concerns shearing between a wall and granular body) is related to the entire granular specimen since the stresses σ_{12} and σ_{22} are constant along the layer length and layer height.

Stresses and wall friction angles show a simple asymptotic behaviour. A critical (residual) state indicated by simultaneously vanishing stress rates and volumetric deformation rate, and a constant void ratio ($\overset{o}{\sigma}_{ij}=0$, $d_{kk}=0$, $\dot{e}=0$) is reached for about γ=80%-100%. The maximum wall friction angle decreases with increasing e_0 and p. The effect of p on φ_w is smaller than in the case of plane strain compression (Section 4.1). The maximum wall friction angle for dense sand is $\varphi_w=43^o$ (e_0=0.60, $p=50$ kPa) and $\varphi_w=41^o$ (e_0=0.60, p=500 kPa), respectively. The residual wall friction angle, φ_w =29.5^o, is independent of the pressure and initial void ratio. It differs from the critical angle of internal friction $\phi_c=30^o$ (Eq.3.90) by 0.5^o. The maximum wall friction angle becomes higher with decreasing dilatancy constraint which is equivalent to a decrease of pressure. With a significant increase of pressure, the results approach the results with constant volume deformation. The shear deformation related to the peak of the wall friction angle becomes larger with increasing pressure. During simple shearing, the Lode angle (Eq.3.89) in the residual state is 30° (referred to triaxial compression) and the parameter a_1^{-1}=3.8 (Eq.3.86).

Dense sand (curves ‘1’ and ‘4’ in Fig.4.18b) experiences dilatancy after an initial densification until the residual state is reached. The amount of dilatancy expressed by an increase of the void ratio grows with a decrease of both initial void ratio and pressure. In turn, loose sand (curve ‘3’ in Fig.4.18b) undergoes only densification which decreases with increasing pressure. The stress ratios σ_{11}/σ_{22} and σ_{11}/σ_{33} in the residual state are 1. Thus, due to shearing, the sand specimen is homogenised and behaves like a viscous fluid. With $\sigma_{11}=\sigma_{22}$, the residual angle of internal friction, ϕ calculated from the Mohr’s formula (Eq.4.1) is equal to 34.5^o. It is about 30% higher than the wall friction angle. This relationship between ϕ and φ_w is in accordance with the experimental data (Herle 1997).

The micro-polar FE calculations of shearing have been made for a initially dense sand strip (e_0=0.60, d_{50}=0.5 mm, h=50 mm, h/d_{50}=100) between two very rough boundaries with free dilatancy under constant vertical normal stress p=500 kPa. The sand specimen was subject first to compression by pressure p and then to shearing with free dilatancy. Fig.4.19 shows the evolution of normalised stress components σ_{ij}/h_s at the mid-point of the strip and at the wall, the evolution of the normalised couple stress $m_2/(h_s d_{50})$ at the wall, the evolution of the mobilised wall friction angle $\varphi_w = arctan(\sigma_{12}/\sigma_{22})$, the void ratio e and the pressure rati σ_{11}/σ_{22} at the mid-point of the strip with the normalised horizontal displacement at the top u_1^t/h (h=50 mm – layer height).

As in the case of non-polar calculations, the wall friction angle φ_w is related to the entire granular layer (the stresses σ_{12} and σ_{22} are constant along the layer height and layer length). Fig.4.20 presents the distribution of the normalised horizontal displacement u_1/h, the Cosserat rotation ω^c, the normalised stresses σ_{ij}/h_s and the couple stress $m_2/(h_s d_{50})$ along the normalised height x_2/d_{50} at the residual state. A deformed FE mesh for the residual state is shown in Fig.4.21. Darker region indicates higher

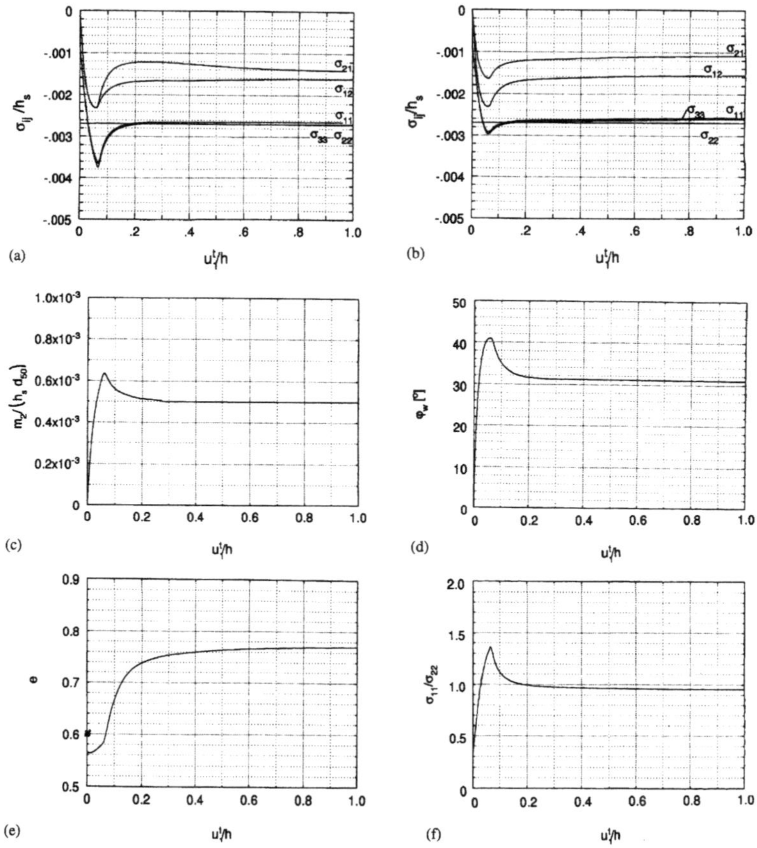

Fig. 4.19. Shearing within a micro-polar continuum with dense sand (e_0=0.60, p=500 kPa, d_{50}=0.5 mm, h=50 mm): a) evolution of normalised stresses σ_{ij}/h_s at the mid-point versus shear deformation u_1^t/h, b) evolution of normalised stresses σ_{ij}/h_s at the wall versus u_1^t/h, c) evolution of normalised wall couple stress $m_2/(h_s d_{50})$ versus u_1^t/h, d) evolution of wall friction angle φ_w versus u_1^t/h, e) evolution of void ratio e in the middle of the layer versus u_1^t/h, and f) evolution of stress ratio σ_{11}/σ_{22} in the middle of the layer versus u_1^t/h

void ratio. The presented quantities were taken as the mean values in each quadrilateral element.

All state variables (stress, couple stress, void ratio and wall friction angle) tend to asymptotic values. The evolution of stresses and wall friction angle during shearing is similar as in a non-polar continuum. The polar stress σ_{12} is higher than the non-polar one. The maximum wall friction angle is the same as in a non-polar continuum, but the residual wall friction angle is by 2° higher.

The Cosserat rotation, curvature, couple stress and non-symmetry of the stress tensor ($\sigma_{12} \neq \sigma_{21}$) are noticeable during shearing. In the middle of the layer, a shear zone is characterised by the appearance of the Cosserat rotation and a strong increase of the

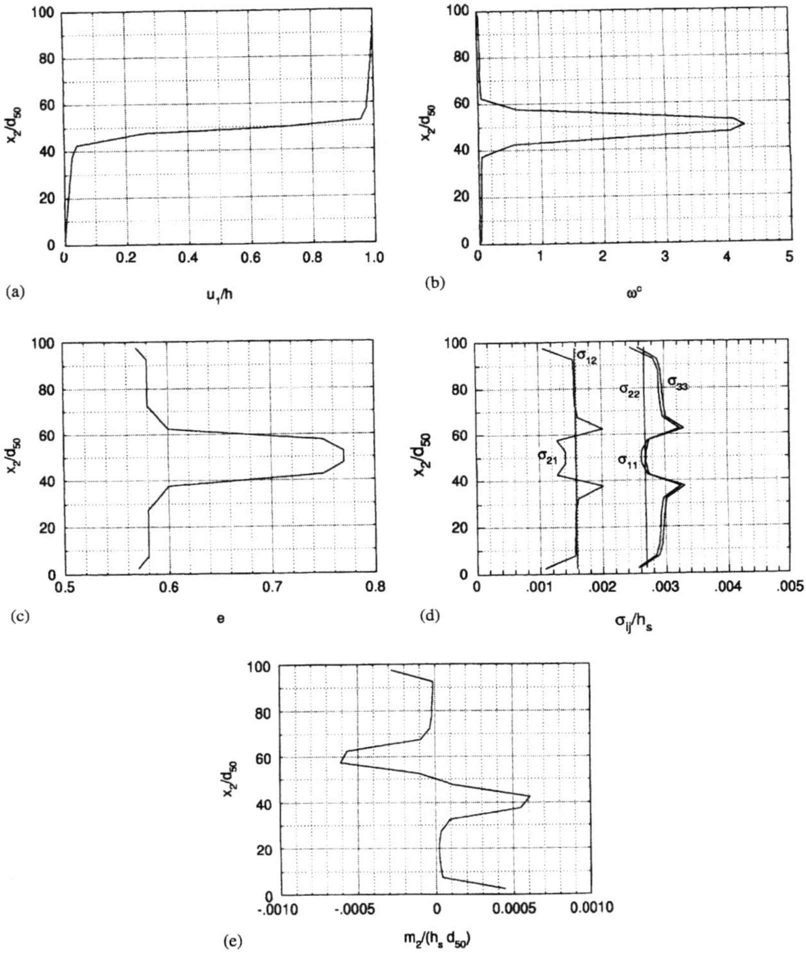

Fig. 4.20. Shearing within a micro-polar continuum with dense sand at $u_1^t/h=1.0$ ($e_0=0.60$, $p=500$ kPa, $d_{50}=0.5$ mm, $h=50$ mm): a) distribution of normalised lateral displacement u_1/h, b) Cosserat rotation ω^c, c) void ratio e, d) normalised stresses σ_{ij}/h_s, and e) normalised wall couple stress $m_2/(h_s d_{50})$ across the normalised height x_2/d_{50}

void ratio. At the base and top of the shear zone, a strong jump of the horizontal displacement, curvature, stresses and couple stress takes place.

The thickness of the shear zone, as visible from the Cosserat rotation and the stress jump at the shear zone edges, is about $20\times d_{50}$. The calculated value exceeds the one calculated from a plane strain compression test, $13\times d_{50}$ (Section 4.1), with the same initial void ratio and mean grain diameter of sand, and the similar pressure level and number of finite elements across the shear zone.

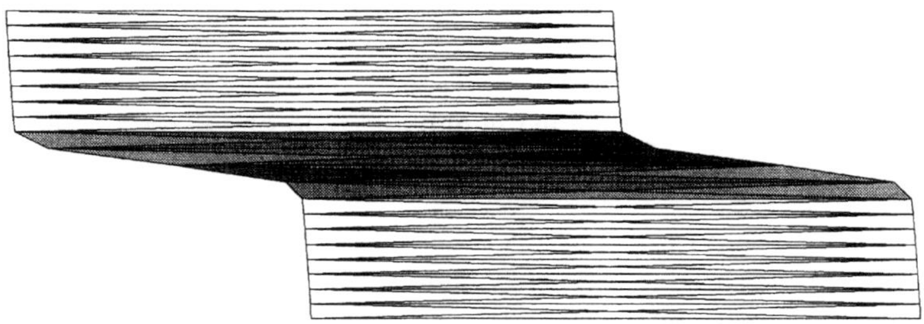

Fig. 4.21. Deformed FE mesh with distribution of void ratio e for dense sand (e_o=0.60, p=500 kPa, d_{50}=0.5 mm, h=50 mm)

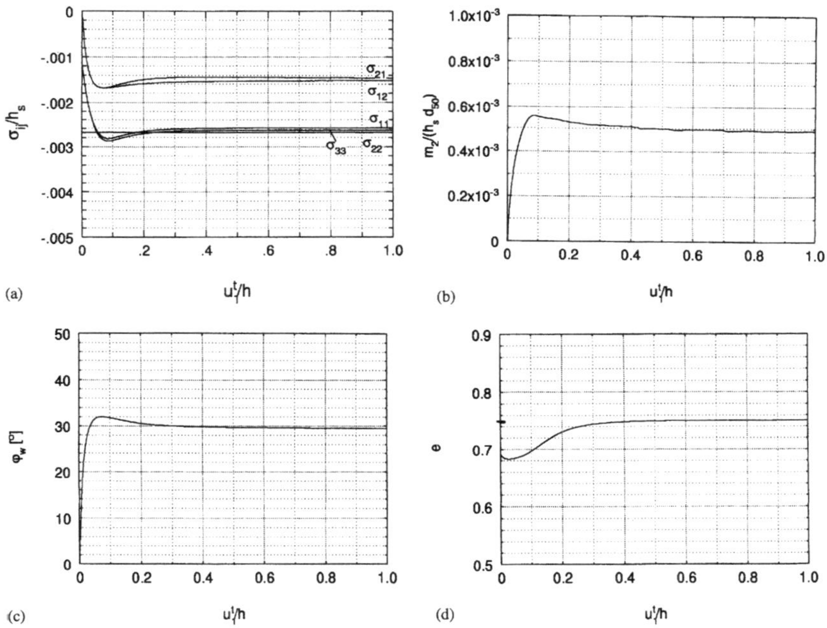

Fig. 4.22. Shearing within a micro-polar continuum with medium dense sand (e_0=0.75, p=500 kPa, d_{50}=0.5 mm, h=50 mm): a) evolution of normalised stresses σ_{ij}/h_s at the mid-point versus shear deformation u_1^t/h, b) evolution of normalised wall couple stress $m_2/(h_s d_{50})$ versus u_1^t/h, d) evolution of wall friction angle φ_w versus u_1^t/h, and e) evolution of void ratio e in the middle of the layer versus u_1^t/h

Thus, the thickness of shear zones is also influenced by the direction of the deformation and the way of the loading. The distribution of stresses σ_{11}, σ_{33} and σ_{21} and the horizontal displacement u_1 across the shear zone is strongly non-linear. The stresses σ_{11}, σ_{33} and σ_{21} in the shear zone have a parabolic distribution. The stresses σ_{11} and σ_{33} have their minimum and the stress σ_{21} its maximum in the middle of the shear zone.

The stress ratios σ_{11}/σ_{22} and σ_{11}/σ_{33} are at residual state almost equal to 1 in the shear zone.

The void ratio in the middle of the shear zone is e_c=0.77. Outside the shear zone is about 0.57-0.58. It is smaller than the initial void ratio of e_0=0.60 due to the initial contractancy. At the shear zone edges, the void ratio changes from 0.58 to 0.60.

Influence of initial void ratio

The results of simple shearing with medium dense sand (e_0=0.75) and very loose sand (e_0=0.90) under constant vertical normal stress p=500 kPa (d_{50}=0.5 mm, h/d_{50}=100) are presented in Figs.4.22 and 4.23, and Figs.4.24 and 4.25, respectively. Comparison with Figs.4.20 and 4.21 shows that the thickness of the shear zone increases with increasing initial void ratio e_0. It is about 30×d_{50} for medium dense sand. For very loose sand, it is equal to the layer height, i.e. 100×d_{50}. In this last case, the Cosserat rotation

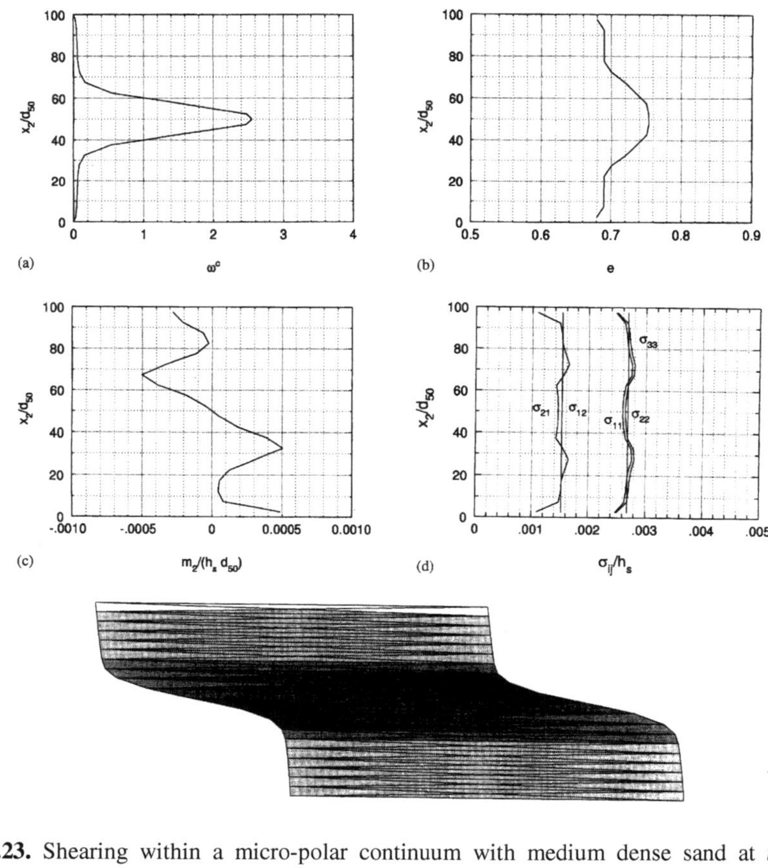

Fig. 4.23. Shearing within a micro-polar continuum with medium dense sand at u_1^t/h=1.0 (e_0=0.75, σ_{ij}=500 kPa, d_{50}=0.5 mm, h=50 mm): a) distribution of Cosserat rotation ω^c, b) void ratio e, c) normalised wall couple stress $m_2/(h_s d_{50})$ and d) normalised stresses σ_{ij}/h_s across the normalised height x_2/d_{50}, and e) deformed FE mesh with distribution of void ratio e

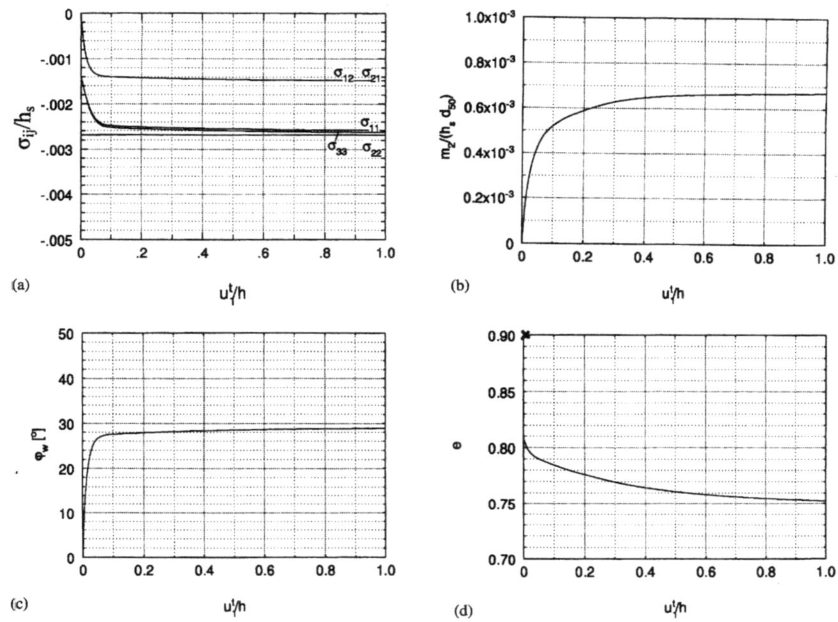

Fig. 4.24. Shearing within a micro-polar continuum with loose sand (e_0=0.90, p=500 kPa, d_{50}=0.5 mm, h=50 mm): a) evolution of normalised stresses σ_{ij}/h_s at the mid-point versus shear deformation u_1^t/h, b) evolution of normalised wall couple stress $m_2/(h_s d_{50})$ versus u_1^t/h, c) evolution of wall friction angle φ_w versus u_1^t/h, and d) evolution of void ratio e in the middle of the layer versus u_1^t/h

spreads over the entire layer height. For a lightly densified sand (e_0=0.80, d_{50}=0.5 mm), the thickness of the shear zone is about 40×d_{50}. The thickness of the shear zone with $e_0 \geq e_c$ is always equal to the layer height.

The residual wall friction angle decreases with increasing initial void ratio: φ_w=31° (e_0=0.60), φ_w=30° (e_0=0.75) and φ_w=29.5° (e_0=0.90). For very loose sand, it is the same as for a non-polar continuum. The maximum Cosserat rotation also decreases with increasing initial void ratio. With an increase of the initial void ratio, the distribution of stresses becomes more uniform across the layer, and the non-symmetry of the stress tensor significantly decreases. The stress gradient at the edges of the shear zone is smaller. The distribution of stresses in loose sand is practically uniform across the entire layer height. The wall couple stress m_2 slightly increases with increasing e_0. The normalised wall couple stress $\hat{m}_2 = m_2/(\sigma_{kk} d_{50})$ is 0.06 with e_0=0.60 and 0.09 with e_0=0.90 in the residual state ($\sigma_{kk} \cong$1500 kPa).

Influence of pressure level

Results with dense sand (e_0=0.60, d_{50}=0.5 mm, h/d_{50}=100) under constant vertical normal stress p=50 kPa and p=5000 kPa (instead of p=500 kPa as before) are presented in Figs.4.26 and Fig.4.27, respectively. They show a reduction of the wall

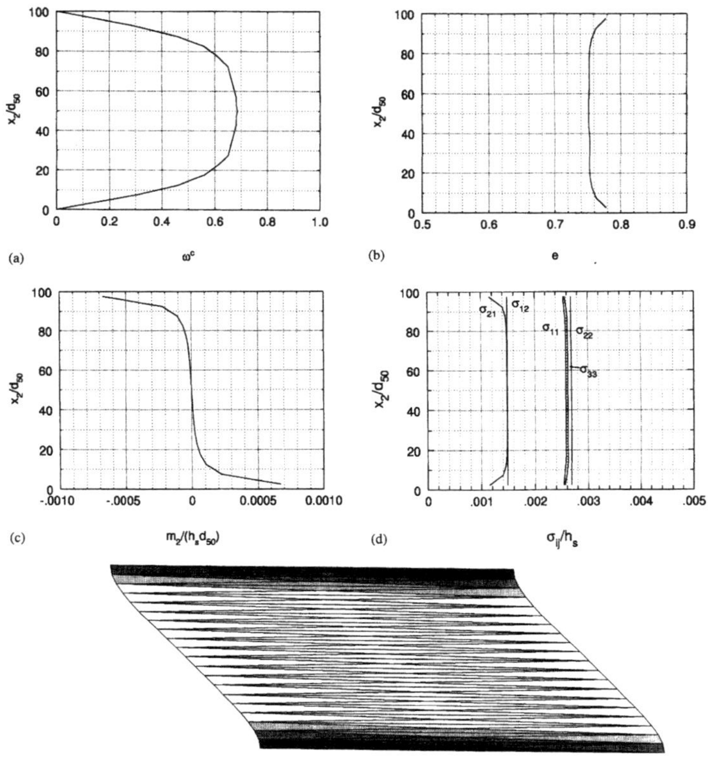

Fig. 4.25. Shearing within a micro-polar continuum with loose sand at u_1^t/h=1.0 (e_0=0.90, p=500 kPa, d_{50}=0.5 mm, h=50 mm): a) distribution of Cosserat rotation ω^c, b) void ratio e, c) normalised wall couple stress $m_2/(h_s d_{50})$ and d) normalised stresses σ_{ij}/h_s across the normalised height x_2/d_{50}, and e) deformed FE mesh with distribution of void ratio e

friction angle at peak and in the residual state with pressure. The wall friction angles at peak and in the residual state are: 35° and 29° (p=5000 kPa), 41° and 31° (p=500 kPa), and 43° and 31° (p=50 kPa), respectively. The difference between the shear stresses in the shear zone becomes smaller with increasing pressure.

The thickness of the shear zone taken from the Cosserat rotation and the stress jumps grows with pressure from about (18-20)×d_{50} (p=50 kPa) and 20×d_{50} (p=500 kPa) up to 25×d_{50} (p=5000 kPa). The effect of pressure on the shear zone thickness is less pronounced than in the case of a plane strain compression test (Section 4.1). The Cosserat rotation decreases with increasing pressure.

The normalised shear zone thickness t_s/d_{50} is depicted versus the relative void ratio

$$r_e = \frac{e - e_d}{e_c - e_d}, \tag{4.5}$$

in Fig.4.28, wherein e_d and e_c are calculated with Eqs.4.36 and 4.37, and e is the initial void ratio (before the shearing but after the introduction of the vertical pressure).

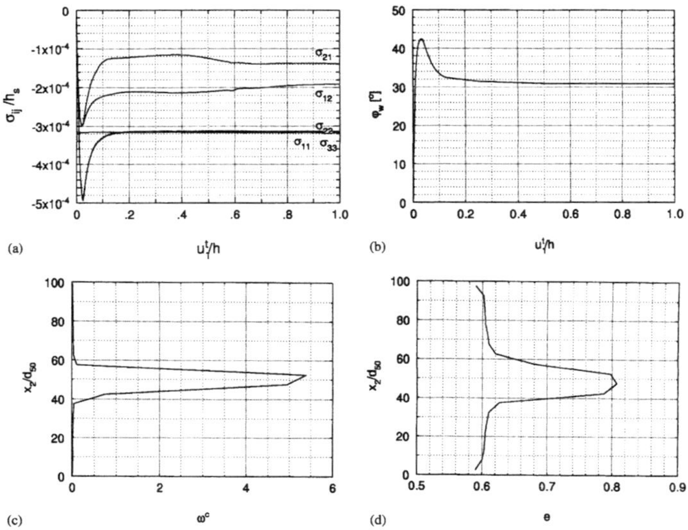

Fig. 4.26. Shearing within a micro-polar continuum with dense sand (e_0=0.60, p=50 kPa, d_{50}=0.5 mm, h=50 mm): a) evolution of normalised stresses σ_{ij}/h_s at the mid-point versus shear deformation u_1^t/h, b) evolution of wall friction angle φ_w versus u_1^t/h, c) distribution of Cosserat rotation ω^c and d) distribution of void ratio e across the normalised height x_2/d_{50} at u_1^t/h=1.0

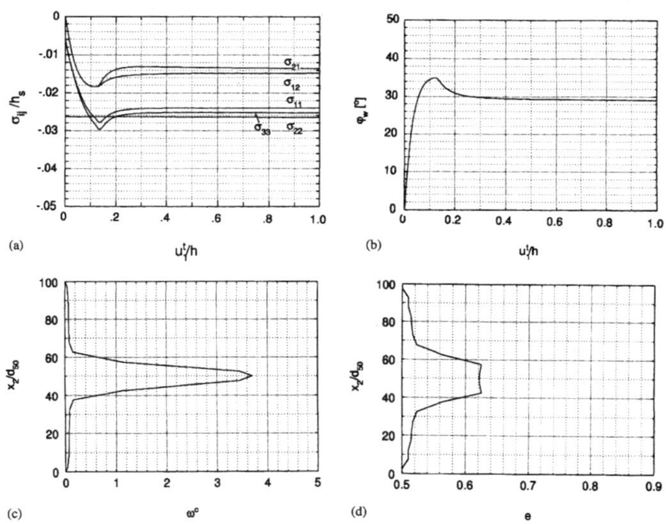

Fig. 4.27. Shearing within a micro-polar continuum with dense sand (e_0=0.60, p=5000 kPa, d_{50}=0.5 mm, h=50 mm): a) evolution of normalised stresses σ_{ij}/h_s at the mid-point versus shear deformation u_1^t/h, b) evolution of wall friction angle φ_w versus u_1^t/h, c) distribution of Cosserat rotation ω^c and d) distribution of void ratio e across the normalised height x_2/d_{50} at u_1^t/h=1.0

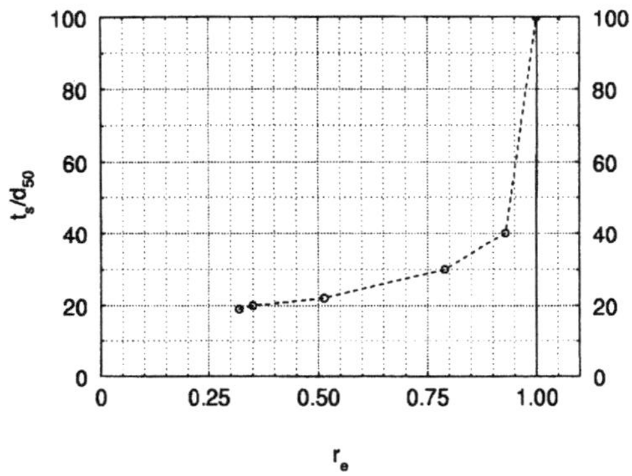

Fig. 4.28. Normalised shear zone thickness t_s/d_{50} versus relative void ratio r_e (Eq.4.5) during shearing of a narrow granular layer with polar quantities (layer height h=50 mm, mean grain diameter d_{50}=0.5 mm)

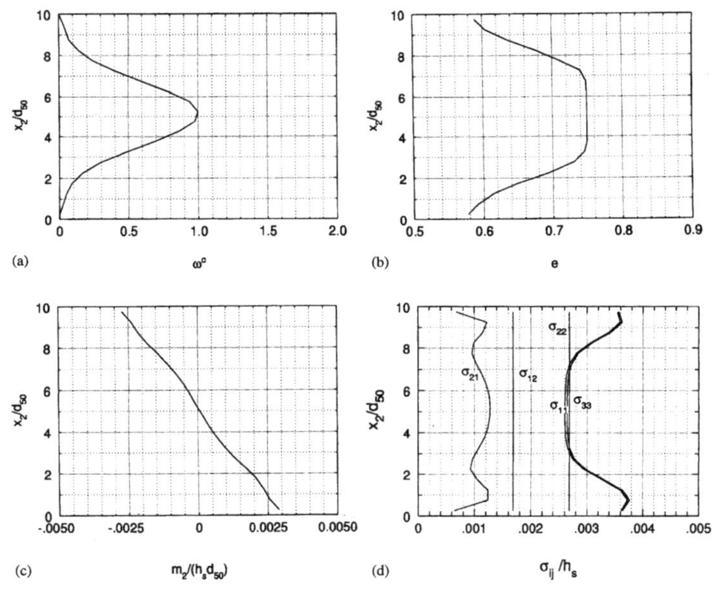

Fig. 4.29. Shearing within a micro-polar continuum with dense sand at u_1^t/h=1.0 (e_0=0.60, p=500 kPa, d_{50}=5 mm, h=50 mm): a) distribution of Cosserat rotation ω^c, b) void ratio e, c) normalised wall couple stress $m_2/(h_s d_{50})$ and d) normalised stresses σ_{ij}/h_s across the normalised height x_2/d_{50}

The result of Fig.4.28 demonstrates that the normalised shear zone thickness t_s/d_{50} is an unique and nearly exponential function of r_e (Gudehus 1998). If the void ratio is equal or exceeds the pressure-dependent critical value, i.e. r_e=1, the shear zone reaches the size of the granular body (in this case $100 \times d_{50}$).

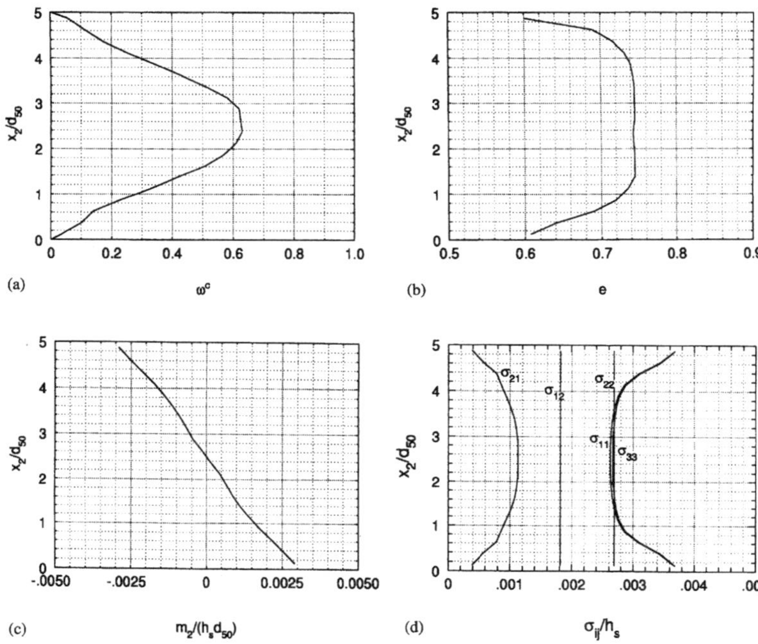

Fig. 4.30. Shearing within a micro-polar continuum with dense sand at u_1^t/h=1.0 (e_0=0.60, p=500 kPa, d_{50}=10 mm, h=50 mm): a) distribution of Cosserat rotation ω^c, b) void ratio e, c) normalised wall couple stress $m_2/(h_s d_{50})$ and d) normalised stresses σ_{ij}/h_s across the normalised height x_2/d_{50}

Influence of mean grain diameter

The effect of the mean grain diameter on shearing of initially dense sand (e_0=0.60, p=500 kPa) is demonstrated in Figs.4.29 (d_{50}=5 mm, h/d_{50}=10) and 4.30 (d_{50}=10 mm, h/d_{50}=5), respectively. The results reveal that the increase of the mean grain diameter increases the stress σ_{12}, stresses σ_{11} and σ_{22} at the shear zone edges, non-symmetry of the stress tensor in the shear zone, wall friction angle at peak and in the residual state, wall couple stress, shear zone thickness and decreases the maximum Cosserat rotation and maximum void ratio. The mean grain diameter d_{50} influences results as a polar granular body is stiffer than a non-polar one using the same constants (Eq.4.22): the stiffness increases with increasing mean grain diameter. The wall friction angles at peak and in the residual state are: φ_w=41° and φ_w=31° (d_{50}=0.5 mm), φ_w=42° and φ_w=32° (d_{50}=5 mm) and φ_w=45° and φ_w=34° (d_{50}=10 mm), respectively. In turn, the thickness of the shear zone is about 10 mm (20×d_{50}) with d_{50}=0.5 mm, 40 mm (8×d_{50}) with d_{50}=5 mm, and 50 mm (5×d_{50}) with d_{50}=10 mm on the basis of the Cosserat rotation and stress jump, respectively. The quantity of 8×d_{50} was also obtained for the same ratio of h/d_{50}=10 but with h=5 mm and d_{50}=0.5 mm (Tejchman 2000). For $d_{50}\geq$1 mm, the thickness of the shear zone was always equal to the layer height of h=5 mm (Tejchman 2000). In this case, the Cosserat rotation spreads across the entire layer and no stress and couple stress jumps are obtained. The distribution of the stresses, couple stress and Cosserat rotation is then similar as in the case of initially loose sand.

The normalised wall couple stress $\hat{m}_2 = m_2/(\sigma_{kk} d_{50})$ is about 0.37 for d_{50}=5-10 mm in the residual state. For the case if the thickness of the shear zone is approximately equal to the layer height, $\hat{m}_2$ slightly decreases with increasing void ratio. It is 0.27 (e_0=0.75, d_{50}=5 mm) and 0.22 (e_0=0.90, d_{50}=5 mm).

Influence of grain roughness

The effect of the grain roughness, expressed by the micro-polar constant a_c (Eq.4.41), is presented in Fig.4.31. The calculations were performed with initially dense sand (e_0=0.60, p=500 kPa, d_{50}=5 mm, h/d_{50}=10) assuming $a_c=0.5a_1^{-1}$. The parameter a_1^{-1} (Eq.4.43) lies between 4.5 (at peak) and 3.7 (at residual state). A decrease of a_c causes a greater influence of Cosserat quantities on the material behaviour by diminishing the effect of the function N_i^c in Eq.4.41, which is equivalent to the growth of the grain roughness.

The lower the micro-polar constant a_c, the higher are σ_{12} across the layer, σ_{21}, σ_{11} and σ_{33} at the shear zone edges, wall couple stress m_2, wall friction angle at peak and residual state, and non-symmetry of the stress tensor in the shear zone. The wall friction angle is higher by 1° at peak and by 3° in the residual state with a larger grain roughness. The thickness of the shear zone and the maximum Cosserat rotation are practically the same.

The lower the micro-polar constant a_c, the higher are the void ratio changes. The void ratio in the shear zone can be even higher than e_c (Fig.4.31b). In the case of loose

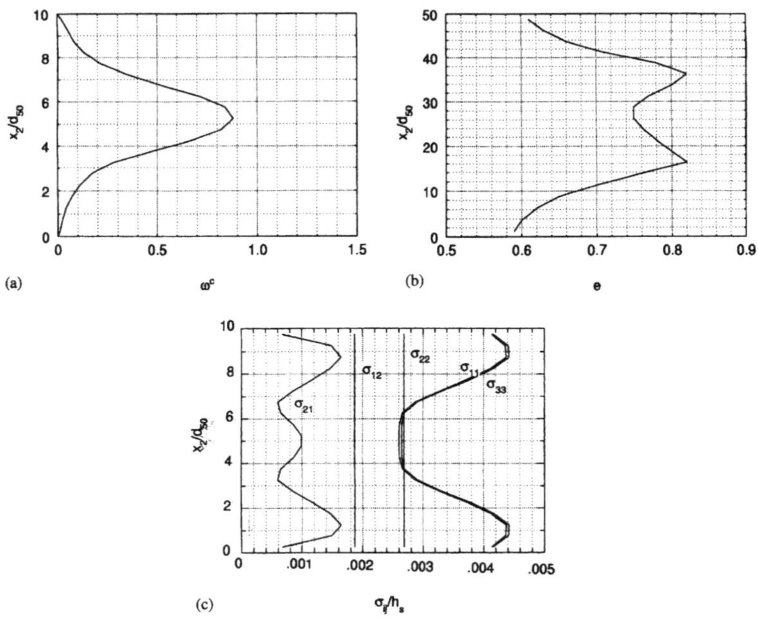

Fig. 4.31. Shearing within a micro-polar continuum with dense sand at u_1^t/h=1.0 (e_0=0.60, p=500 kPa, d_{50}=5 mm, h=50 mm, $a_c=0.5a_1^{-1}$): a) distribution of Cosserat rotation ω^c, b) void ratio e, c) normalised stresses σ_{ij}/h_s across the normalised height x_2/d_{50}

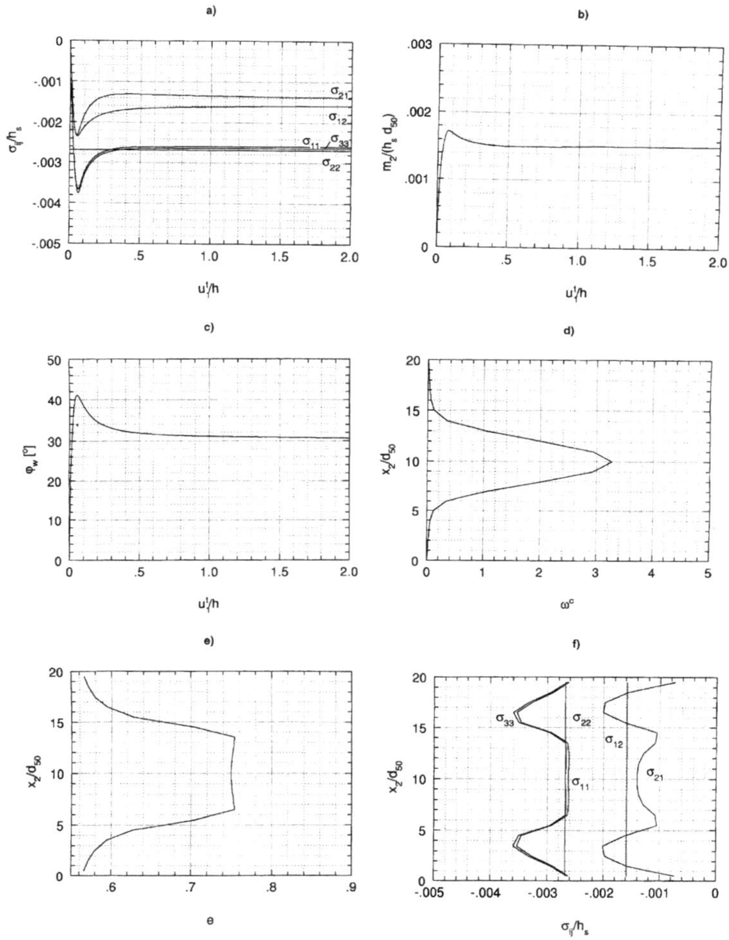

Fig. 4.32. Shearing within a polar continuum with dense sand (e_0=0.60, p=500 kPa, d_{50}=0.5 mm, h=10 mm): a) evolution of normalised stresses σ_{ij}/h_s at the mid-point versus shear deformation u_1^t/h, b) evolution of normalised wall couple stress $m_2/(h_s d_{50})$ versus u_1^t/h, c) evolution of wall friction angle φ_w versus u_1^t/h, d), e) and f) distribution of Cosserat rotation ω^c, void ratio e and normalised stresses σ_{ij}/h_s across the normalised height x_2/d_{50} (u_1^t/h=2.0)

sand with e_0=0.90 ($a_c=0.5a_1^{-1}$), the calculations show that the material close to the wall undergoes an initial contractancy and dilates with further shearing.

Influence of the layer height

The results for the layer height of h=10 mm are shown in Fig.4.32 (e_o=0.60, d_{50}=0.5 mmm, p=500 kPa).

The effect of h on φ_w is rather insignificant. The stresses σ_{11} and σ_{33} at the shear zone edges, stress σ_{12}, and wall couple stress m_2 increase with decreasing layer height.

The stress σ_{21} at the shear zone edges and the Cosserat rotation decrease with decreasing layer height.

In turn, the thickness of the shear zone decreases with decreasing h. It is about $6\times d_{50}$ (h=5 mm), $10\times d_{50}$ (h=10 mm) and $20\times d_{50}$ for h=50 mm.

Influence of dilatancy constraint

The effect of the dilatancy constraint is presented in Fig.4.33. The stiffness of vertical truss elements prescribed to the top was so chosen that pressure was augmented approximately twice as compared to calculations with free dilatancy. The results indicate that an increase of the dilatancy constraint (which is equivalent to a continous pressure growth) increases the thickness of the shear zone almost up to the layer height. Thus, the growth of the thickness of the shear zone under increasing pressure is stronger than in the case of the constant vertical load. It shows that the load history also influences the thickness of shear zones.

The following conclusions can be drawn on the basis of the performed FE studies on an induced shear zone:

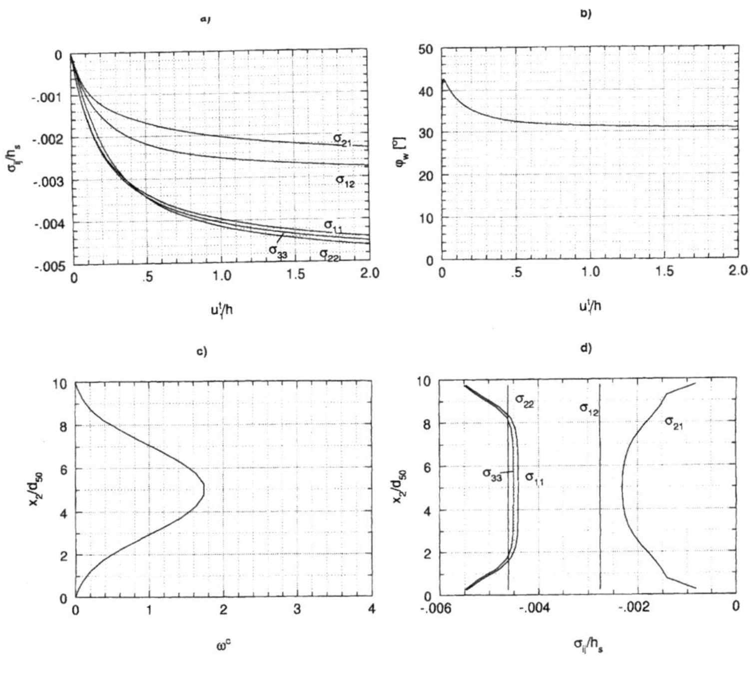

Fig. 4.33. Shearing within a polar continuum with dense sand under dilatancy constraint (e_0=0.60, d_{50}=0.5 mm, h=5 mm): a) evolution of normalised stresses σ_{ij}/h_s at the mid-point versus shear deformation u_1^t/h, b) evolution of wall friction angle φ_w versus u_1^t/h, c) and d) distribution of Cosserat rotation ω^c and normalised stresses σ_{ij}/h_s across the normalised height x_2/d_{50} (u_1^t/h=2.0)

- The micro-polar quantities become noticeable by shearing. The Cosserat rotation, the increasing void ratio and the non-symmetry of the stress tensor in the shear zone, and high gradients of curvatures, stresses and couple stresses at the shear zone edges can be used to identify shear zones. The lateral displacement and stress distribution are non-uniform in the shear zone.
- The thickness of shear zones increases with increasing initial void ratio, pressure level, layer height and mean grain diameter. The thickness related to the mean grain diameter grows with the relative void ratio. If the initial void ratio approaches or exceeds the pressure-dependent critical void ratio, the shear zone reaches the size of the granular body.
- The non-symmetry of the stress tensor, the non-uniformity of the stress distribution in the shear zone, and the stress and couple stress jump at the shear zone edges decrease with increasing initial void ratio, pressure level and layer height, and with decreasing mean grain diameter.
- The residual normal stresses at the shear zone edges increase with decreasing void ratio and layer height, and increasing mean grain diameter and grain roughness. The shear stress at peak along a localized zone increases with an increase of the mean grain diameter and grain roughness, and a decrease of the initial void ratio and layer height. At residual state, it increases with an increase of the mean grain diameter and grain roughness, and a decrease of the layer height.
- The couple stresses in the shear zone grows with increasing mean grain diameter and grain roughness. And decreasing initial void ratio and layer height. The maximum Cosserat rotation increases with decreasing shear zone thickness.
- The residual wall friction angle increases with decreasing pressure level and initial void ratio, and increasing mean grain diameter and grain roughness. The void ratio in the shear zone can be greater than its critical value, in particular for rough grains.

A similar FE analysis of shearing of an infinite narrow granular layer between two very rough boundaries boundaries was performed by Huang and Bauer (2003) with a micro-polar hypoplastic law using Eqs.4.50 and 4.51. The shear zone thickness was found to increase linearly with increasing mean grain diameter, to increase non-linearly with increasing initial void ratio and pressure, and to decrease non-linearly with increasing micro-polar constant (decreasing grain roughness).

4.3 Cyclic Shearing of an Infinite Layer

There are few experimental studies in the literature on cyclic shearing of granular materials. Tests with sand specimens were performed with the so-called simple shear devices (Youd 1971, Wood and Budhu 1980, Pradhan and Tatsuoka 1989) where global quantities were measured. They indicate that the evolution of an overall mean void ratio within a sand specimen strongly depends on the magnitude of a cyclic shear amplitude. For medium dense sand and a small cyclic shear amplitude, the mean value of the void ratio becomes lower while a global dilatant behaviour can be observed for a larger shear amplitude.

A similar macroscopic behaviour can be reproduced theoretically in calculations of so-called element tests using a non-polar hypoplastic constitutive law (Bauer 1996, 2000, Osinov 2001, Gudehus et al. 2001). However, a simple shear device is not appropriate as an element test for granular materials due to that the observed strain field within the specimen is non-uniform and stress concentrations occur in the sharp corners at specimen ends (Section 4.7). The orientation of zones of shear deformation is not parallel to the direction of shearing and it is strongly influenced by the ratio of the height to the length of the granular specimen. This shortcoming can be overcome during shearing of an infinite granular layer, which can be experimentally almost realised during torsional shearing of hollow cylindrical specimens (Pradhan and Tatsuoka 1989) or using a ring shear apparatus (Löffelmann 1989, Garga and Infante Sedano 2002). In both cases, the behaviour of the material is independent of a co-ordinate in the direction of shearing. The effect of micro-polar properties on the evolution of shear localizations under cyclic shearing was rarely investigated experimentally. Since shear localization is a precursor of failure of cohesionless soils, the effect of cyclic shearing on their strength and deformation is important during earthquake and dynamic impacts resulting from machines.

The calculations were performed for an infinite granular strip with a height of h=20 mm (Tejchman and Bauer 2004). The calculations were performed with a section of an infinite shear layer with a width of b=10 cm, discretised by 20 quadrilateral elements. The height of the elements was $5\times d_{50}$ for the mean grain diameter d_{50}=0.5 mm. The behaviour of an infinite shear layer was modelled by lateral boundary conditions, i.e. displacements and rotations along both sides of the column were constrained by the same amount (Section 4.2). The calculations were carried out with large deformations and curvatures (Section 3.3). The influence of the gravity was neglected. Both bottom and top surfaces were assumed to be very rough, i.e. sliding and rotation of particles against the bounding surface were excluded. The boundary conditions were along the bottom: $u_1=0$, $u_2=0$ and $\omega^c=0$, and along the top: $u_1=n\Delta u$, $\omega^c=0$, and $\sigma_{22}=p$ (Section 6.2).

Initially, the granular layer was compressed under the pressure p and then subject to shearing in one direction up to an almost stationary state at $u_1^t/h=1$ (u_1^t denotes the horizontal displacement of the top and h denotes the height of the layer). Afterwards, the direction of shearing was repeatedly changed by applying a large ($u_1^t/h=\pm2$), medium ($u_1^t/h=\pm0.2$) and small ($u_1^t/h=\pm0.02$) cyclic shear amplitude. Totally, the calculations were carried out with six full shear cycles.

The FE investigations were performed with two different initial void ratios of sand: low (e_0=0.60) and high (e_0=0.90). The vertical pressures was assumed to be p=500 kPa and the mean grain diameter was equal to d_{50}=0.5 mm (h/d_{50}=40). In addition, the effect of a random distribution of the initial void ratio e_o across the height of the shear layer was investigated. In this case, e_o was determined by means of a random generator in such a way that the initial void ratio was increased in every element layer by the value $a\times r$, where a=0.1 and r is a random number within the range of (0.01, 0.99). The assumed void ratio distribution was arbitrary to show only qualitatively the effect of imperfections (inherently presented in granular bodies) on shear localization.

The global stiffness matrix (calculated with only two first terms of the constitutive equations (Eqs.3.137-3.140) was updated every 10 steps.

The FE analyses of wall shearing were carried out with the following material constants for Karlsruhe sand: e_{i0}=1.3, e_{d0}=0.51, e_{c0}=0.82, ϕ_c=30°, h_s=190 MPa, α=0.3, β=1, n=0.5, d_{50}=0.5 mm and $a_c=a_1^{-1}$.

Figs.4.34-4.39 show results for an initially dense sand strip (h=20 mm, e_o=0.60, d_{50}=0.5 mm, p=500 kPa) subject to symmetric cyclic shearing with a large magnitude of the shear amplitude of $u_1^t/h=\pm 2$. Fig.4.34 presents the evolution of normalized stress components σ_{ij}/h_s (Fig.4.3) (h_s – granular hardness, σ_{33} – perpendicular to the plane of deformation) at the mid-point of the strip against the normalised horizontal displacement of the top u_1^t/h with the normal stress components. In Fig.4.35 the evolution of the normalized vertical couple stress $m_2/(h_s d_{50})$ at the layer boundary and the evolution of the mobilized friction angle $\varphi=arctan(\sigma_{12}/\sigma_{22})$ versus u_1^t/h is shown. The mobilised friction angle φ is related to the entire granular layer since the stresses σ_{12} and σ_{22} are independent of both the height and length of the layer. Fig.4.36 demonstrates the evolution of the void ratio e in 20 elements from the bottom (x_2/d_{50}=1) up to mid-height ($x_2/d_{50}=19$) of the layer versus u_1^t/h during initial compression and following cycle shearing. Figs.4.37 and 4.38 present the distribution of the normalised stresses σ_{ij}/h_s, couple stress $m_2/(h_s d_{50})$, Cosserat rotation ω^c and void ratio e across the normalised height x_2/d_{50} after the initial shearing and after six shear cycles (absolute values of σ_{ij}, m_2 and ω^c were shown). The deformed FE mesh for different states of shearing is shown in Fig.4.39. A darker region indicates a higher void ratio. The entire range of void ratio was divided into 10 different shadows. The presented quantities were taken as the mean values in each quadrilateral element.

All state variables (stress, couple stress and void ratio) tend to asymptotic values. The shear stresses insignificantly decrease during cyclic shearing. The shear stress σ_{12} is slightly larger than σ_{21} in the middle of the layer. During each reversal shearing, the normal stresses σ_{11} and σ_{33} diminish by 25%. The behaviour of the couple stress m_2 is similar as for the shear stresses. The maximum friction angle is 42^0 (obtained during initial shearing). The residual friction angle is about 30^0 and almost independent of the number of shear cycles. The void ratio close to the boundaries ($x_2/d_{50}\leq 9$ and $x_2/d_{50}\geq 31$) continuously decreases. The void ratio in the middle of the shear layer ($x_2/d_{50}\geq 13$ and $x_2/d_{50}\leq 27$) increases globally during each cycle and tends towards the pressure-dependent critical value ($e=e_c$=0.75). During each reversal shearing, contractancy occurs in the entire shear layer. After this small compaction, the void ratio in the middle of the shear zone reaches again its critical value. Since the densification of the material at the bottom and top boundary of the shear layer is less pronounced than the loosening in the middle of the layer, the mean value of the void ratio within the specimen increases. This value is influenced by the height of the shear layer.

The increase of the mean value of void ratio is qualitatively in accordance with cycle shear tests carried out for initially dense sand specimens and large cyclic shear amplitudes (Wood and Budhu 1980). Shear localization is manifested by the appearance of a Cosserat rotation (Fig.4.38a). At the boundaries of the shear zone, a strong jump of stresses σ_{11},σ_{33} and σ_{21}, (Figs.4.37a and 4.37b) and couple stress m_2 (Fig.4.37c) takes place. The thickness of the shear zone, is about $14\times d_{50}$ after the initial shearing and $18\times d_{50}$ after six full shear cycles. Thus, during the cyclic shearing, the thickness of the shear zone grows. The increase of the thickness is pronounced within the first three shear cycles.

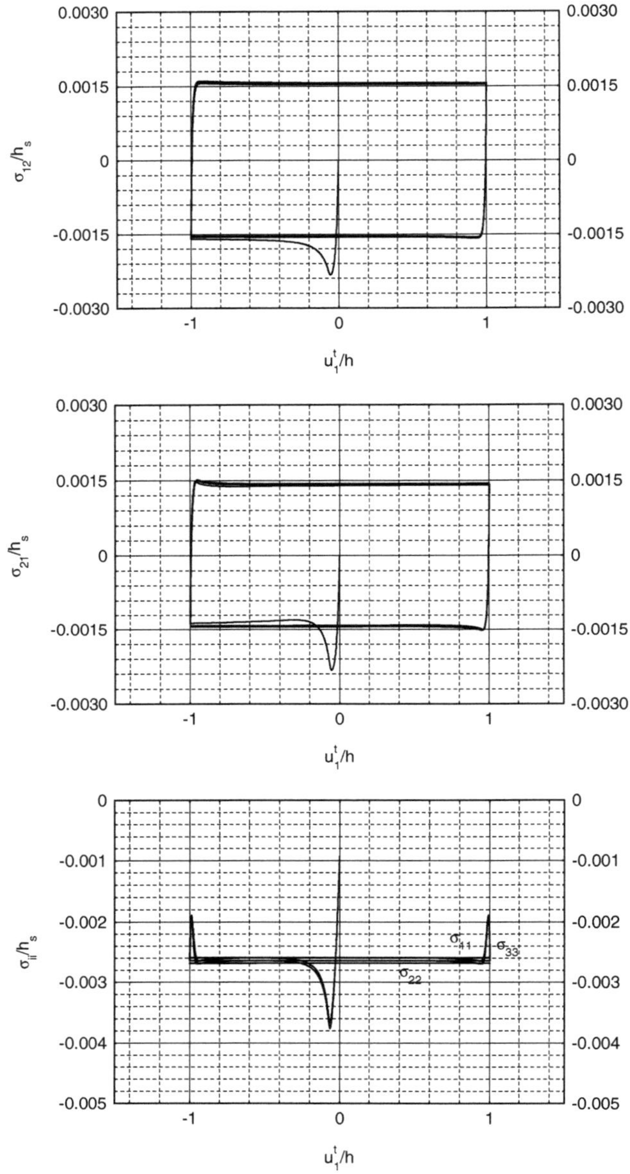

Fig. 4.34. Cyclic shearing with initially dense sand: evolution of normalised stresses σ_{ij}/h_s at the mid-point versus u_1^t/h

After the transition of the top of the granular layer by its initial position (u_1^t/h=0), the distribution of horizontal displacements is not uniform, i.e. a zig-zag occurs across the height of the localized zone (Fig.4.39b). With continuing shearing, the

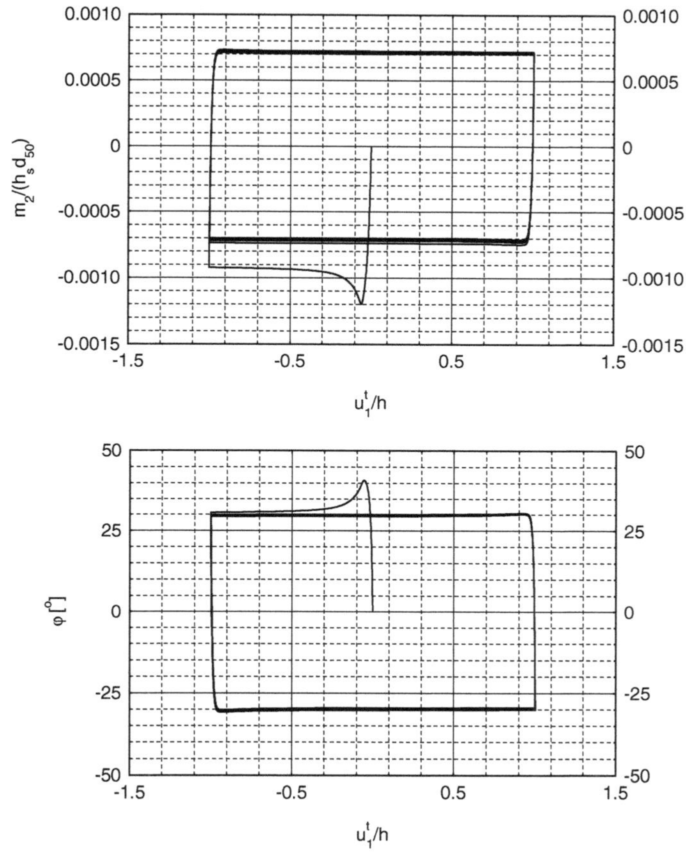

Fig. 4.35. Cyclic shearing with initially dense sand: evolution of normalised wall couple stress $m_2/(h_s d_{50})$ and evolution of mobilized friction angle $\varphi=\arctan(\sigma_{12}/\sigma_{22})$ versus u_1^t/h

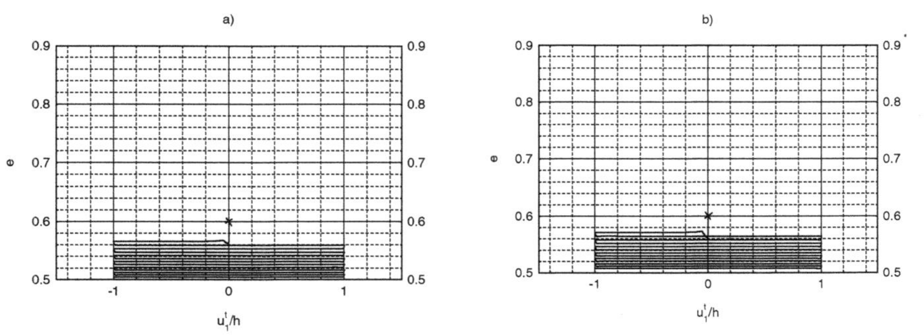

Fig. 4.36. Cyclic shearing with initially dense sand: evolution of void ratio e across the layer height versus u_1^t/h at: a) $x_2/d_{50}=1$, b) $x_2/d_{50}=3$, c) $x_2/d_{50}=5$, d) $x_2/d_{50}=7$, e) $x_2/d_{50}=9$, f) $x_2/d_{50}=11$, g) $x_2/d_{50}=13$,h) $x_2/d_{50}=15$, i) $x_2/d_{50}=17$, j) $x_2/d_{50}=19$

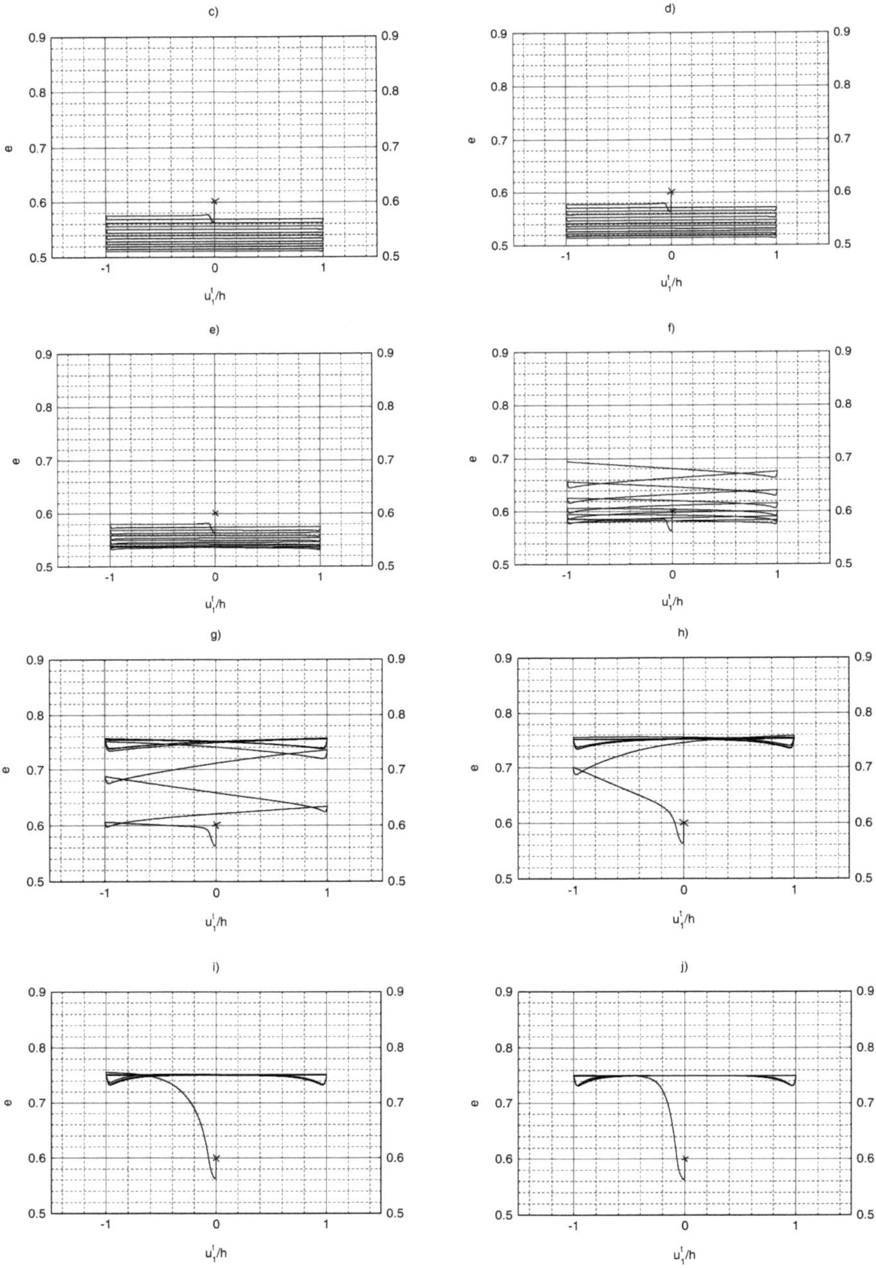

Fig. 4.36. (*continued*)

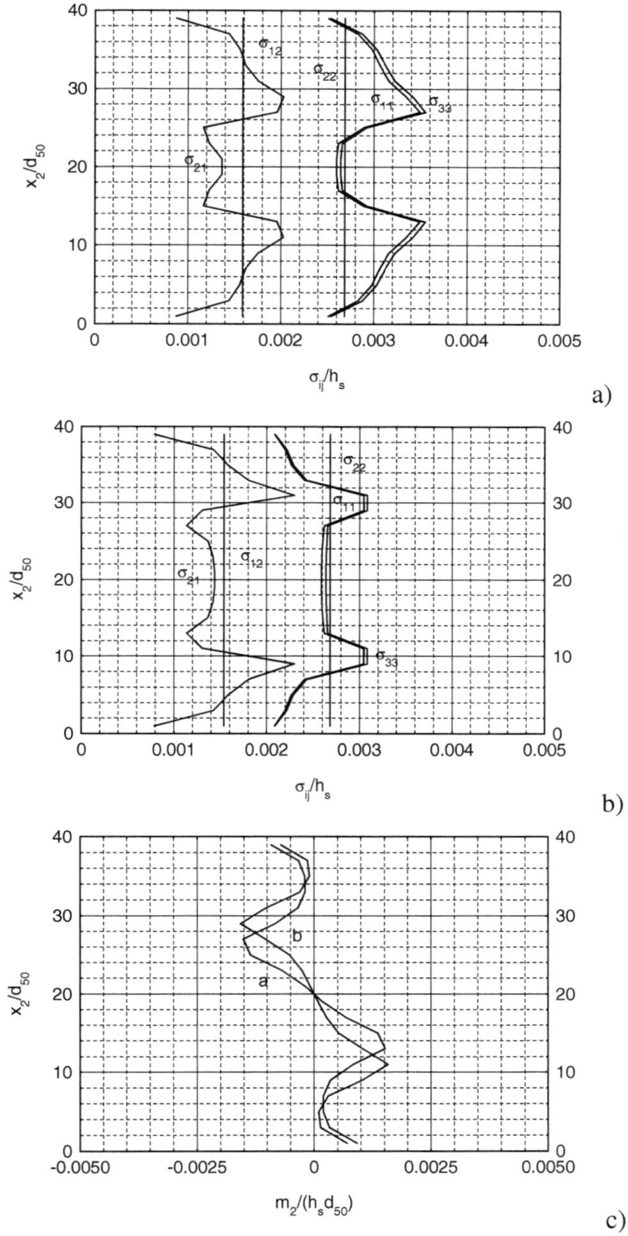

Fig. 4.37. Cyclic shearing with initially dense sand: a) distribution of normalised stresses σ_{ij}/h_s after the initial shearing, b) distribution of normalised stresses σ_{ij}/h_s after the sixth shear cycle, c) normalised wall couple stress $m_2/(h_s d_{50})$ across the normalised height x_2/d_{50} (a - after the initial shearing, b - after the sixth shear cycle)

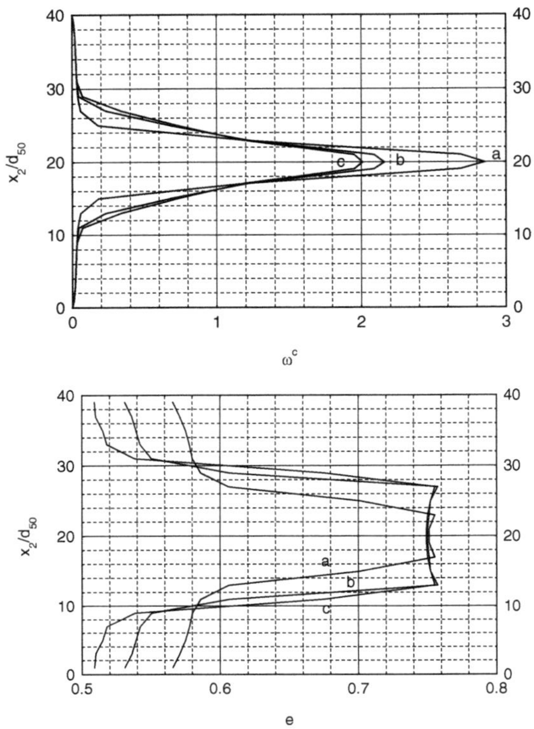

Fig. 4.38. Cyclic shearing with initially dense sand: distribution of Cosserat rotation ω^c and void ratio e across the normalised height x_2/d_{50} (a - after the initial shearing, b - after the third shear cycle, c - after the sixth full shear cycle)

displacement field becomes s-shaped similar to that after initial shearing which is in accordance with the displacement field observed in a sand specimen in a torsional ring shear apparatus (Garga and Infante Sedano 2002). The distribution of stresses σ_{11}, σ_{33} and σ_{21} across the shear zone is strongly non-linear (Figs.4.38a and 4.38b). In the middle of the shear zone, the stresses σ_{11} and σ_{33} show their minimum and the stress σ_{21} its maximum. In a stationary state, the stress ratios σ_{11}/σ_{22} and σ_{11}/σ_{33} become equal to 1 (Section 4.2).

The FE results with an initially loose sand strip (h=20 mm, e_o=0.90, d_{50}=0.5 mm, p=500 kPa) are shown in Figs.4.40-4.45. The evolution of stresses in the middle of the strip versus u_1^t/h is similar as for the initially dense material with the exception of the initial shearing where no pronounced stress peak appears. The wall couple stress is slightly higher for the initially dense material. Close to stationary states, the friction angle is similar as for the dense specimen. It remains almost unchanged during further cyclic shearing. The granular material globally undergoes contractancy which is mainly pronounced within the first shearing (Fig.4.44). The void ratio close to the mid-point of the layer ($x_2/d_{50}\geq 7$ and $x_2/d_{50}\leq 33$) decreases and reaches the pressure-dependent critical value (e=e_c=0.75).

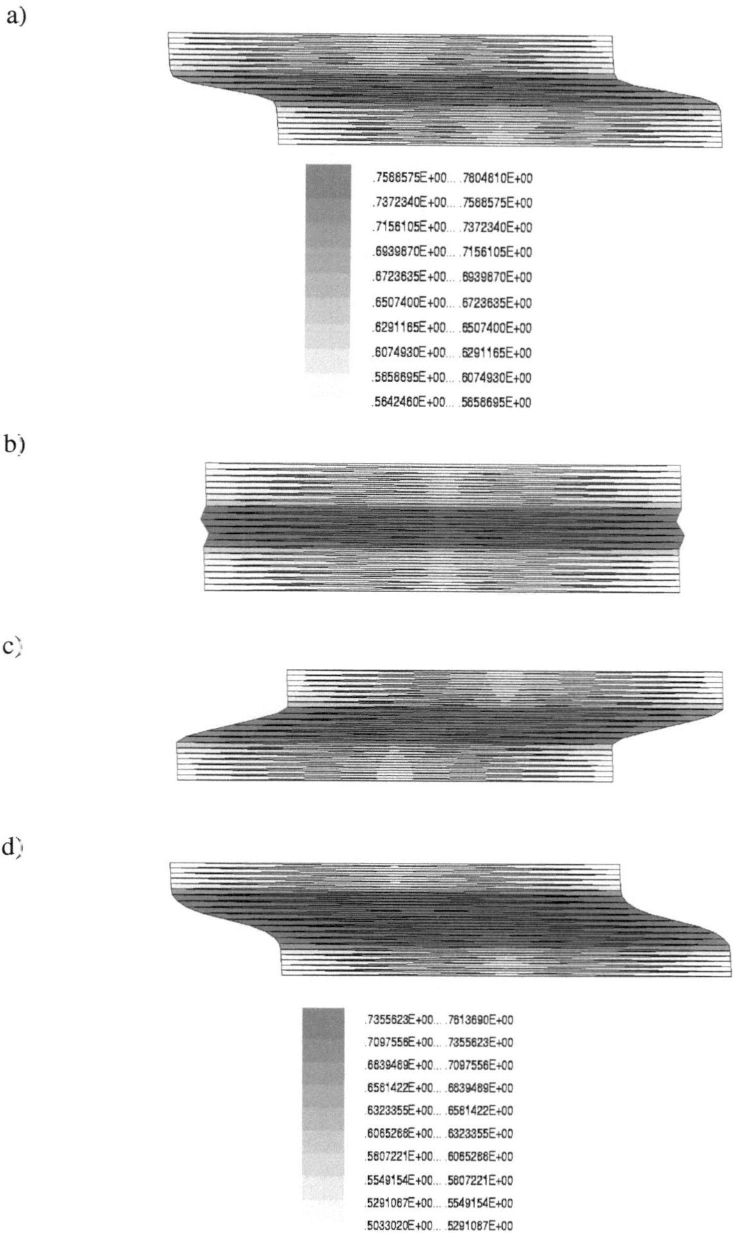

Fig. 4.39. Deformed FE mesh with distribution of void ratio e for dense sand: a) after the initial shearing, b) after the reversed shearing u_1^t/h=1, c) after the reversed shearing u_1^t/h=2, d) after the sixth shear cycle

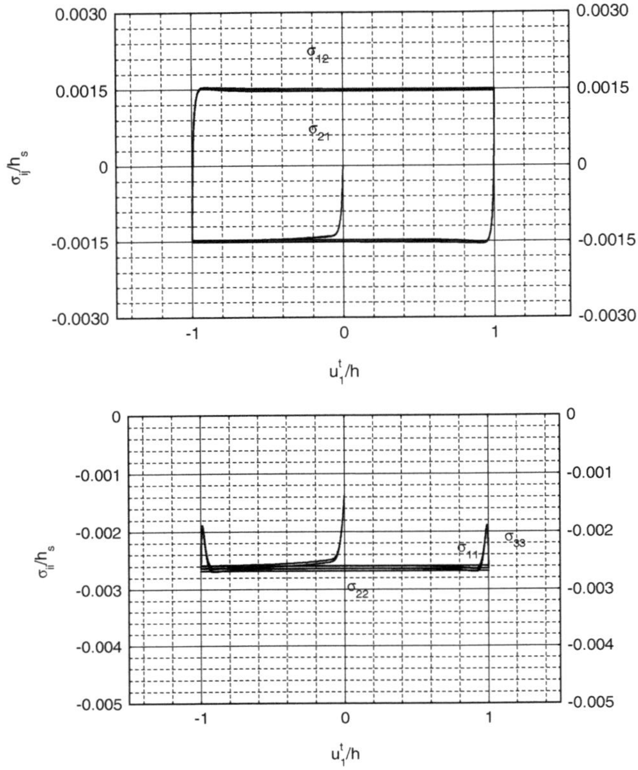

Fig. 4.40. Cyclic shearing with initially loose sand: evolution of normalised stresses σ_{ij} /h_s at the mid-point versus u_1^t/h

The results show that the critical void ratio is independent of the initial void ratio as it is assumed in the concept of critical sate soil mechanics. At the beginning of each reversal shearing, an additional compaction takes place in the entire layer (Fig.4.43). After this compaction, the void ratio in the middle of the shear zone increases with continuous shearing and reaches again a stationary state.

At the end of the initial shearing, the thickness of the localized zone is about 40×d_{50}, which is equal to the height of the layer (Section 4.2). As a result of cyclic shearing, the thickness of the shear zone decreases and after six shear cycles is about 30×d_{50}. The distribution of stresses in the shear zone is more uniform (Fig.4.43) and the stress gradient at the boundaries of the shear zone is smaller than in dense specimen. In contrast to the dense specimen the maximum Cosserat rotation is smaller and it increases with the number of shear cycles (Fig.4.44a).

The influence of an initially inhomogeneous distribution of the void ratio in a dense specimen is depicted in Fig.4.46. The initial void ratio was distributed stochastically across the height of the shear layer with a random generator, i.e. e_o=0.6+0.1r. (0<r<1).

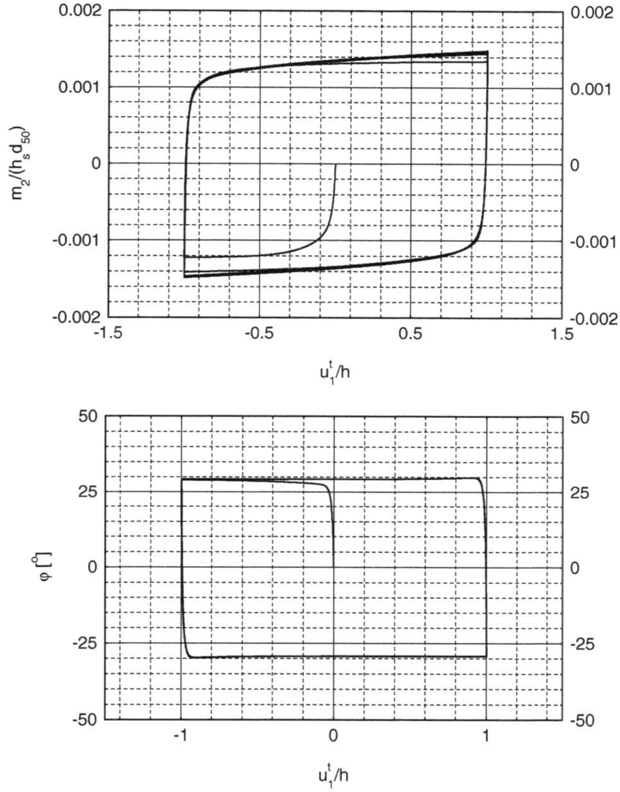

Fig. 4.41. Cyclic shearing with initially loose sand: evolution of normalised wall couple stress $m_2/(h_s d_{50})$ and evolution of mobilized friction angle $\varphi=\arctan(\sigma_{12}/\sigma_{22})$ versus u_1^t/h

The thickness of the shear zone increases again during cyclic shearing, however, the location of the localized zone is different, i.e. the shear zone does not appear in the middle of the layer as in the previous cases with an initially homogeneous void ratio distribution. The evolution of the state quantities within the localized zone and the growth of its thickness is similar as in a symmetric shear zone.

The effect of the symmetric cyclic shear amplitude on the behaviour of initially dense sand is shown in Figs.4.47-4.50.

The thickness of the shear zone does not change for both a medium (u_1^t/h=0.2) and small shear amplitude (u_1^t/h=0.02). The range of the Cosserat rotation remains the same after each full shear cycle. The distribution of the void ratio at the mid-height of the layer is the same after the initial shearing and after 6 shear cycles (medium shear amplitude). For a small shear amplitude, the behaviour of the material in the entire layer is only contractant. During each change of the shear direction with a medium shear amplitude, small contractancy first occurs and then dilatancy takes place again in the elements near the mid-height (as for a large shear amplitude). For a small shear

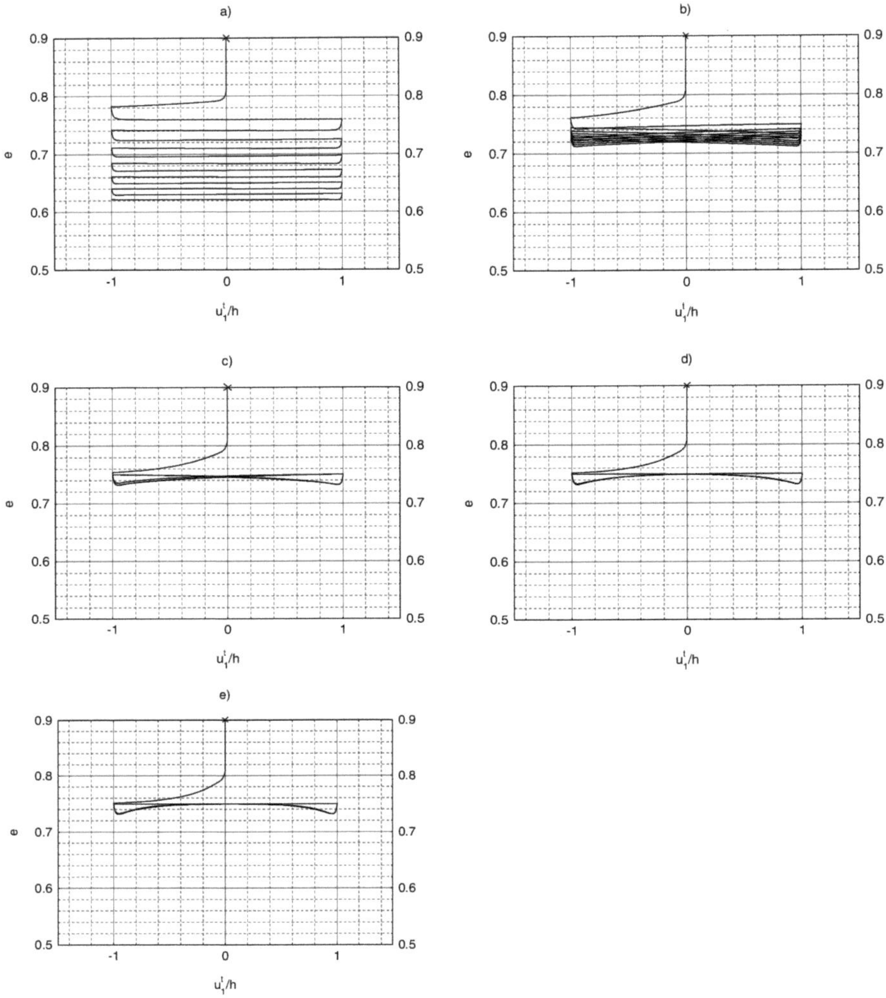

Fig. 4.42. Cyclic shearing with initially loose sand: evolution of void ratio e across the layer height versus u_1^t/h at: a) x_2/d_{50}=1, b) x_2/d_{50}=5, c) x_2/d_{50}=9, d) x_2/d_{50}=13, e) x_2/d_{50}=17

amplitude, the material is only contractant at the change of the shear direction. The void ratio at the boundaries tends to the value of e_d (pressure-dependent minimum void ratio), Eq.4.36.

Figs.4.51 and 4.52 present the results with loose sand at a small symmetric shear cyclic amplitude (u_1^t/h=0.02). The material is subject to continuous contractancy with the number of shear cycles. The void ratio decreases uniformly across the layer. The behaviour of stresses and void ratio is similar as for an initially dense specimen with a small cyclic amplitude.

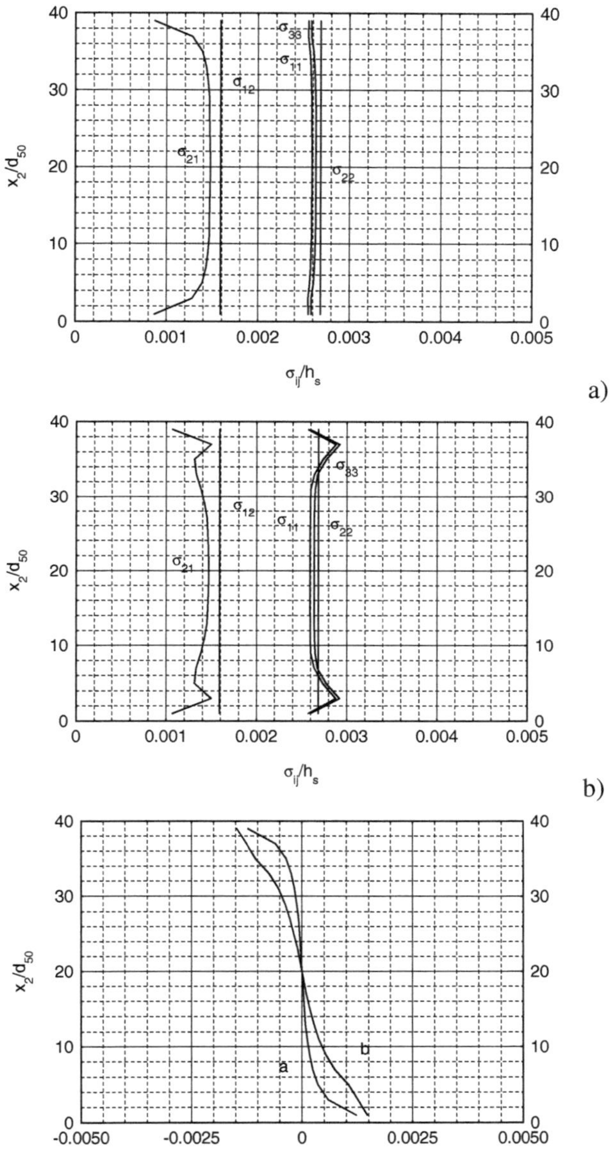

Fig. 4.43. Cyclic shearing with loose sand: a) distribution of normalised stresses σ_{ij}/h_s after the initial shearing, b) distribution of normalised stresses σ_{ij}/h_s after the sixth shear cycle, c) normalised wall couple stress $m_2/(h_s d_{50})$ across the normalised height x_2/d_{50} (a - after the initial shearing, b - after the sixth shear cycle)

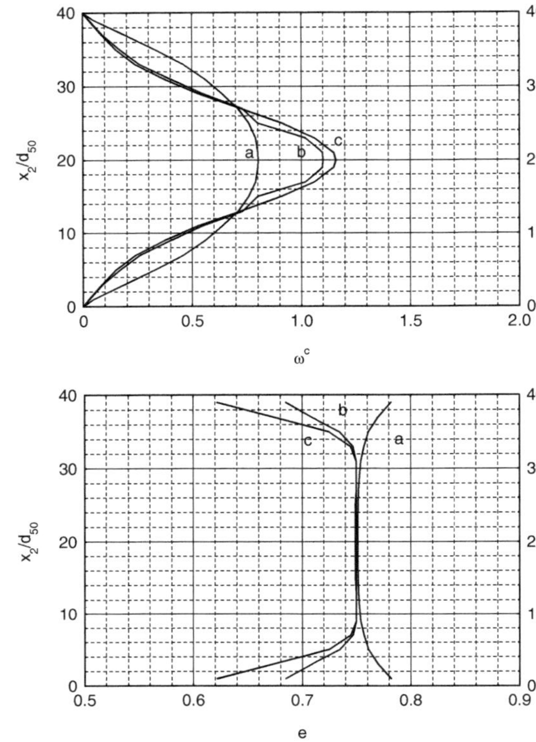

Fig. 4.44. Cyclic shearing with initially loose sand: distribution of Cosserat rotation ω^c and void ratio e across the normalised height x_2/d_{50} (a - after the initial shearing, b - after the third full shear cycle, c - after the sixth shear cycle)

a)

b)

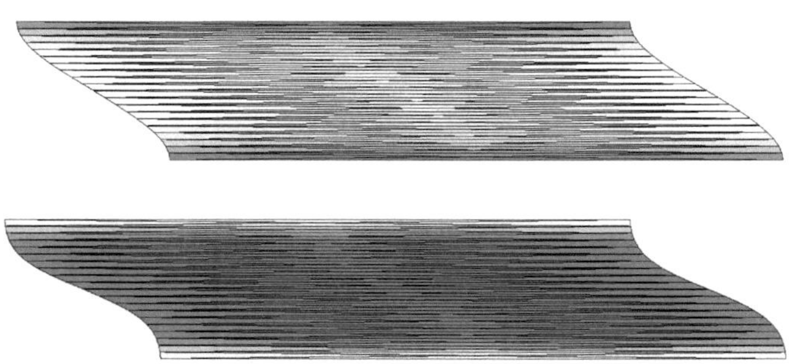

Fig. 4.45. Deformed FE mesh with distribution of void ratio e for initially loose sand: a) after the initial shearing, b) after the sixth shear cycle

a)

b)

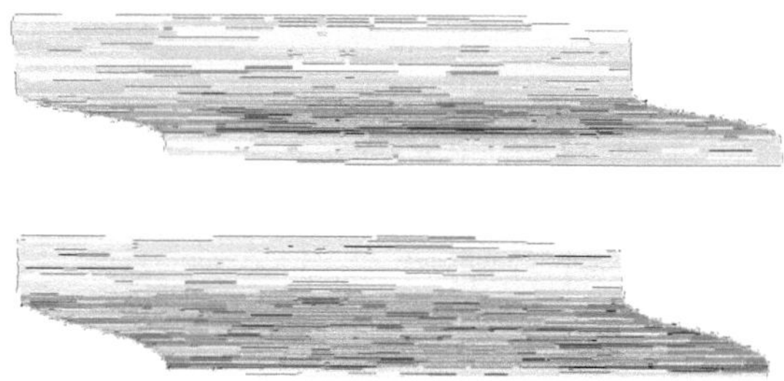

Fig. 4.46. Deformed FE mesh with distribution of void ratio e for initially dense sand with e_o=0.60+0.1r (a - after the initial shearing, b - after the sixth shear cycle)

From the presented numerical investigations on cyclic shearing of an infinite granular strip, the following conclusions can be drawn:

- Cyclic shearing with large cyclic amplitudes influences the evolution of the thickness of a localised zone (in contrast to the results for small cyclic shear amplitudes). With an increase of the number of cycles, the thickness of the localised zone increases with an initially dense specimen, and it decreases with an initially loose specimen.
- If the shear layer is higher than the thickness of the localized zone, the void ratio beyond the localized zone decreases with each shear cycle. Within the localized zone, the void ratio always decreases after changing the shear direction and then increases tending to the critical void value (large and medium shear amplitude), and always decreases after changing the shear direction (small shear amplitude). The critical void ratio depends on the mean pressure but it is independent of the initial void ratio.
- A random distribution of the initial void ratio influences the location of the localized shear zone.

4.4 Biaxial Compression

The FE calculations of quasi-static biaxial tests (Tejchman 2002a) were performed for a rectangular sand specimen with a height of h=50 mm and a length of l=100 mm. Totally, 800 quadrilateral elements with 3200 triangular elements were used. The height and the width of the quadrilateral elements was always 2.5 mm (5×d_{50}). The calculations were carried out with large deformations and curvatures (Section 3.5).

As the initial stress state in the granular specimen, a K_o-state without polar quantities ($\sigma_{22}=\gamma_d x_2$, $\sigma_{11}=\sigma_{33}=K_o\gamma_d x_2$, $\sigma_{12}=\sigma_{21}=m_1=m_2=0$) was assumed ($K_0$=0.40 – pressure coefficient at rest).

Two different sets of conditions along boundaries of the sand specimen were assumed: one with three non-deforming rigid boundaries and one free boundary, and second with four non-deforming rigid boundaries. In the first case, the granular specimen was placed on the smooth fixed bottom, its smooth top was subject to the uniform vertical pressure p,

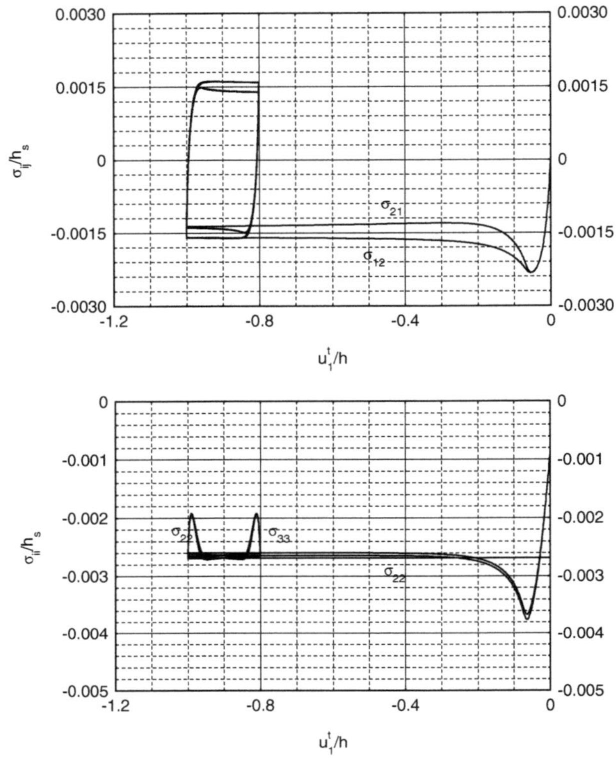

Fig. 4.47. Shearing with initially dense sand (medium shear amplitude): evolution of normalised stresses σ_{ij}/h_s at the mid-point versus shear deformation u_1^t/h

and its vertical smooth sides were subject to equal horizontal displacement increments directed to the specimen inside (Fig.4.53a). The boundary conditions were along the bottom: u_2=0, σ_{12}=0 and m_1=0, along the top: $\sigma_{22}=p$, σ_{12}=0 and m_1=0, along the left side: $u_1=n\Delta u$, σ_{21}=0 and m_2=0, and along the right side: $u_1=-n\Delta u$, σ_{21}=0 and m_2=0. In the second case, the granular specimen was also placed on the smooth fixed bottom. The vertical smooth sides were subject to equal horizontal displacement increments directed to the specimen inside and the smooth top was subject to uniform vertical displacement increments directed to the specimen outside (Fig.4.53b). The vertical displacement increments were equal to the horizontal ones. The boundary conditions were along the bottom: $u_2=0$, $\sigma_{12}=0$ and $m_1=0$, along the top: $u_2=n\Delta u$, $\sigma_{12}=0$ and $m_1=0$, along the left side: $u_1=n\Delta u$, $\sigma_{21}=0$ and $m_2=0$, and along the right side: $u_1=-n\Delta u$, $\sigma_{21}=0$ and $m_2=0$. n denotes the number of the time steps and Δu is the constant displacement increment in one step. The displacement increments were chosen as $\Delta u/h$=0.0001. About 2000-4000 steps were performed.

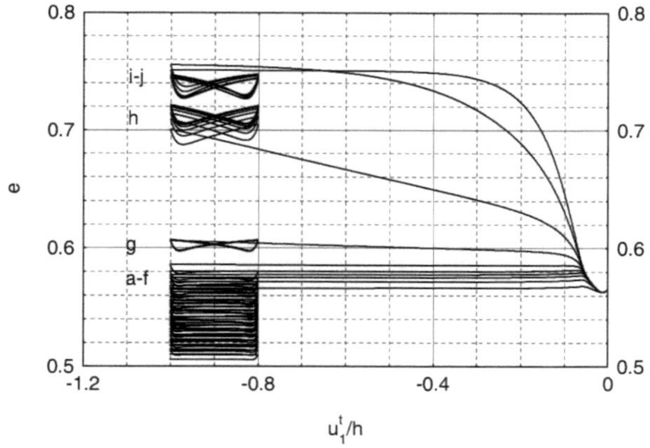

Fig. 4.48. Shearing with initially dense sand (medium shear amplitude): evolution of void ratio e across the layer height versus $u_1{}^t/h$ at x_2/d_{50}: a) 1, b) 3, c) 5, d) 7, e) 9, f) 11, g) 13 h) 15, i) 17, and j) 19

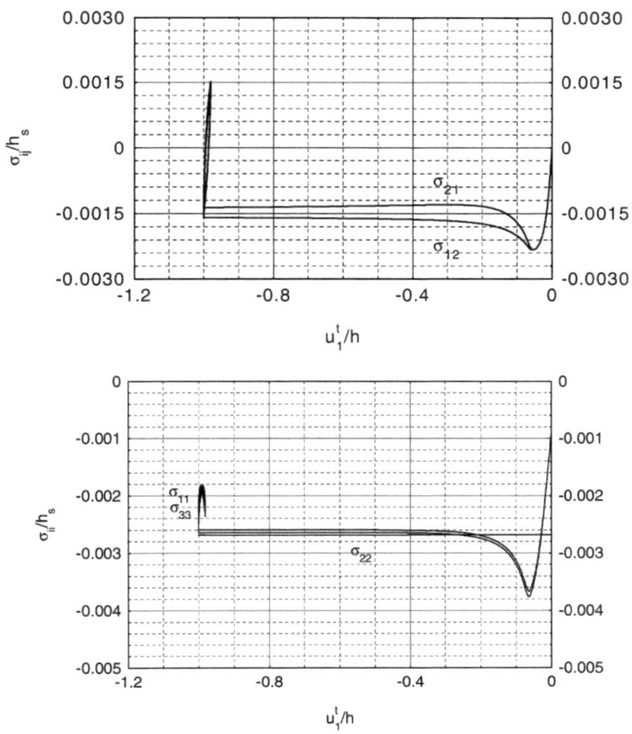

Fig. 4.49. Shearing with initially dense sand (small shear amplitude): evolution of normalised stresses σ_{ij}/h_s at the mid-point versus shear deformation $u_1{}^t/h$

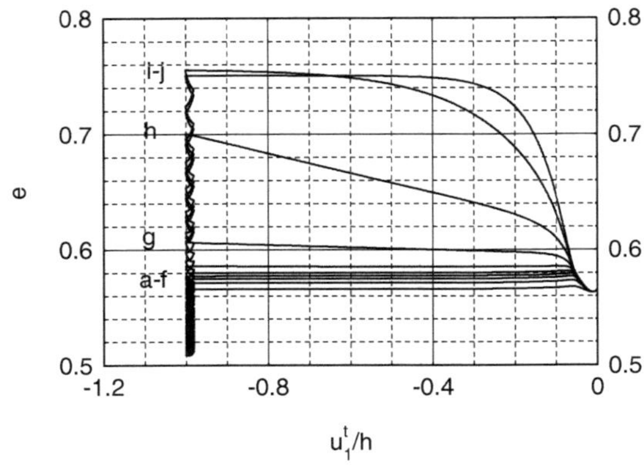

Fig. 4.50. Shearing with initially dense sand (small shear amplitude): evolution of void ratio e across the layer height versus u_1^t/h at x_2/d_{50}: a) 1, b) 3, c) 5, d) 7, e) 9, f) 11, g) 13, h) 15, i) 17 and j) 19

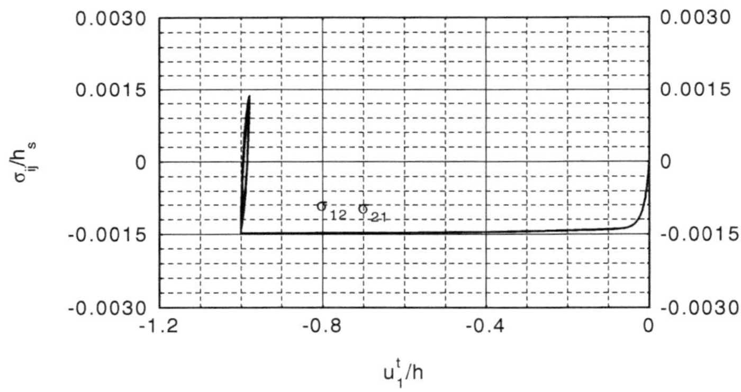

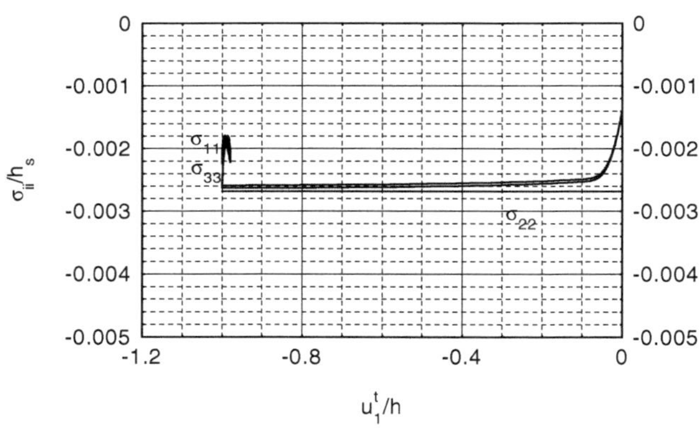

Fig. 4.51. Shearing with initially loose sand (small shear amplitude): evolution of normalised stresses σ_{ij}/h_s at the mid-point versus shear deformation u_1^t/h

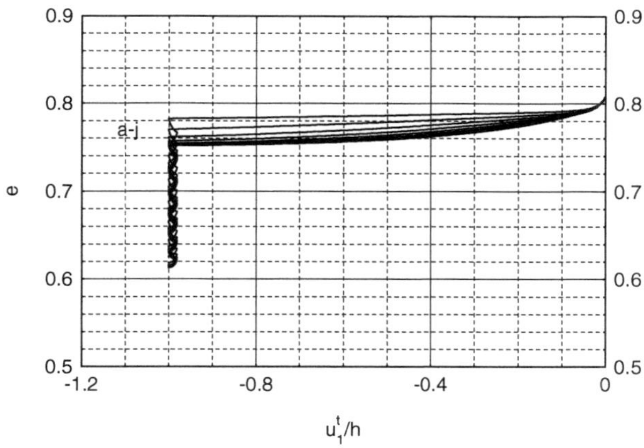

Fig. 4.52. Shearing with loose sand (small shear amplitude): evolution of void ratio e across the layer height versus u_1^t/h at x_2/d_{50}: a) 1, b) 3, c) 5, d) 7, e) 9, f) 11, g) 13, h) 15, i) 17 and j) 19

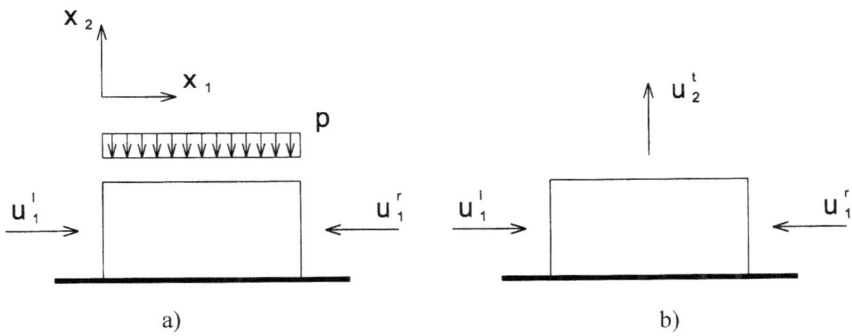

Fig. 4.53. Biaxial test: a) deformation produced by horizontal displacements of both sides ($u_1^l=u_1^r$) and vertical pressure on the top p, b) deformation produced by horizontal displacements of both sides ($u_1^l=u_1^r$), and vertical displacement of the top ($u_1^t=u_1^l=u_1^r$); l –left, r – right, t -top

The FE calculations of a plane strain compression test with the boundary conditions of Fig.4.53a, the K_o-initial stress state (K_o=0.4) and one weak element inserted in the middle of the left side are shown in Figs.4.54-4.60. Figs.4.54 and 4.55 show the deformed FE mesh at residual state, the evolution of the resultant horizontal force P_1 acting on the sides versus the horizontal displacement of both sides u_1, and the deformed mesh with the distribution of the Cosserat rotation ω^c and void ratio e at residual state for dense sand (e_0=0.60, d_{50}=0.5 mm, p=100 kPa). The Cosserat rotation is marked by circles with a maximum diameter corresponding to the maximum rotation in the given step. In turn, the grade of an increase of the void ratio is marked by a dark region. Darker region indicates higher void ratio.

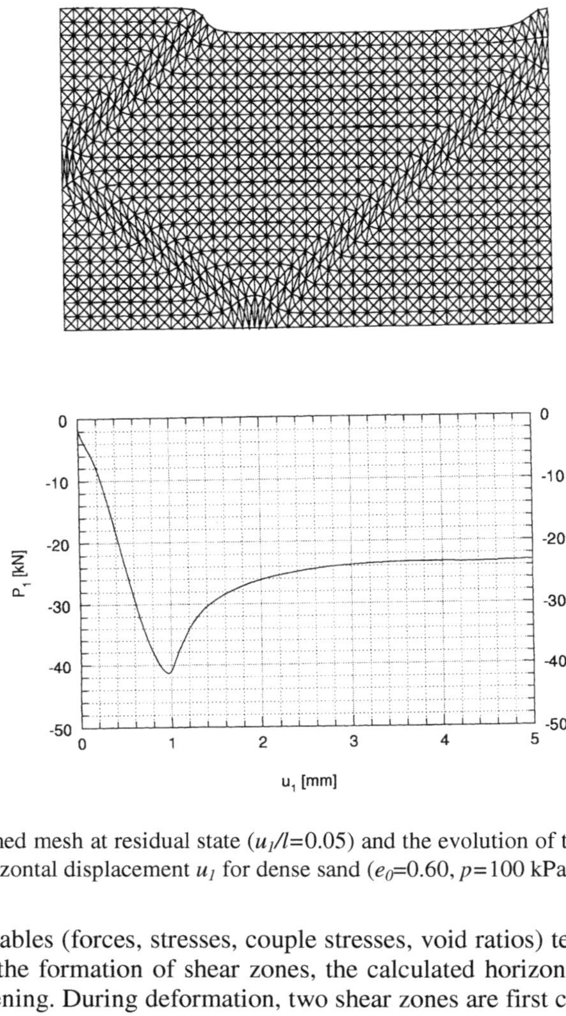

Fig. 4.54. Deformed mesh at residual state (u_1/l=0.05) and the evolution of the horizontal force P_1 versus the horizontal displacement u_1 for dense sand (e_0=0.60, p=100 kPa, d_{50}=0.5 mm)

All state variables (forces, stresses, couple stresses, void ratios) tend to asymptotic values. Due to the formation of shear zones, the calculated horizontal force P_1 indicates large softening. During deformation, two shear zones are first created expanding outward from the weakest element in the middle of the left side. They propagate at the same time towards both the fixed bottom and free top. The shear zone propagating to the fixed bottom, reaches it, reflects from it and propagates again upwards. It reaches the right rigid side, reflects from it and moves towards the free top. Thus, four shear zones are created.

The shear zones are marked out by the concentration of shear deformations and Cosserat rotations and by an increase of the void ratio. The thickness of the shear zones on the basis of displacements and Cosserat rotations is about t=13×d_{50}, and the distance between two inclined shear zones s=45 mm=90×d_{50}. The distance s is approximately equal to the specimen height h=50 mm. The thickness of the shear zones on the basis of an increase of the void ratio is larger since dense granular material dilates before the shear zone is created (Herle 1997). The Cosserat rotations are only noticeable in the shear zone. Outside the shear zone, they are negligible. The void

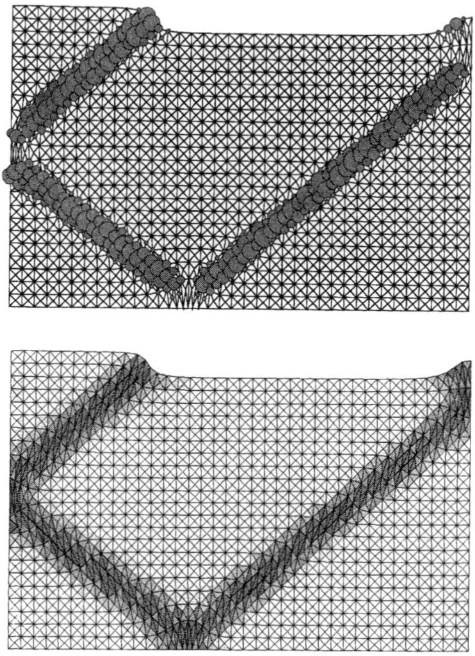

Fig. 4.55. Deformed mesh with the distribution of the Cosserat rotation and void ratio for dense sand specimen (e_0=0.60, p=100 kPa, d_{50}=0.5 mm) at residual state (u_1/l=0.05)

ratio changes across the shear zone from 0.65-0.79. Outside the shear zone, the void ratio is e=0.55, and is lower than its initial value of e_0=0.60 since each granular material undergoes contractancy at the beginning of shearing (Herle 1997). The largest void ratio in the shear zone corresponds approximately to the pressure-dependent critical value e_c (Eq.4.37). The inclination of shear zones is about 40° and is different than the mesh alignment of 45°.

The effect of the initial void ratio e_0, vertical pressure p and the mean grain diameter d_{50} is presented in Figs.4.54-4.58. The following results of t and s were obtained: t=25×d_{50} and s=85×d_{50} (e_0=0.75, d_{50}=0.5 mm, p=100 kPa), t=9×d_{50} and s=45×d_{50} (e_0=0.60, d_{50}=1.0 mm, p=100 kPa), and t=15×d_{50} and s=100×d_{50} (e_0=0.60, d_{50}=0.5 mm, p=1000 kPa). The results show that the thickness of shear zones increases with increasing initial void ratio, mean grain diameter and vertical pressure. The distance between the inclined zones decreases with increasing e_0 and decreasing p. For e_0>e_c, the thickness of shear zones almost reaches the size of the granular body during deformation.

Fig.4.55 presents the evolution of the Cosserat rotation during deformation in loose sand (e_0=0.90). The maximum Cosserat rotation is 0.046, 0.15 and 0.26 for u_1/l=0.025, u_1/l=0.05 and u_1/l=0.075, respectively. The thickness of the shear zone grows during deformation, and for u_1/l=0.075 is equal to the specimen height. If the linearly increasing vertical pressure p (changing from 0 kPa to 1000 kPa) is prescribed, the increase of the thickness of shear zones with increasing pressure is

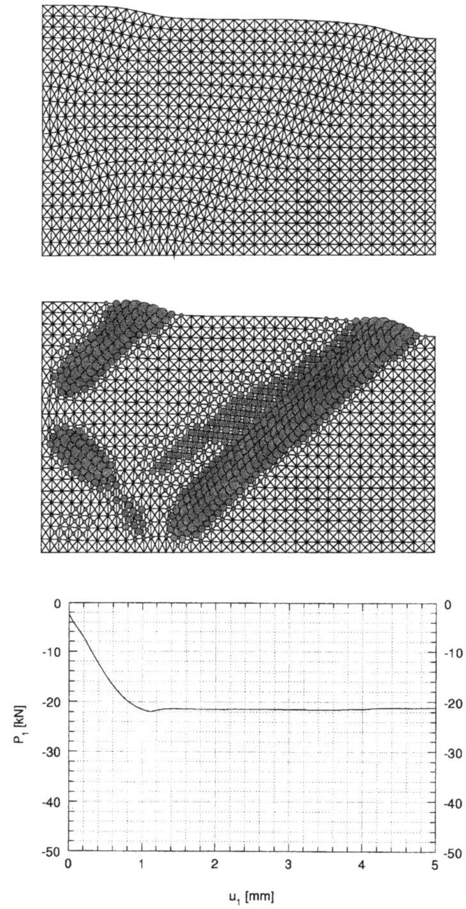

Fig. 4.56. Deformed mesh and distribution of the Cosserat rotation at residual state (u_1/l=0.05) and load-displacement curve for medium dense sand specimen (e_0=0.75, p=100 kPa, d_{50}=0.5 mm)

more pronounced. The results show than in contrast to free boundaries which cannot reflect moving shear zones, both a moving rigid or a fixed boundary are able to reflect each shear zone.

All load-displacement curves for $e_0 \leq 0.75$ show a pronounced peak, drop later and reach almost the same residual state. The maximum normalised horizontal force P_1 increases with decreasing e_0 and p and increasing d_{50}. With an increase of the initial void ratio and vertical pressure, the horizontal displacement of sides u_1 related to the peak of P_1 becomes larger.

The increase of the mean grain diameter increases the shear zone thickness and the maximum horizontal force as a polar granular body is stiffer than a non-polar one using the same constants (Eq.4.22).

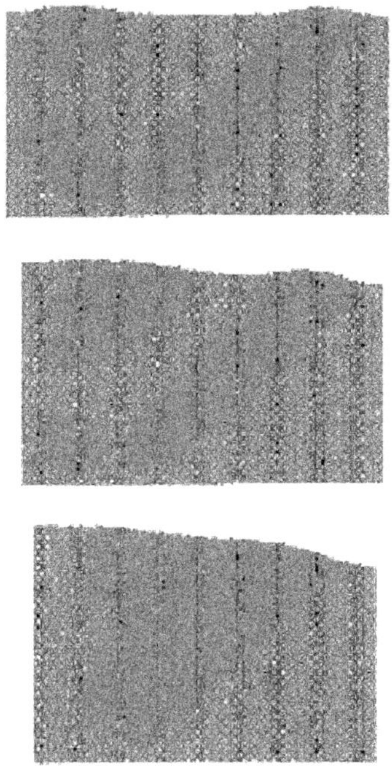

Fig. 4.57. Deformed mesh with the distribution of the Cosserat rotation at residual state (u_1/l=0.05) for loose sand specimen (e_0=0.90, p=100 kPa, d_{50}=0.5 mm) a) u_1/l=0.025, b) u_1/l=0.05, c) u_1/l=0.075

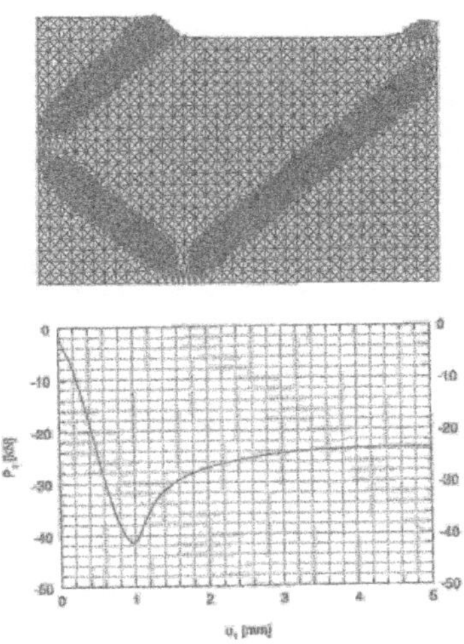

Fig. 4.58. Deformed mesh with the distribution of the Cosserat rotation at residual state (u1/l=0.05), and load-displacement curve for dense sand specimen (e_0=0.60, p=100 kPa, d_{50}=1.0 mm)

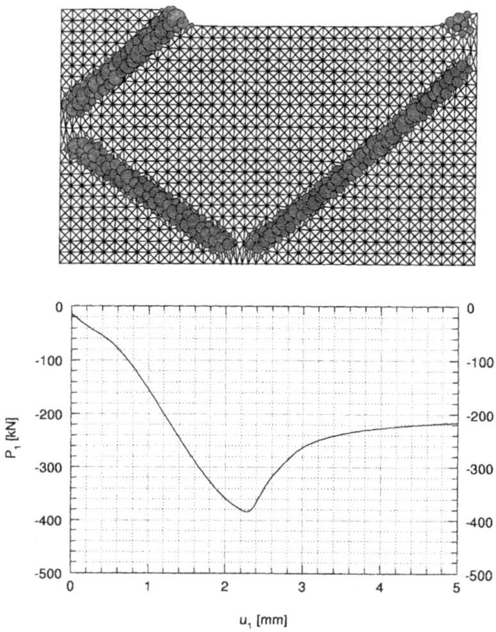

Fig. 4.59. Deformed mesh with the distribution of the Cosserat rotation at residual state (u_1/l=0.05), and load-displacement curve for dense sand specimen (e_0=0.60, p=1000 kPa, d_{50}=0.5 mm)

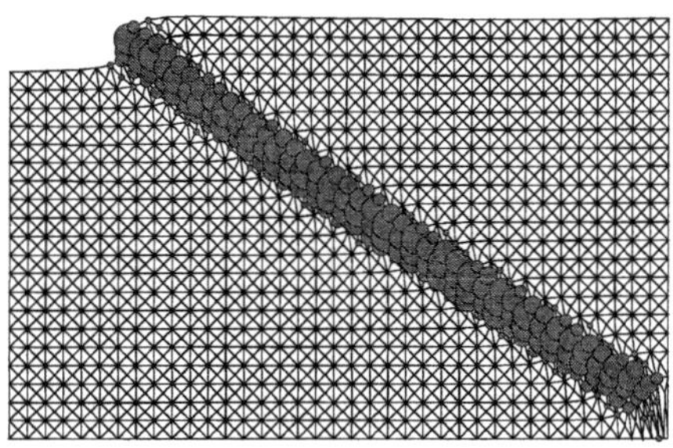

Fig. 4.60. Distribution of the Cosserat rotation at residual state (u_1/l=0.05) for dense specimen (e_0=0.60, p=100 kPa, d_{50}=0.5 mm) with the imperfection located in the middle of the specimen or in the left top corner

The location of the imperfection in sand determines the geometry of shear zones. When the imperfection is assumed in the middle of the specimen or in the left top corner only one shear zone is obtained (Fig.4.60). In this case, the shear zone is not able to reflect from the bottom or sides since it moves to the right bottom corner. The shape of this zone is slightly curved.

The initial stress state has also a certain effect on the geometry of shear zones. The FE calculations were carried out with the K_o-initial stress state but assuming K_o=1.0. In this case, the same amount of shear zones was created as in Fig.4.53 but the distance between the inclined shear zones was slightly larger (s=100×d_{50}) due to a different zone inclination.

The numerical results with the boundary conditions of Fig.4.53a, the K_o-initial stress state (K_o=0.40) and a stochastic distribution of the initial void ratio in the specimen are shown in Figs.4.61 and 4.62. In the calculations, the vertical pressure and the initial void ratio were varied.

The results show that the geometry of shear localization is strongly influenced by the location of the first shear zone implied by the distribution of imperfections, and the magnitude of e_0 and p. For the small vertical pressure of p=10 kPa (Fig.4.61a), the number of shear zones (4) is similar as in the case of FE calculations with one weak element (Figs.4.53 and 4.54) since the first two shear zones occurred at the similar place (mid-point of the left side). Their thickness (t=10×d_{50}) is slightly smaller due to the smaller pressure. The distance between the inclined zones (s=120×d_{50}) is larger.

If the uniform vertical pressure becomes greater (p=100 kPa), only one shear zone is created in the middle of the specimen since it has not a possibility to be reflected from rigid boundaries. The shape of this zone is slightly curved again. However, if the prescribed vertical pressure increases linearly from 0 kPa to 100 kPa, three shear zones appear with a thickness of t=16×d_{50} (Fig.4.62a). The distance between the

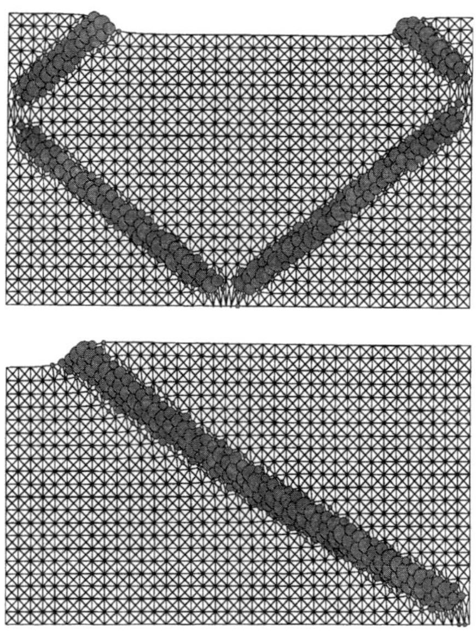

Fig. 4.61. Deformed mesh with the distribution of the Cosserat rotation at residual state (u_1/l=0.05) for dense specimen (e_0=0.60+0.05r, d_{50}=0.5 mm): p=10 kPa and p=100 kPa

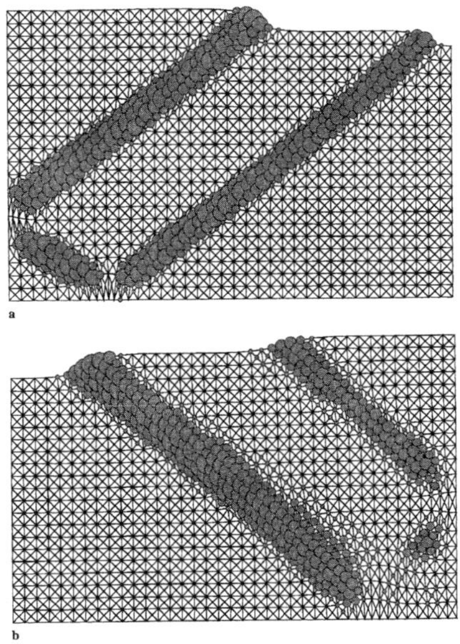

Fig. 4.62. Deformed mesh with the distribution of the Cosserat rotation at residual state (u_1/l=0.05) for dense specimen (d_{50}=0.5 mm): a) e_0=0.60+0.05r, p=0-100 kPa, b) e_0=0.70+0.05r, p=100 kPa

zones, $s=50\times d_{50}$, is significantly smaller than in Fig.4.54. For the medium dense specimen (Fig.4.62b), again three shear zones appear with a thickness of $t=20\times d_{50}$ and a distance of $s=60\times d_{50}$. Thus, the calculations indicate that the distance between the shear zones is influenced by reflection positions of shear zones from the fixed bottom and rigid sides implied by the position of the first created shear zone.

The FE results with boundary conditions of Fig.4.53b, the K_o-initial stress state (K_o=0.40) and one weak element inserted in the middle of the left side are shown in Figs.4.63 and 4.64. A different pattern of shear zones is obtained as compared to calculations with a free boundary since all boundaries can now reflect them. The thickness of shear zones is $t=18\times d_{50}$ and the spacing $s=(80\text{-}90)\times d_{50}$.

The FE calculations of a plane strain compression test with boundary conditions of Fig.4.53b, the K_o-initial stress state (K_o=0.40) and the stochastic distribution of the initial void ratio in the specimen are shown in Figs.4.65 and 4.66. The effect of the initial void ratio e_0 and its distribution deviation a was studied. The calculations show that the geometry of shear zones is strongly influenced by factors e_0 and a. For a large deviation of the initial void ratio, different patterns with many intersecting shear zones are obtained with dense and medium dense sand. Their thickness and distance on the basis of the Cosserat rotation are variable. In the case of dense sand and a small deviation of the initial void ratio (e_0=0.60+0.005r), only one very wide shear zone

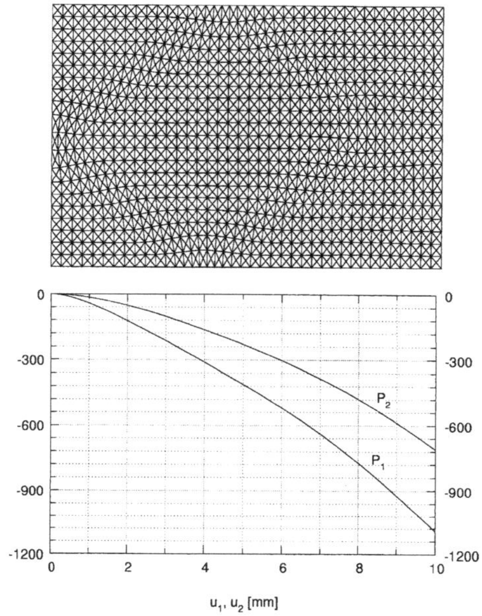

Fig. 4.63. Deformed mesh at residual state (u_1/l=0.10) and the evolution of the horizontal P_1 and vertical force P_2 versus the horizontal u_1 and vertical displacement u_2 for dense sand (e_0=0.60, d_{50}=0.5 mm)

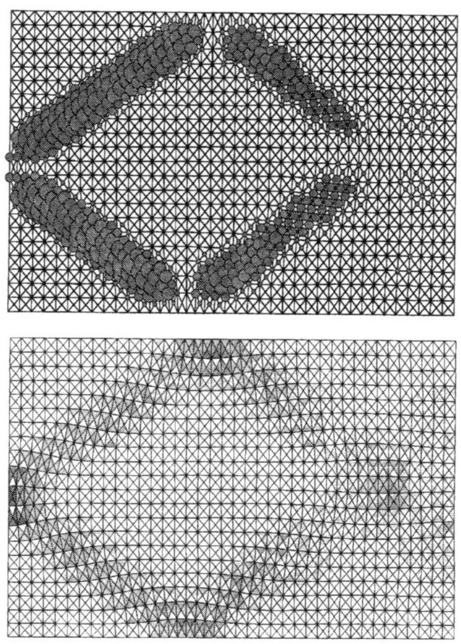

Fig. 4.64. Deformed mesh with the distribution of the Cosserat rotation and void ratio for dense sand specimen (e_0=0.60, d_{50}=0.5 mm) at residual state (u_1/l=0.10)

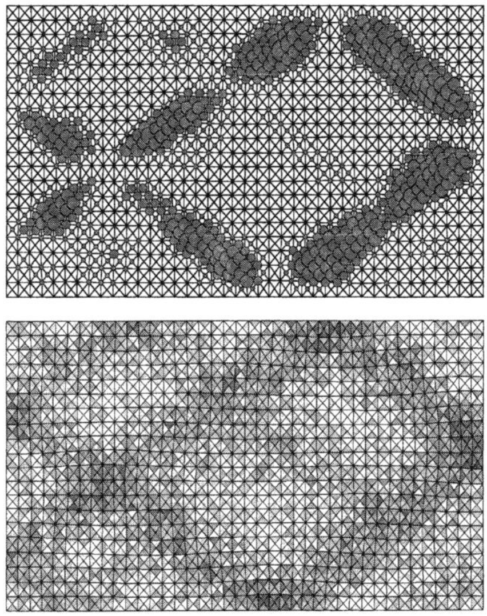

Fig. 4.65. Deformed mesh with the distribution of the Cosserat rotation and void ratio for dense specimen (e_0=0.60+0.05r, d_{50}=0.5 mm) at residual state (u_1/l=0.10)

(due to high pressure) occurs (t=30×d_{50}), Fig.4.66b. The shear zone moves from left bottom corner to the right top corner, and cannot be reflected from boundaries.

If medium dense sand with a large deviation of the initial void ratio is used (e_0=0.70+0.05r), two pronounced shear zones are obtained; one very wide with a thickness of t=25-30×d_{50}, and one with a thickness of t=15×d_{50} (Fig.4.65c). The results demonstrate that not only the geometry of shear zones but also their thickness depend on the boundary conditions of the entire system.

The following conclusions can be drawn on the basis of the performed FE studies:

- Shear zones have a tendency for reflection only from fixed or moving rigid boundaries.
- The geometry of shear zones depends mainly on the type of the boundary along the specimen (deforming or non-deforming boundary), specimen form and size, and the location of the first shear zone implied by the distribution and kind of imperfections. The distance between shear zones increases also with decreasing initial void ratio.
- The thickness of shear zones increases with increasing initial void ratio, pressure level and mean grain diameter. If the initial void ratio approaches or exceeds the pressure-dependent critical void ratio, the shear zone reaches the size of the granular body. The thickness is also dependent upon the boundary conditions of the entire system.

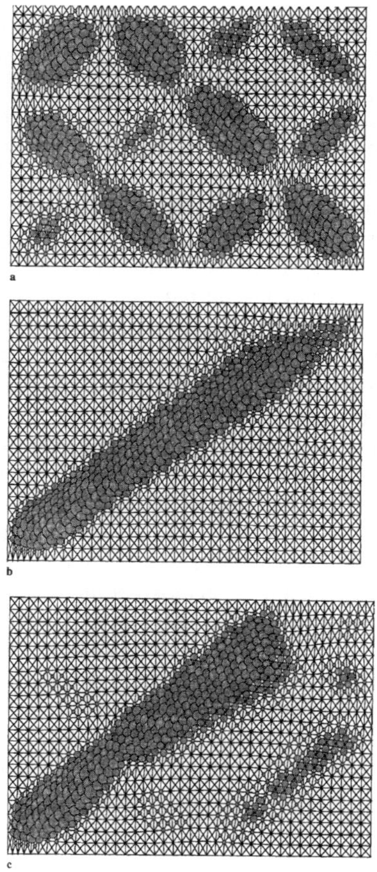

Fig. 4.66. Deformed mesh with the distribution of the Cosserat rotation for sand specimen (d_{50}=0.5 mm) at residual state (u_1/l=0.10): a) e_0=0.65+0.05r, b) e_0=0.60+0.005r, c) e_0=0.70+0.05r

4.5 Strip Foundation

A normalized bearing capacity of model footings on granular material cannot be directly transferred to large prototype footings due to scale effects caused by grain size and pressure level (Steenfelt 1979, Bätcke 1982, Tatsuoka et al. 1991, 1997). An increasing mean grain diameter and decreasing pressure level increase, namely, a normalized bearing capacity. Thus, model tests overestimate the bearing capacity of large footings. The grain size influences the shear resistance of granular materials which increases with increasing mean grain diameter (Section 4.2). In turn, the pressure level has an influence both on the internal friction angle and dilatancy angle of the granulate. With an increase of the pressure level, the maximum internal friction angle and the maximum dilatancy angle become smaller. As a result, the shear resistance decreases (Herle 1997). In this way, a realistic prediction of the bearing capacity of footings should be achieved with a constitutive model taking into account pressure level and mean grain diameter. Both of them influence also the thickness of a shear zone occurring in granular material under the footing during its loading. This thickness is linked with the bearing capacity of footings (Tejchman 1997).

In this Section, results of the numerical modelling of the bearing capacity of model footings under plane strain conditions are presented (Tejchman and Herle 1999). They are compared with results of experiments carried out at the Tokyo University by Tatsuoka et al. (1997). During experiments, the footings were subjected to vertical central loading, and they were placed on dry dense Silver Leighton Buzzard sand. The presented FE calculations belong to the so-called class '*A*' prediction since they were finished before the model experiments started.

Fig.4.67 shows the effects of the particle size and the pressure level on the normalised bearing capacity factor $N_b=2P_{max}/(\gamma_d b^2 l)$ of some model footings on dense sand from *1g* and cetrifuge tests (Bätcke 1982, Tatsuoka et al. 1991, 1997). The foundations were vertically and centrally loaded. P_{max} denotes the maximum measured vertical force on the footing, γ_d is the initial volume weight of sand, b is the footing width and l denotes the footing length. In the case of the centrifuge tests, the footing width b was multiplied with a factor n of the increase of the field of gravity. Sand used by Bätcke (1982) had a mean grain diameter of d_{50}=0.45 mm and an initial void ratio of e_0=0.6. In the experiments by Tatsuoka et al. (1991, 1997), the first sand (so-called Toyoura sand) had d_{50}=0.16 mm and e_0=0.665 (Tatsuoka et al. 1991), and the second sand (so-called Silver Leighton Buzzard sand) d_{50}=0.60 mm and e_0=0.55 (Tatsuoka et al. 1997).

The results of the *1g* tests for different footing widths (Fig.4.67) evidently show a scale effect (an increase of the normalised bearing capacity factor with decreasing foundation width) due to the pressure level induced by the footing width. In turn, the normalised bearing capacity factor is greater in centrifuge tests as in corresponding *1g* tests (both tests have the same pressure level) that indicates a scale effect due to the mean grain diameter (an increase of the normalised bearing capacity factor with increasing mean particle size).

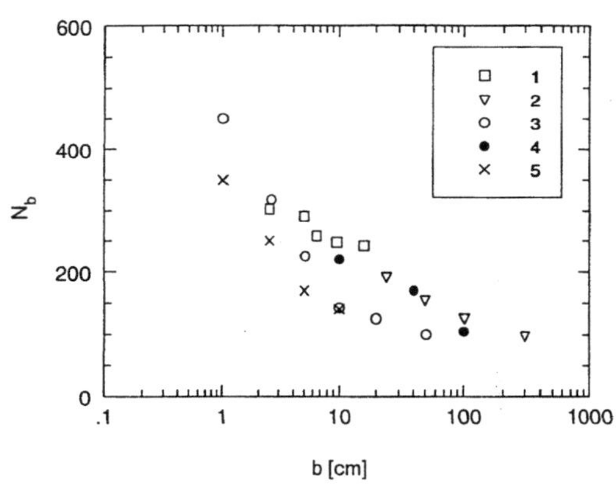

Fig. 4.67. Relationship between the normalised bearing capacity factor $N_b=2P_{max}/(\gamma_d b^2 l)$ and the footing width b from *1g* and cetrifuge model tests: 1. *1g* tests (Bätcke 1982), 2. centrifuge tests (Tatsuoka et al. 1991) 3. *1g* tests (Tatsuoka et al. 1991), 4. centrifuge tests (Tatsuoka et al. 1997), 5. *1g* tests (Tatsuoka et al. 1997)

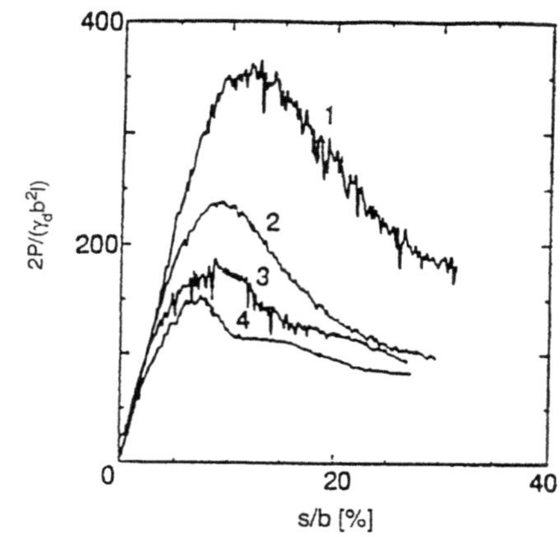

Fig. 4.68. Test results with SLB sand (Tatsuoka et al. 1997): relationship between the normalised vertical force $2P/(\gamma_d b^2 l)$ and the normalised vertical displacement s/b from $1g$ tests: 1. b=1.0 cm, 2. b=2.5 cm, 3. b=5 cm, 4. b=10 cm

Thus, results from model tests cannot be simply transferred to full-scale foundations. The experiments by Tatsuoka et al. (1991, 1997) show that a scale effect due to the particle size becomes insignificant for $b/d_{50}>500$ using Toyoura sand and $b/d_{50}>100$ using SLB sand. However, a scale effect due to the pressure level always exists. In turn, in other model tests with sand, a scale effect due to the particle size occurred for the ratio $b/d_{50}<33$ (Bätcke 1982) or $b/d_{50}<100$ (Steenfelt 1979, Jarzombek 1989).

The relationship between the normalised vertical force on the footing $2P/(\gamma_d b^2 l)$ versus the normalised vertical displacement of the footing s/b for SLB sand in $1g$ tests (Tatsuoka et al. 1997) is presented in Fig.4.68. The maximum and the residual normalised vertical force increase with decreasing footing width. The normalised vertical displacement s_p/b of the footing corresponding to the maximum force increases also with decreasing b. The measured thickness of the shear zone directly under the footing was found to be in all tests about 6 mm ($10\times d_{50}$) (Fig.2.26).

In addition, Fig.4.69 shows the effect of the initial sand density expressed by the internal friction angle at peak on the bearing capacity factor from different 1g model tests. The maximum internal friction angle was determined with triaxial compression tests. The experiments indicate that the bearing capacity of foundations strongly grows with increasing initial density of sand.

Identification of parameters for SLB sand

The laboratory tests on dense SLB sand by Tatsuoka et al. (1994) were the only source of the data for the determination of the constants for the hypoplastic constitutive model. Unfortunately, no experiments on loose sand were at disposal. Thus, the standard procedure (Section 4) for the determination of the constants h_s and n from an oedometric test with loose sand and of the constant φ from the angle of

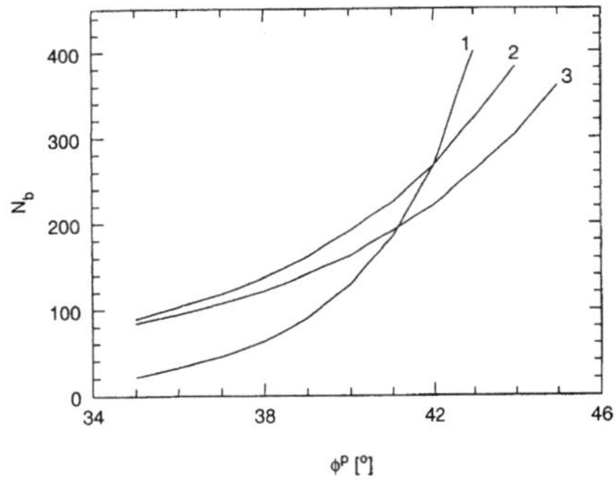

Fig. 4.69. Relationship between the normalised bearing capacity factor N_b and internal friction angle at peak ϕ^p from different *1g* model tests: 1. b=5 cm, d_{50}=0.45 mm (Bätcke 1982), 2. b=6.5-9 cm, d_{50}=0.8 mm (Jarzombek 1989), 3. b=9-15 cm, d_{50}=0.19 mm (de Beer 1970)

repose was not possible. The critical friction angle ϕ_c was evaluated from the stress ratio at the residual state in plane strain tests. The exponent α and the granulate hardness h_s were found by fitting the peak stress values in plane strain and triaxial tests for an assumed value of the exponent n. The characteristic void ratios and the mean grain diameter were taken from the given standard index tests. In this way, a set of the following parameters used in the FE calculations of footings on sand was obtained: ϕ=29^0, h_s=300 MPa, α=0.16, n=0.4, e_{i0}=0.86, e_{d0}=0.51, e_{c0}=0.55 and d_{50}=0.60 mm. The micro-polar constant was a_c=1.0. The calculations were also performed with a_c=2.5.

Figs.4.70 and 4.71 present the measured and calculated results of plane strain compression tests with e_o=0.55 for different confining pressures (σ_{22} – vertical normal stress, σ_{11} –horizontal normal stress equivalent to the confining pressure). The calculation of element tests was performed by the numerical integration of all relevant stress components using corresponding boundary conditions. Only homogeneous deformation was assumed. Consequently, the micro-polar terms were not considered. The influence of the confining pressure on the internal friction angle and the shear deformation at peak from the tests and the calculations is summarised in Figs.4.72 and 4.73. The calculated and measured results from triaxial compression tests are depicted in Fig.4.74.

For plane strain compression, the effect of the stress level on the internal friction angle at peak (Fig.4.72) is fairly described with a non-polar hypoplastic law, although it overestimates the internal friction angle for low pressures and underestimates it for high ones. The calculated dilatancy angle (the inclination of the curve ε_v versus γ) at the beginning of compression (Fig.4.71), and the calculated shear strain at peak (Fig.4.73) are smaller than the measured values (Fig.4.70). In turn, the calculated dilatancy for large values of γ (Fig.4.71) is too large because strain localization is not taken into account by the non-polar hypoplastic model. The inclination of the calculated curve σ_{22}/σ_{11} versus the shear strain γ (Fig.4.71) is too small at the beginning of compression, and too large near the peak as compared to the experiments.

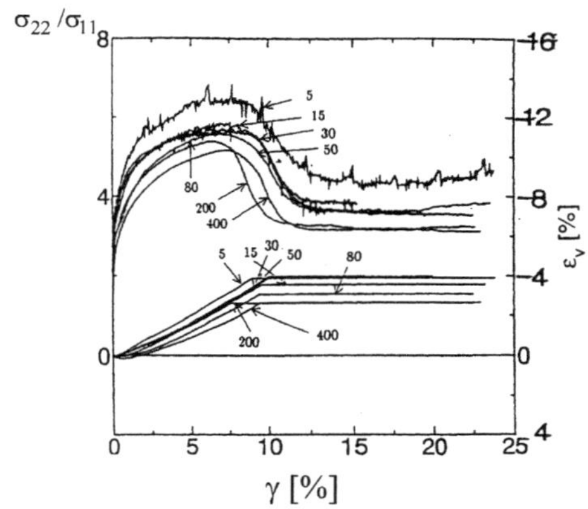

Fig. 4.70. Relationship between the stress ratio σ_{22}/σ_{11} and the average shear strain $\gamma=\varepsilon_{22}-\varepsilon_{11}$, and between the average volumetric strain $\varepsilon_v=\varepsilon_{11}+\varepsilon_{22}$ and the average shear strain γ from plane strain compression tests (Tatsuoka et al. 1994). σ_{22}, σ_{11} - vertical and horizontal normal stress, ε_{22}, ε_{11}- vertical and horizontal strain (numbers at curves correspond to σ_{11} in kPa)

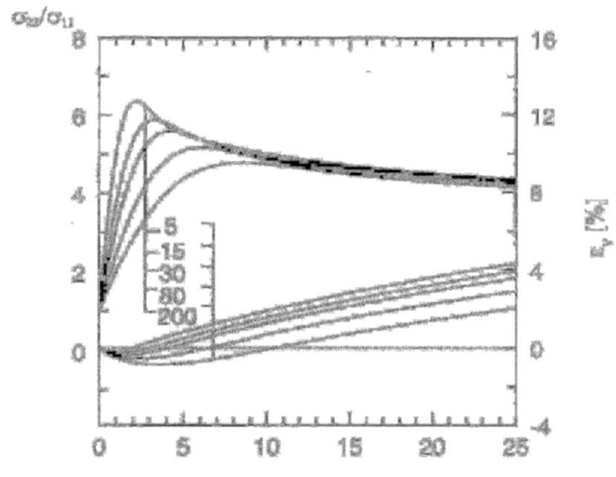

Fig. 4.71. Element test calculations for plane strain compression for different confining pressures in kPa: relations between the stress ratio σ_{22}/σ_{11} and the shear strain $\gamma=\varepsilon_{22}-\varepsilon_{11}$, and between the volumetric strain $\varepsilon_v=\varepsilon_{11}+\varepsilon_{22}$ and the shear strain $\gamma=\varepsilon_{22}-\varepsilon_{11}$. σ_{22}, σ_{11} - vertical and horizontal normal stress, ε_{22}, ε_{11}- vertical and horizontal strain (numbers at curves correspond to σ_{11} in kPa)

The calculated curve σ_{22}/σ_{11} versus the axial strain ε_{22} during triaxial compression (Fig.4.71) results in a stiffer behaviour of sand shortly before the peak is reached as compared to the experiments. The onset of the calculated volumetric strain curve versus the axial strain corresponds well to the experimental ones which were registered only up to ε_{22}=4%.

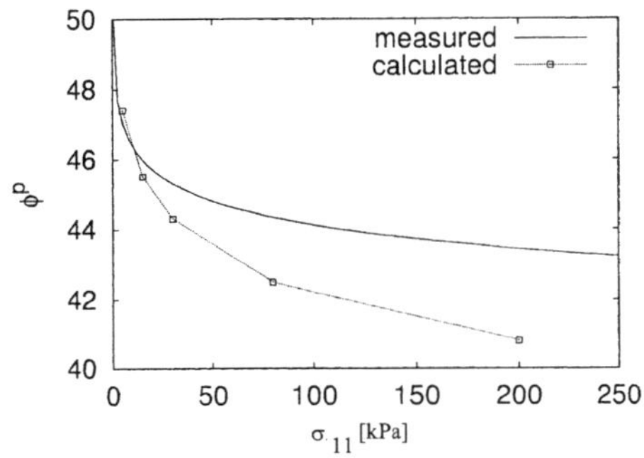

Fig. 4.72. Measured and calculated relationship between the internal friction angle at peak φ^p and the confining pressure σ_{11} during plane strain compression

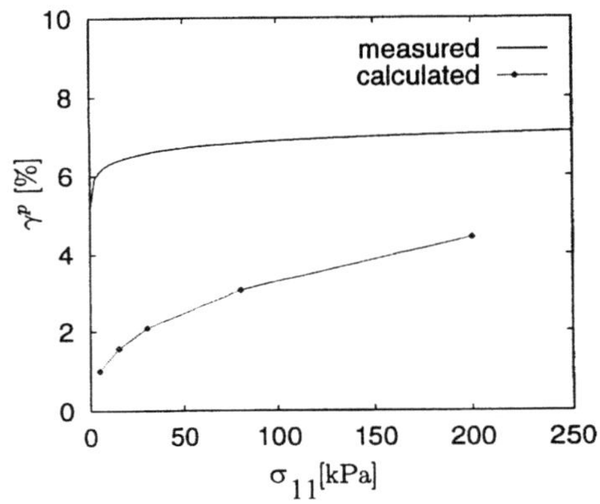

Fig. 4.73. Measured and calculated relationship between the shear deformation at peak γ^p and the confining pressure σ_{11} during plane strain compression

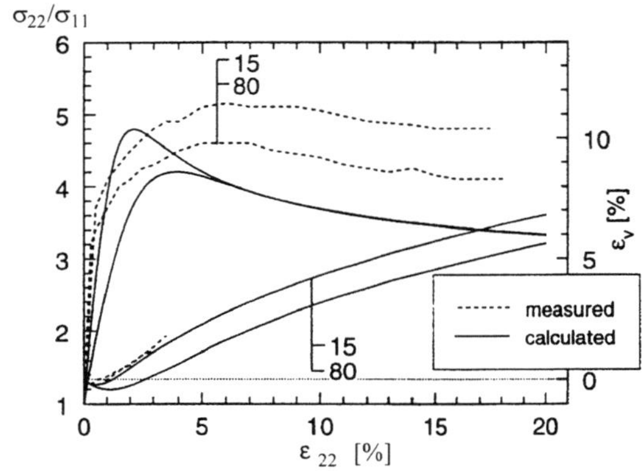

Fig. 4.74. Measured and calculated relationship between the stress ratio σ_{22}/σ_{11} and the axial strain ε_{22}, and between the average volumetric strain $\varepsilon_v=\varepsilon_{22}+2\varepsilon_{11}$ and the axial strain ε_{22} from triaxial compression tests (numbers at curves correspond to σ_{11} in kPa)

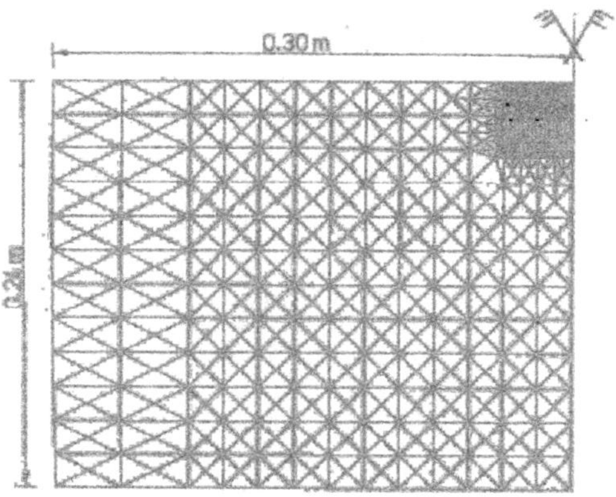

Fig. 4.75. FE mesh used for numerical analyses of footings on SLB sand

In general, the non-polar hypoplastic model shows a satisfactory agreement with the experiments when taking into account the simplicity of the determination of the material constants. The FE calculations of plane strain compression tests considering the formation of shear zones would allow for a more accurate calibration (Tejchman et al. 1999, Section 4.1).

The FE calculations of the bearing capacity of plane strain model foundations placed on dense SLB sand were carried out with the mesh of Fig.4.75 (Tejchman and Herle 1999). The width of the foundation was assumed to be b=1.0, 2.5 and 5.0 cm. The dimensions of the sand specimen surrounding the foundations were 24 cm (height) and 60 cm (width), and corresponded to the small-size sand box tests (Tatsuoka et al. 1997). Symmetry with respect to the centreline was taken into account during FE calculations. In total, 1700 triangular elements were used. The dimensions of finite elements directly under the foundation were approximately $5\times d_{50}$. The calculations were carried out for large deformations and curvatures (Section 3.5).

As the initial stress state, the K_0-state with $\sigma_{22}=\gamma_d x_2$, $\sigma_{11}=K_0\sigma_{22}$, $\sigma_{33}=\sigma_{11}$, $\sigma_{12}=\sigma_{21}=0$ (γ_d=17 kN/m^3, K_0=0.4) and e_o=0.55 were assumed in the sand specimen, where x_2 is the vertical coordinate measured from the top of the sand box and γ_d denotes the initial volume weight.

The quasi-static deformation in sand was initiated through a constant vertical displacement increment Δs ($\Delta s/b$=0.0002) prescribed at the nodes along the foundation. The boundary conditions along a very rough foundation were: $u_1=0$, $u_2=0$, $\omega^c=0$. They correspond to full shearing of sand (without slip) along the foundation surface. The boundary conditions of the sand specimen were along the top traction and moment free. They were along a smooth bottom: $u_2=0$, $\sigma_{12}=0$, $m_2=0$; along smooth sides: $u_1=0$, $\sigma_{21}=0$, $m_1=0$ and in the symmetry axis: $u_1=0$, $\omega^c=0$, $\sigma_{21}=0$. Two nodes at the bottom ends were kept as fixed: $u_1=0$, $u_2=0$, $\omega^c=0$. The global stiffness matrix calculated with only two first terms of the constitutive equations (Eq.4.92-4.95) was updated every 50 steps.

Fig.4.76 shows the calculated normalised vertical force $2q/(\gamma_d b)=2P/(\gamma_d b^2 l)$ versus the normalised vertical displacement of the plane strain footing s/b for three various footing widths (q – mean vertical normal stress, l=1.0 m). The vertical force P was

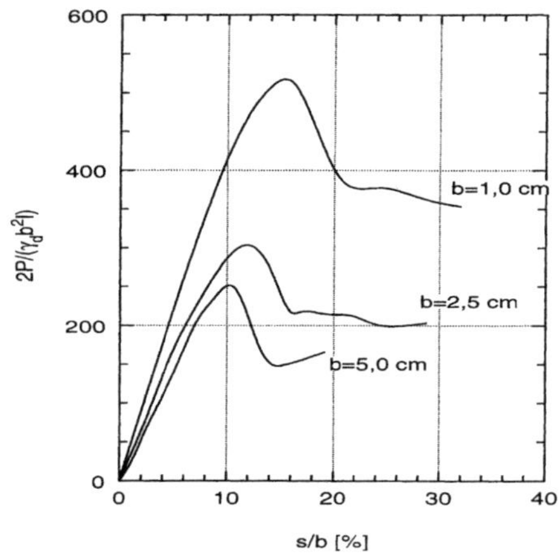

Fig. 4.76. Relationship between the normalised vertical force $2P/(\gamma_d b^2 l)$ and the normalised vertical displacement *s/b* from the FE calculations

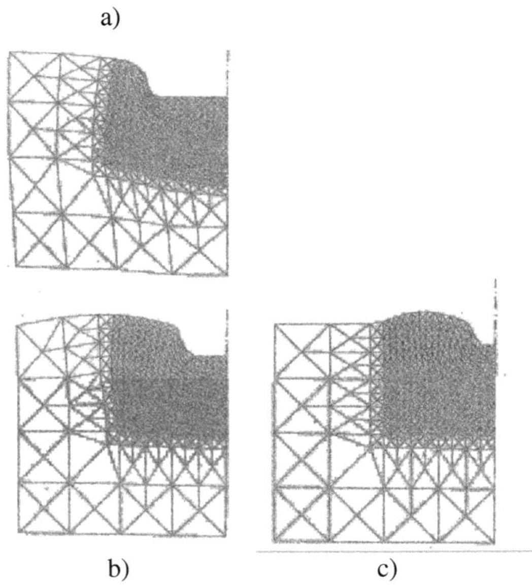

Fig. 4.77. Calculated displacements in sand close to the footing: a) *b*=5 cm, *s*=1.6 cm, b) *b*=2.5 cm, *s*=1.0 cm, c) *b*=1.0 cm, *s*=1.0 cm

calculated as the total sum of vertical forces at the nodes along the footing. The deformed meshes 8×8 cm^2 of the sand specimen close to the footing are presented in Fig.4.77. The results of the Cosserat rotation ω^c and the void ratio *e* under the footing with *b*=5 cm (*s*=1.5 cm) are shown in Fig.4.78. The distribution of the vertical normal stress directly under the footing is depicted as well. The Cosserat rotation

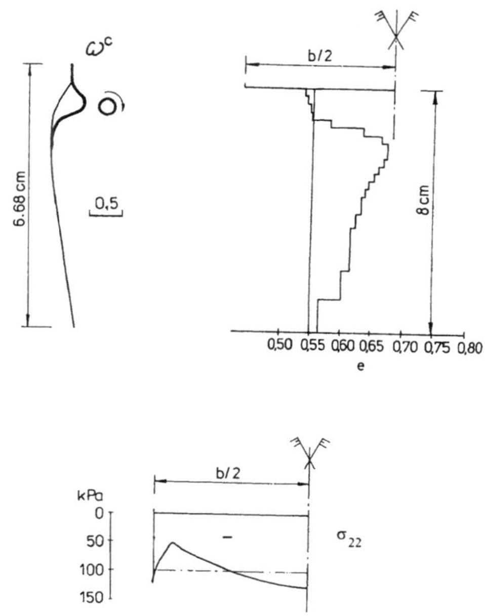

Fig. 4.78. Results of the Cosserat rotation ω^c along a cross-section in the deformed state, the void ratio e along a cross-section in the initial state and the vertical normal stress σ_{22} under the footing from the FE analyses (b=5 cm, s=1.5 cm)

is shown along a vertical cross-section under the footing in a displaced configuration (Fig.4.77a) and the void ratio along the same cross-section in an initial configuration (Fig.4.75). The position of the cross-section was at *b/4* (Fig.4.78b). In addition, the direction of the Cosserat rotation in the shear zone is indicated in Fig.4.78a.

The maximum normalised vertical force and the corresponding normalised settlement *s/b* increase with a decrease of the footing width. During the footing displacement, an almost stiff wedge of sand is created directly underneath. The material in the wedge undergoes slight densification. Below the edge, a single shear zone occurs, manifested by the concentration of both large shear deformations and Cosserat rotations, and a strong increase of the void ratio *e*. Thus, the Cosserat rotation ω^c and a strong increase of the void ratio *e* seem to be suitable criteria to detect shear zones.

The calculated displacements are similar to those in the tests (Fig.2.26). However, the secondary shear zones are not numerically obtained under the footing due to that the FE calculations were carried out with consideration of a symmetry axis. The thickness of the calculated shear zone is found to be about 10 mm ($14\times d_{50}$) on the basis of the displacements and is slightly greater than the experimental value. To obtain a better agreement, a micro-polar constant should be increased (Tejchman 1997). However, judging from the Cosserat rotation, the thickness of the shear zone is slightly greater, i.e. 12.5 mm ($20\times d_{50}$). In turn, the thickness of the shear zone estimated on the basis of the increase of the void ratio is significantly larger because dense sand starts to dilate before the shear

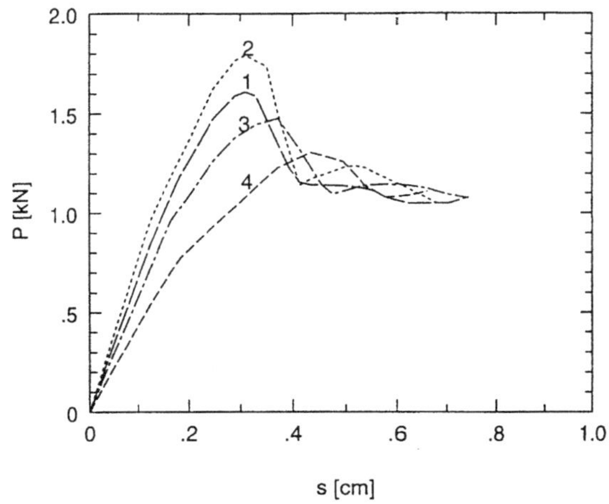

Fig. 4.79. Calculated load-displacement diagrams (b=2.5 cm, d50=0.6 mm, e0=0.55): 1. α=0.16, hs=300 MPa, n=0.4, 2. α=0.17, hs=300 MPa, n=0.4, 3. α=0.16, hs=200 MPa, n=0.4, 4. α=0.16, hs=300 MPa, n=0.35

zone is created. The void ratio in the shear zone (whose thickness was estimated with the Cosserat rotation) changes from 0.65 to 0.675 for s/b=0.3 (b=5.0 cm).

The distribution of the vertical normal stress σ_{22} under the foundation is non-uniform. The mean inclination of the sand slope close to the foundation is about 50° and is higher than the angle of internal friction of sand at peak, ϕ=44°. The mesh along the upper boundary is slightly distorted due to the occurrence of tensile stresses in some elements (which could not be avoided at the assumed mesh). To overcome this disadvantage and to obtain a more realistic slope inclination, the mesh along the free boundary near the footing should be finer. However, the preliminary calculations showed that this slight distortion of the mesh along the free boundary did not influence the bearing capacity of the footings.

The influence of the assumed material constants: h_s, n and α on the load-displacement curve for b=2.5 cm is presented in Fig.4.79. An increase of α corresponds to an increase of the initial density of sand (curves '1' and '2'). An increase of h_s and n is related to an increase of the stiffness of sand (curves '1', '3' and '4'). The FE results show that the greater α, the larger both the maximum vertical force on the footing P_{max} and the softening regime. The vertical displacement s_p related to P_{max} remains the same. In turn, the smaller h_s and n, the smaller P_{max} and the larger s_p.

Fig.4.80 describes the effect of the mean grain diameter and the initial void ratio on the load-displacement curve for b=2.5 cm. In addition, the results for small deformations and curvatures are shown. In this case, the calculations were performed without taking into account both of the Jaumann terms and the change of the element configuration and element volume. The maximum vertical force increases with increasing mean grain diameter and decreasing initial void ratio. The increase of d_{50} by 100% (from d_{50}=0.4 mm to d_{50}=0.8 mm) makes the force P_{max} larger by about 10% (curves '1' and '3'), and the increase of e_0 by about 20% (from e_0=0.55 to e_0=0.65) makes

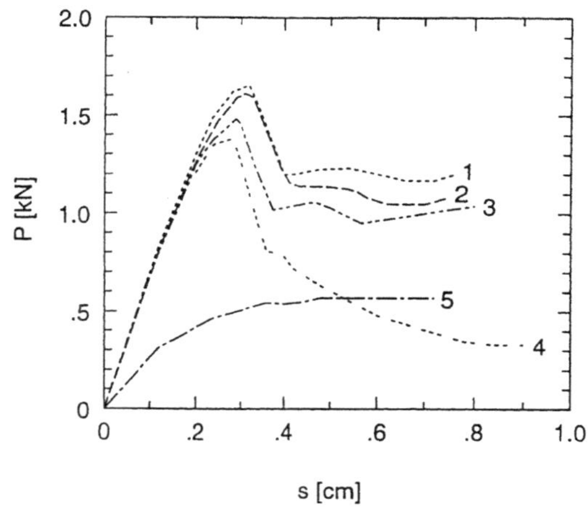

Fig. 4.80. Calculated load-displacements diagrams (b=2.5 cm): 1. d_{50}=0.8 mm, e_0=0.55, 2. d_{50}=0.6 mm, e_0=0.55, 3. d_{50}=0.4 mm, e_0=0.55, 4. d_{50}=0.6 mm, e_0=0.55 (small deformations and curvatures), 5. d_{50}=0.6 mm, e_0=0.65

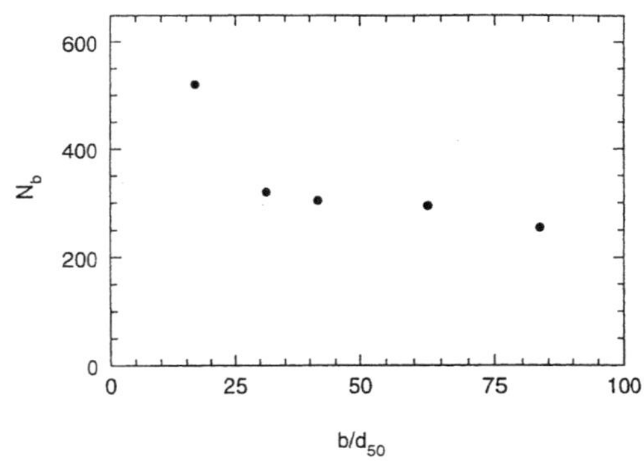

Fig. 4.81. Relationship between the normalised bearing capacity factor $N_b=2P_{max}/(\gamma_d b^2 l)$ and the ratio b/d_{50} from the FE analyses

P_{max} approximately 3 times smaller (curves '2' and '5'). In the calculations with small deformations and curvatures, P_{max} is smaller by about 30% and P_{cr} by about 300% (curves '2' and '4').

A decrease of the effect of the Cosserat quantities on the material behaviour through an increase of the polar constant a_c from 1 to 2.5 decreases the maximum normalised vertical force by about 5%. The thickness of the shear zone is slightly diminished as well.

The effect of the ratio between the footing width and the mean grain diameter on the normalised bearing capacity factor on the basis of the FE calculations is summarised in Fig.4.81. The influence of the ratio b/d_{50} on the normalised bearing capacity of plane strain footings becomes negligible for b/d_{50}>33.

Comparing the calculated load-displacements curves of Fig.4.76 with the experimental curves of Fig.4.68, it follows that the effect of the foundation width on the normalised bearing capacity is realistically described. The shape of the experimental and the predicted curves is similar. The calculated normalised vertical displacements (s_p/b=0.1-0.15) differ not much from the measured ones (s_p/b=0.09-0.12). However, there are some discrepancies in the magnitude of the normalised bearing capacity. The calculations result, namely, in the normalised bearing capacity larger by about 30-35%. There are two reasons for it which have been already mentioned. First, the calibration of the material constants was not exactly performed after the standard procedure (Section 4.2). Probably, the value of the exponent n was set too high. This exponent has, namely, a crucial influence on the dependence of the incremental stiffness $\dot{\sigma}_{ij}/d_{ij}$ on the pressure level σ_{kk}. Consequently, the stiffness may be overestimated at lower pressures and underestimated at higher pressures (if the parameter n is too high). Moreover, the maximum friction angle is also influenced by the parameter n. For a higher n, the relative void ratio $e_r=(e-e_d)/(e_c-e_d)$ is too low at low pressures and too high at high pressures, respectively. As a result, the calculated friction angle is too large in the first case and too small in the second case, respectively. This agrees with the comparison between the predicted and measured results where the largest difference in the normalised vertical force was obtained for the smallest footing, i.e. for very low pressures. The second reason for the discrepancies between the calculated and measured results is related to the fact that the calibration procedure was performed for a homogeneous specimen. It did not take into account a shear zone which took place in dense sand during a plane strain compression test before the peak on the load-displacement curve occurred. The calculations show, namely, that the internal friction angle at peak is smaller for a non-homogeneous specimen than for a homogeneous one (Tejchman 1997).

The FE calculations demonstrate that:

- The normalised bearing capacity of footings on sand increases with decreasing footing width, increasing initial density of sand, and increasing mean grain diameter of sand.
- The Cosserat effect manifested by rotations is noticeable in shear zones and its influence on the results is substantial. Outside the shear zone this effect becomes negligible.

The calculations of footings on sand were also carried out by Nübel (2002) using a micro-polar hypoplastic constitutive law. In the calculations, the entire footing was taken into account.. The footing was modelled as an elastic beam. The initial void ratio in a dense specimen was distributed stochastically with an exponential frequency function. A small wall friction angle was assumed between the footing and soil. The FE results were in good accordance with experiments by Vesic (1973) and Tatsuoka et al. (1997).

4.6 Earth Pressure

The earth pressure on retaining walls belongs to a classical problem of soil mechanics. In spite of intense theoretical and experimental research on this problem over more than 200 years, there are still large discrepancies between theoretical solutions and experimental results due to the complexity of the deformation field in granular bodies near the wall caused by localization of deformations. Coulomb (1775) indicated for the first time the occurrence of shear zones during earth pressure tests. Darwin (1883) showed with model tests that the explanation of the behaviour of granular bodies during earth pressure is not possible without taking into account shear localization and procedure of filling (called by him a "historical effect"). Comprehensive experimental studies on earth pressure in sand have been carried out at Cambridge University and Karlsruhe University (Section 2.1, Figs.2.17-2.19).

Earth pressures on a retaining wall are calculated within a theory of elasticity and plasticity. Within plastic limit states, there are generally two approaches: static and kinematic. Within a first approach assuming the material yielding behind the wall (according to the Mohr-Coulomb law), one can obtain mathematically closed solutions of pressures for simple boundary conditions (Caquot and Kerisel 1948, Negre 1959, Dembicki 1979) or numerical solutions using a method of stress and velocity characteristics by Sokolovsky (1965) for cases with complex boundary and load conditions (Roscoe 1970, James and Bransby 1971, Szczepiński 1974, Bransby and Milligan 1975, Houlsby and Wroth 1982, Milligan 1983). In turn, within a kinematic approach, different failure mechanisms consisting of slip surfaces are assumed. From the equilibrium of forces on sliding wedges, a resultant total earth pressure is calculated (Coulomb 1775, Terzaghi 1951, Gudehus 1978, Wang 2000, Leśniewska and Mróz 2000, 2001). All theoretical solutions are very sensitive to the angle of internal friction and wall friction angle. They are not able to predict consistently deformations observed in experiments.

Finite element calculations are more realistic than analytical solutions since first, they take into account advanced constitutive laws describing the granular material behaviour and second, they can predict the evolution of localization of deformation. For FE analyses of earth pressures in granular soils, a perfect plastic (Nakai 1985), an elasto-plastic (Simpson 1972, Simpson and Wroth 1974, Christian et al. 1977, Potts and Fourie 1984, Fourie and Potts 1989, Leśniewska and Mróz 2003), an elasto-plastic with remeshing (Hicks et al. 2001), an elasto-plastic with embedded strong discontinuity (Lai et al. 2003), a micro-polar elasto-plastic (Kitsabunnarat et al 2008), a hypoplastic (Ziegler 1986), and a micro-polar hypoplastic constitutive law (Tejchman and Dembicki 2001, Nübel 2002) were used. A characteristic length of micro-structure was not taken into account in the analyses, except of the calculations with a micro-polar hypoplastic and elsto-plastic constitutive law. A pattern of shear zones calculated by Nübel (2002) for the case of an active dredged test with flexible walls was in good accordance with experiments (Milligan 1974). In this analysis, a fluctuation of the initial void ratio with exponential distribution was attributed to the granular body. The numerical calculations by Leśniewska and Mróz (2001, 2003) revealed that the pattern of shear zones was easier to observe when a non-associated Coulomb-Mohr model without dilatancy was assumed. A FE study carried out by Tejchman (2002b) with an uniform distribution of the initial void ratio in sand showed that the calculated deformation field with a wall rotating around the bottom (passive and active case) was different than the experimental one (Bransby 1968, Smith 1972). In addition, this study showed that the geometry of calculated shear zones was influenced by the size of the sand specimen and its initial void ratio. The

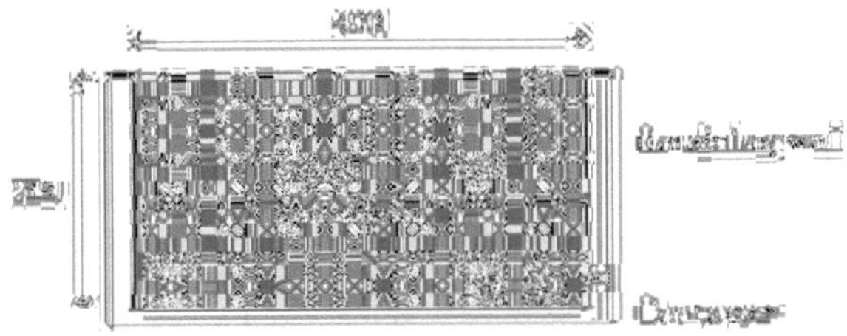

Fig. 4.82. Initial dimension of the sand specimen (units in mm)

pattern of induced shear zones was created also in the case of loose sand continuously subject to contractancy.

The calculations were performed with a sand body of a height of H=200 mm and length of L=400 mm (models test at Karlsruhe University, Gudehus 1986, Gudehus and Schwing 1986), Fig.4.82. Totally, 3200 triangular elements were used. The height of the retaining wall located at the right side of the sand body was assumed to be h=170 mm (h/H=0.85). The calculations were carried out with large deformations and curvatures (Section 5.3).

The initial stresses were generated using a K_o-state without polar quantities: $\sigma_{22}=\gamma x_2$, $\sigma_{11}=\sigma_{33}=K_0\gamma x_2$, $\sigma_{12}=\sigma_{21}=m_1=m_2=0$ (Fig.4.3). The pressure coefficient at rest was assumed for dense sand as K_0=0.47 on the basis of a so-called element test for oedometric compression (Herle 1997). Two sides and the bottom of the sand specimen were assumed to be very rough: u_1=0, u_2=0 and ω^c=0. The top of the sand specimen was traction and moment free. The retaining wall was assumed to be stiff and very rough (u_2=0 and ω^c=0). Three different wall modes were assumed in passive and active tests: uniform horizontal translation, rotation around the wall bottom and rotation around the wall top. The maximum horizontal displacement increments were chosen as $\Delta u/h$=0.00002 (passive mode – wall moves against the backfill) and $\Delta u/h$=0.000004 (active mode – wall moves away from the backfill)).

The calculations were carried out with a random distribution of the initial void ratio to enhance and promote the whole process of the shear zone formation. The non-homogenous distribution of void ratio is an inherent property of each granulate. The fluctuations of the initial void ratio are usually more marked for loose than for dense skeletons (Nübel 2002). The greater the fluctuations of void ratio, the bigger are the stress changes (Behringer and Miller 1997, Geng et al. 2003). The initial void ratio e_o was distributed non-uniformly in elements of the sand body by means of a random generator in such a way that the initial void ratio e_0=0.60 was increased in every element by the value 0.05×r (e_o=0.60+0.05r), where r is a random number within the range of (0.01, 0.99).

The FE analyses were carried out with the following material constants for fine Karlsruhe sand: e_{i0}=1.3, e_{d0}=0.51, e_{c0}=0.82, ϕ_c=30°, h_s=190 MPa, α=0.3, n=0.5, β=1, d_{50}=0.5 mm and $a_c=a_1^{-1}$.

Passive case

The FE results of a plane strain passive earth pressure problem for dense sand (e_o=0.60+0.05r, γ=17.0 kN/m^3) are shown in Figs.4.83-4.88 (Tejchman 2004a, Tejchman et al. 2007). Fig.4.83 presents the evolution of the normalized horizontal

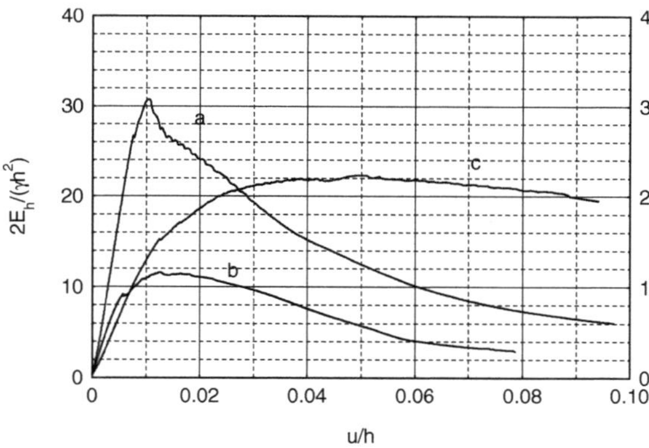

Fig. 4.83. Resultant normalized earth pressure force $2E_h/(\gamma h^2)$ versus normalized wall displacement u/h (passive case): a) translating wall, b) wall rotating around its top, c) wall rotating around its bottom

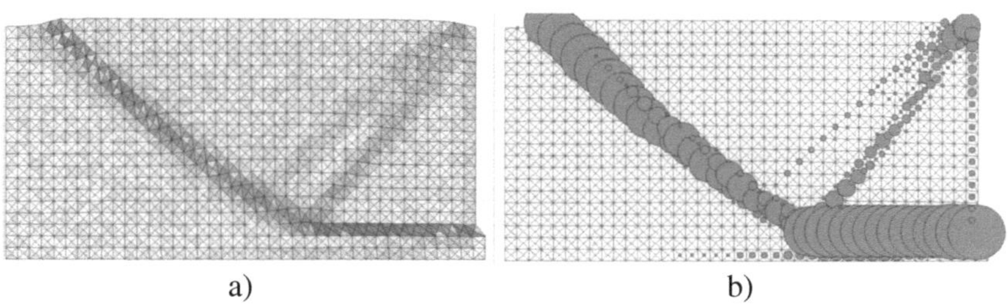

Fig. 4.84. Deformed FE meshes with distribution of void ratio (a) and Cosserat rotation(b) for dense sand during passive earth pressure with translating wall (u/h=0.05)

earth pressure force $2E_h/(\gamma h^2)$ versus the normalized horizontal wall displacement u/h for three different wall movements. In the case of a rotating wall, the horizontal displacement u is related to the wall displacement of the bottom point (wall rotating about the top) or top point (wall rotating around the bottom). The force E_h was calculated as the integral of mean horizontal normal stresses σ_{11} from quadrilateral elements along the retaining wall. In Figs.4.84-4.86, the deformed meshes with the distribution of the void ratio and Cosserat rotation in the residual state are shown. The darker region indicates the higher void ratio. The Cosserat rotation is marked by circles with a diameter corresponding to the magnitude of the rotation in the given step. The distribution of the Cosserat rotation at the beginning of the passive wall translation is described in Fig.4.87. The distribution of Cosserat rotation ω^c and void ratio e across the normalized initial specimen height x_2/d_{50} in two sections at x_1=0.15 m and x_1=0.30 m (measured from the left side) at residual state is presented in Fig.4.88 (during passive wall translation). The positive Cosserat rotation corresponds to that of Fig.4.1.

The evolution of curves of the horizontal earth pressure $2E_h/(\gamma h^2)$ is similar in all three cases (Fig.4.83). The forces increase, reach a maximum for about u/h=1-5%,

a) b)

Fig. 4.85. Deformed FE meshes with distribution of void ratio (a) and Cosserat rotation (b) for dense sand during passive earth pressure with wall rotating around its top (u/h=0.06)

a) b)

Fig. 4.86. Deformed FE meshes with distribution of void ratio and Cosserat rotation for dense sand during passive earth pressure with wall rotating around its bottom (u/h=0.07)

next show softening and tend to asymptotic values for about u/h=7-9%. For the wall rotation about the bottom, a decrease of the curve is smaller (in the considered range of u/h). The maximum horizontal force on the wall is the highest for the wall translation, and the lowest for the wall rotation about the top. The maximum normalized horizontal earth pressure forces are high ($2E_h/\gamma h^2$=12-31) due to the high initial void ratio of sand, large wall roughness, high relationship between the mean grain diameter and wall height and low initial stress level. The calculated minimum (residual) earth pressure coefficients are about 3.0-6.0 (Fig.4.83).

Shear localization which is characterized in dense granulates among others by the appearance of the Cosserat rotation and a strong increase of the void ratio is strongly dependent upon the type of the wall mode (Figs.4.84-4.86). For the wall translation (Figs.4.84 and 4.87), five shear zones are obtained: one vertical along the very rough retaining wall, one zone projecting horizontally from the wall base, one inclined (slightly curved) zone spreading between the wall bottom and free boundary, and two radial oriented shear zones starting to form at the wall top. The inclined shear zone becomes dominant in the course of deformation. The horizontal shear zone develops only at the beginning of the wall translation (Fig.4.87). The material starts to generate the Cosserat rotation and to dilate at the same time at three different places (wall bottom, wall top and free boundary at the left side), Fig.4.87. Next, the shear zone (starting from the wall base) curves upwards, becomes straight and reaches the free boundary. Later, it is approached by one radial shear zone. The second radial shear zone is not fully developed at u/h=0.07. The thickness of the dominant shear zone is about

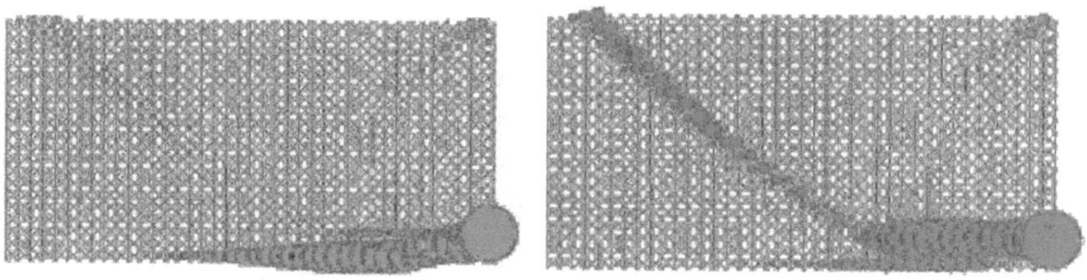

a) b)

Fig. 4.87. Distribution of Cosserat rotation in dense sand at the beginning of passive earth pressure with translating wall: a) u/h=0.01, b) u/h=0.02

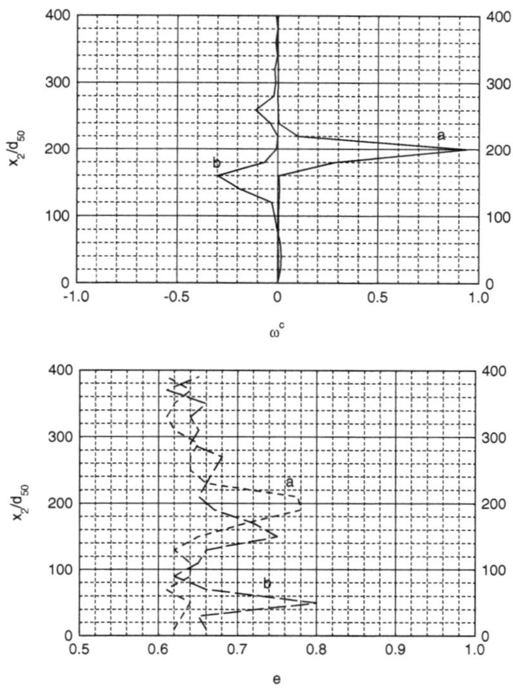

A)

B)

Fig. 4.88. Distribution of Cosserat rotation ω^c (A) and void ratio e (B) across the normalized specimen height x_2/d_{50} at residual state during passive earth pressure with translating wall (u/h=0.08): a) x_1=0.0.15 m, b) x_1=0.30 m

30×d_{50} (on the basis of the Cosserat rotation, Fig.4.88) and its inclination from the bottom is about θ=40°.

The values of void ratio e and mobilized internal friction angle ϕ in the middle of the shear zones are near the peak (at u/h=0.01): e=0.65 and ϕ=45.2^0 (inclined zone at x_1=0.05 m), e=0.68 and ϕ=42.0^0 (inclined zone at x_1=0.10 m), e=0.64 and ϕ=44.8^0 (horizontal zone at x_1=0.15 m), e=0.63 and ϕ=48.8^0 (lower radial zone at x_1=0.15 m), and e=0.63 and ϕ=48.1^0 (upper radial zone at x_1=0.15 m). At u/h=0.08, they are, respectively: e=0.79 and ϕ=34.6^0 (inclined zone at x_1=0.05 m), e=0.78 and ϕ=35.0^0 (inclined zone at x_1=0.10 m), e=0.80 and ϕ=33.5^0 (horizontal zone at x_1=0.15 m), e=0.74 and ϕ=37.2^0 (lower radial zone at x_1=0.15 m), and e=0.67 and ϕ=43.3^0 (upper radial zone at x_1=0.15 m). The angles of internal friction were calculated from the Mohr's formula (Eq.4.1).

The largest void ratios in the inclined and horizontal shear zones at u/h=0.08 are equal to the pressure-dependent critical values e_c (Fig.4.88). The critical void ratios in the radial shear zones have not been obtained yet. Beyond the shear zones, the residual void ratio is equal to 0.61-0.65. The calculated geometry of shear zones is in agreement with experimental observations at Cambridge University by Lucia 1966 (Fig.2.17a) and at Karlsruhe University by Gudehus 1986 and Gudehus and Schwing 1986 (Fig.2.19).

The maximum normalized horizontal earth pressure forces ($2E_h/\gamma h^2$=12-31) are approximately in the range of the usual (engineering) earth pressure coefficients (Dembicki 1979) determined under the assumption of one circular slip line (K_{pr}=11.33-25.80) and three straight slip lines (K_{pt}=13.40-23.70) at $\delta=\phi^p$=40°-45° (δ - wall friction angle, ϕ^p- internal friction angle of dense sand at peak). However, the actual internal friction angles at peak ϕ^p in the shear zones are not known in advance (they depend strongly on the initial and boundary conditions of the entire system). Therefore, it is difficult to obtain realistic earth pressures with a conventional earth pressure theory. In addition (as the numerical calculations show), the different internal friction angles are mobilized in the various shear zones at the same time. The different friction angles occur also along the same shear zone.

In the case of the wall rotation around the top (Fig.4.85), only one shear zone occurs which is more curved than the shear zone during the wall translation. The calculated deformation field is close to the experimental one (Arthur 1962) (Fig.2.17b).

When the retaining wall rotates around the bottom (Fig.4.86), a pattern of curved parallel shear is obtained. This result is approximately in accordance with experiments (Bransby 1968) (Fig.2.17c).

Fig.4.89 presents the evolution of the normalized horizontal earth pressure force versus the normalized horizontal wall displacement for different initial densities and mean grain diameters of sand (Tejchman et al. 2007). The calculations were also carried out for homogeneous initial void ratio (e_o=0.65). The maximum earth pressure coefficient increases with increasing mean grain diameter and decreasing initial void ratio. Finally, Figs.4.90 and 4.91 demonstrate the distribution of the Cosserat rotation ω^c, void ratio e, horizontal normal wall stress σ_{11} and couple stresses m_i along the specimen height. The calculated distribution of the wall horizontal normal stress is almost linear except of the value at the wall bottom (due to the singular point). Small fluctuations of σ_{11} in the upper part of the wall are caused by shear formation of radially oriented shear zones. The distribution of the couple tresses m_i along the wall is almost constant.

Active case

The FE results of a plane strain active earth pressure problem with a dense sand specimen are shown in Figs.4.92-4.95. Fig.4.92 shows the evolution of the normalized

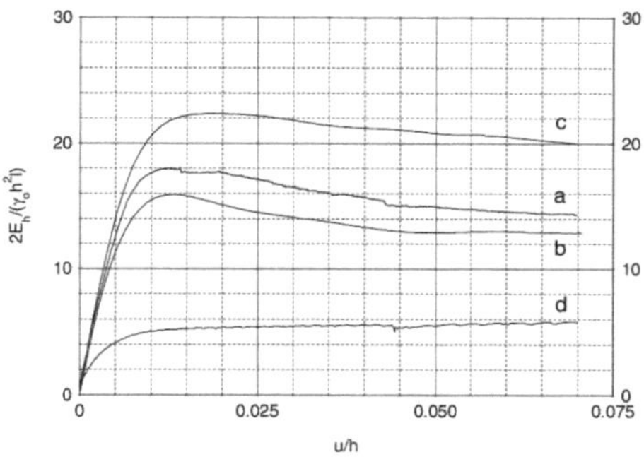

Fig. 4.89. Resultant normalized earth pressure force $2E_h/\left(\gamma_0 h^2 l\right)$ versus normalized wall displacement u/h for: a) homogeneous initial void ratio (e_o=0.65, d_{50}=0.5 mm), b) e_o=0.60+0.1r (d_{50}=0.5 mm), c) e_o=0.60+0.1r (d_{50}=2.0 mm), d) homogeneous initial void ratio (e_o=0.90, d_{50}=0.5 mm)

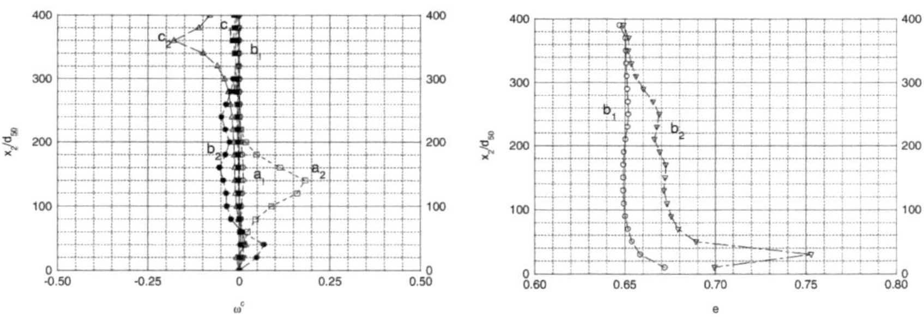

A) B)

Fig. 4.90. Distribution of Cosserat rotation ω^c (A) and void ratio e (B) across the normalized specimen height x_2/d_{50} at: a) x_1=0.20 m, b) x_1=0.30 m, c) x_1=0.38 m, 1) at peak, 2) at residual state (e_o=0.65, d_{50}=0.5 mm)

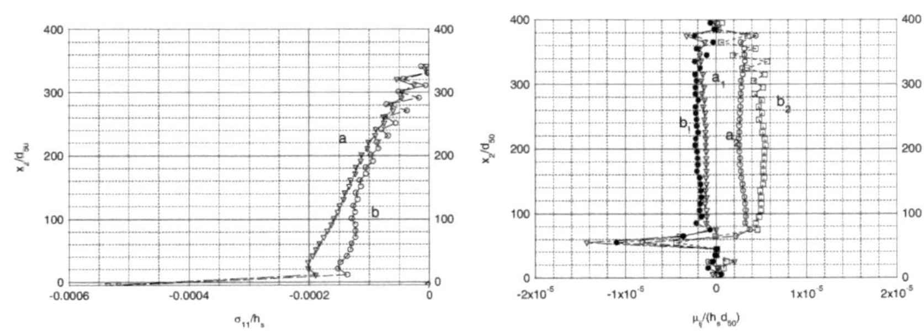

A) B)

Fig. 4.91. Distribution of normalized horizontal normal stress σ_{11} along the normalized wall height x_2/d_{50} (A) and couple stresses m_i (B) along the normalized specimen height x_2/d_{50} (e_o=0.65, d_{50}=0.5 mm) at peak u/h=0.01(a) and residual state u/h=0.07 (b): 1) m_1, 2) m_2

horizontal earth pressure force $2E_h/(\gamma h^2)$ versus the normalized horizontal wall displacement u/h for three different wall movements in the case of dense sand (e_o=0.60+0.05r, γ=17.0 kN/m^3). In Figs.4.93-4.95, the deformed meshes with the distribution of the void ratio and Cosserat rotation in the residual state are demonstrated.

All earth pressure curves drop sharply at the beginning of the wall movement, reach the minimum at u/h=0.001-0.002 and next increase continuously (Fig.4.92). For the case of a wall rotating around its bottom, the residual state was reached for u/h=0.1. The lowest earth pressure force occurs with the wall translation, and the largest with the wall rotation about the top. Thus, the relationship between the minimum active earth pressure and the type of the wall movement is inversed as compared to the maximum passive earth pressure and the type of the wall movement. The minimum normalized earth pressure forces ($2E_h/(\gamma h^2)$=0.10-0.16) are slightly smaller than the usual earth pressure coefficients assuming a circular slip line (K_a=0.16-0.20) or a straight slip line (K_a=0.14-0.16) with $\delta=\varphi^p$ (φ^p=40°-45°).

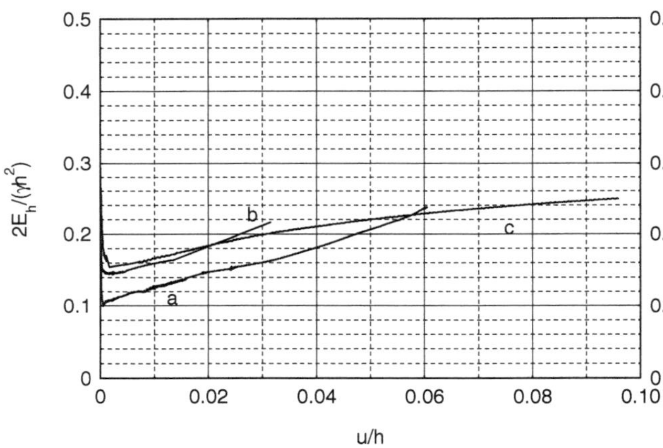

Fig. 4.92. Resultant normalized earth pressure force $2E_h/(\gamma h^2)$ versus normalized wall displacement u/h (active case): a) translating wall, b) wall rotating around its top, c) wall rotating around its bottom

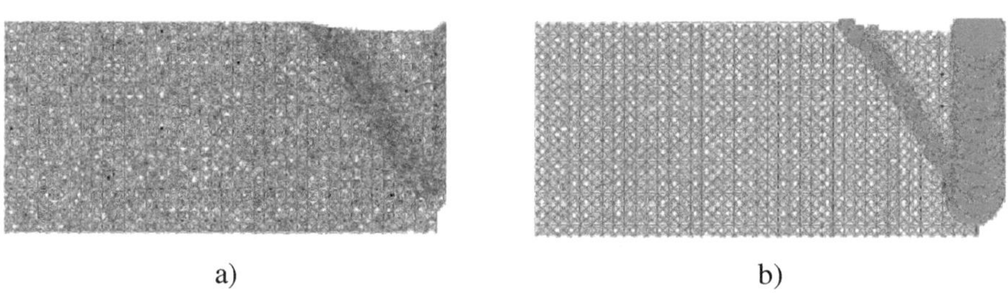

Fig. 4.93. Deformed FE meshes with distribution of void ratio and Cosserat rotation for dense sand during active earth pressure with translating wall (u/h=0.06)

a) b)

Fig. 4.94. Deformed FE meshes with distribution of void ratio and Cosserat rotation for dense sand during active earth pressure with wall rotating around its top (u/h=0.03)

a) b)

Fig. 4.95. Deformed FE meshes with distribution of void ratio and Cosserat rotation for dense sand during active earth pressure with wall rotating around its bottom (u/h=0.09)

The calculated pattern of shear zones depends again on the type of the wall movement. In the case of the wall translation, two pronounced shear zones are obtained (Fig.4.93). A vertical one occurs along the wall, and the second one propagates from the wall bottom up to the free boundary. The internal shear zone is slightly curved with a mean inclination to the bottom of θ=50°. This result is close to the experimental one (Szczepiński 1974). The thickness of the shear zone is again about (30-35)$\times d_{50}$.

When the wall rotates around the top, two shear zones are obtained again: the first along the wall and the second inside of sand starting from the wall bottom (Fig.4.94). The shear zone is strongly curved. The similar result was obtained in the experiment by Lord (1969) (Fig.2.18a).

In the case of a wall rotating about the bottom (Fig.4.95), three shear zones are obtained: one shear zone along the wall and two parallel internal shear zones. It is in accordance with the experiment by (Smith 1972) (Fig.2.18b).

The following conclusions can be drawn on the basis of the performed FE studies on shear localization during plane strain earth pressure problems with very rough and rigid retaining walls:

- The geometry of shear zones depends strongly on the direction and type of the wall movement (passive or active, translation or rotation). The experimental deformation field was realistically reproduced.

- The effect of a non-uniform distribution of the initial void ratio on the geometry of shear zones is of a major importance for the case of a passive and active wall rotation around the bottom.
- The largest passive earth pressures occur with the horizontal translation of the wall, they are smaller with the wall rotation around the bottom and again smaller with the wall rotation around the top. The smallest active earth pressures are created during a wall translation, and the largest during the wall rotation around the top.
- The passive earth pressures increases with increasing mean grain diameter and decreasing initial void ratio.
- The granular material tends to a critical state inside of shear zones.
- The earth pressures significantly change with the wall displacement.
- Conventional earth pressure mechanisms with slip surfaces are roughly reproduced. Realistic earth pressure coefficients can be obtained with actual values of internal friction angles.

4.7 Direct and Simple Shear Test

To investigate the progressive failure of frictional granular materials like soils and powders, their shear resistance in limit states is of particular interest. Herein the maximum shear resistance (or so-called peak internal friction angle) has to be distinguished from the residual shear resistance (or so-called critical internal friction angle). The former is not a material constant and depends strongly on the density and pressure. Beyond the peak state, a reduction of the shear resistance is accompanied by dilatancy of the material and leads to a critical state for large monotonic shearing. As outlined in the critical state concept, the shear resistance in the critical state is independent of the initial density and primarily proportional to the effective mean pressure.

For experimental investigation of the shear resistance, different shear testers are used in soil mechanics and powder technology. The comparison between shear testers reveals, however, significant discrepancies in measured properties of materials even if the same testing procedure, initial density of the solid, consolidation and stress level are used. It is due to different boundary conditions in testers to induce shearing in the material. The size of the specimen, the ratio between the length and height of the specimen, the ratio between the height and the mean grain diameter of the specimen, and the interface behavior between the soil and bounding walls are also of importance (Tejchman 1997, 2000, Tejchman and Gudehus 2001, Bauer and Huang 2001).

The intention of the FE analysis was to show the effect of different boundary conditions during induced shearing on the shear resistance and distribution of stresses and void ratio in a cohesionless granular specimen (Tejchman and Bauer 2005). Numerical calculations were carried out to simulate the material behavior in a direct shear tester (Skempton 1949, Potyondy 1961) and a true simple shear tester (Roscoe 1953, Bjeruum and Landva 1966) under the same initial conditions. Both testers are very popular in soil mechanics, used to determine important properties of granular and cohesive materials such as: evolution of the mobilized friction angle, envelope of maximum peak friction angles, critical friction angle, wall friction angle and cohesion. The built-up of the direct shear tester and the method of measuring the shear resistance are similar as in the Jenike's shear tester (Jenike 1964) which is the most common shear device in the powder technology to determine the flow function of cohesive bulk solids and powders. The only difference between them is that the cross-section of samples is circular at the Jenike's shear tester instead of a rectangular one at the direct shear tester.

Beside the axisymmetric triaxial test, the direct shear box test (Skempton 1949, Potyondy 1961, Wernick 1977, 1978, Kast 1985, Desai et al. 1985, Jewell and Wroth 1987, Tejchman and Wu 1995, Shibuya et al. 1997) is one of the most popular laboratory test in soil mechanics to investigate the shear resistance of granular materials. Its advantages are simple specimen preparation and test procedure, and it can be performed within a short time. However, this test has also disadvantages (Wernick 1977, 1978, Terzaghi and Peck 1948, Morgenstern and Tchalenko 1967, Saada and Townsend 1981, Scarpelli and Wood 1982, Uesugi and Kishida 1986, Chandler and Hamilton 1999, Johanson et al. 2003): deformation and stress fields are strongly non-uniform within the specimen, contact area of the forced shear zone diminishes during shearing, stress distribution is not known, sliding displacement and displacement due to shear deformation cannot be separated and rotation of the top cap takes place which may influence the thickness of a shear zone.

A true simple shear tester (Roscoe 1953, Bjeruum and Landva 1966, Uesugi and Kishida 1986, Budhu 1984, Budhu and Britto 1987) is also used in soil mechanics, however more rarely due to a complexity of the mechanical equipment needed. Its advantages are: constant shear area, simple preparation and procedure. However, stress field is non-uniform due to the increasing length of sides and absence of complementary shear stresses at smooth end walls and stress concentrations occur for large shearing at ends.

In the past, numerical simulations of shear tests in a direct and a true simple box were carried out among other by Potts et al. (1987) using a hardening elasto-plastic continuum model (direct and simple shearing), Budhu and Britto (1987) with a hardening elasto-plastic continuum model (simple shearing), by de Borst and Heeres (2002) using a non-polar hypoplastic model (direct shearing), by Al Hattamleh et al. (2007) applying a second order gradient elasto-plastic multi-slip model (simple shearing), and by Thornton and Zhang (2001, 2003) and Zhang (2003) (direct and simple shearing), Bilgili et al. (2003) (direct shearing), Tykhoniuk et al. (2004) (direct shearing) and Hartl and Ooi (2008) (direct shearing) with a discrete granular dynamic algorithm. A comprehensive FE analysis with an elastic-perfectly plastic Mohr-Coulomb model (Potts et al. 1987) showed that despite of strong non-uniform stresses generated within the direct shear box, the shear resistance was close to the ultimate strength during simple shearing. In addition, the shear resistance was found to be strongly influenced by the initial stress state assumed. The FE results obtained with the classical continuum mechanics depended upon the mesh discretization due to the lack of a characteristic length. In turn, the results of DEM reported by Thornton and Zhang (2001, 2003) and Zhang (2003) showed that the internal friction angle in the localized horizontal shear zone in the central part of the box was slightly smaller than that inferred from boundary global forces during direct shearing and significantly

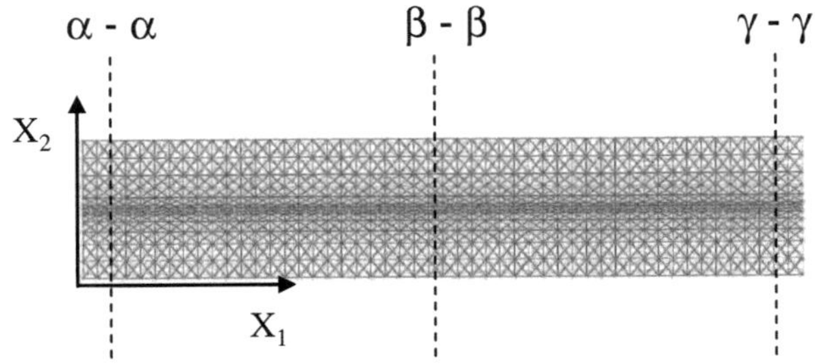

Fig. 4.96. FE mesh used for calculations (with the marked 3 vertical cross-sections)

larger during simple shearing. The Coulomb friction angle in the shear zone was less than the angle of shear resistance defined by Mohr's criterion. The evolution of the average rotation with shear strain was linear. In the critical state (direct and simple shearing), the vertical and horizontal normal stresses in the shear zone were equal and the directions of principal stress and strain rate were coaxial. The pronounced effect of the initial stress state on the shear resistance during simple shearing was also observed. In DEM-calculations by Bilgili et al. (2003), the stress field was found to be strongly non-homogeneous.

For simulations of direct and simple shearing, a specimen with a length l=0.10 m and a height of h=20 mm (l/h=5) was assumed (Wernick 1977, 1978, Kast 1985). The heights of finite elements were equal to (from the bottom up to the top): 2×2.5 mm, 1.75 mm, 1.5 mm, 1 mm, 3×0.5 mm, 1 mm, 1.5 mm, 1.75 mm and 2×2.5 mm (Fig.4.96). The width of elements was 2 mm. Thus, the size of elements in the shear zone (width and height) was not greater than 5×d_{50}.

As the initial stress state in the granular specimen, a K_o-state without polar quantities was assumed ($\sigma_{22}=\gamma_d x_2$, $\sigma_{11}=\sigma_{33}=K_0\gamma_d x_2$, $\sigma_{12}=\sigma_{21}=m_1=m_2=0$), where: γ_d=16.5 kN/m^3 - initial volume weight of sand and K_0=0.47 – pressure coefficient at rest. With respect to an initial void ratio of e_o=0.60 (dense sand), a constant load p=100 kN/m was applied along the entire top boundary. The initial void ratio of sand was assumed to be homogeneous in the specimen.

The bottom of the sand specimen was assumed to be very rough: u_1=0, u_2=0 and ω^c=0. The horizontal displacements of all nodes along the very rough top boundary were tied together in both apparatuses: $u_1=n\Delta u_1$ and ω^c=0. In turn, the vertical displacements along the top boundary were constrained to move by the same amount Δu_2. To simulate a direct shear test, the same horizontal displacement increments were prescribed to the upper part of two smooth side walls: $u_1=n\Delta u_1$, σ_{21}=0 and m_1=0 (Δu_1 – horizontal displacement increment of the top boundary, n – step number). Along the lower part of side walls, the boundary conditions were: u_1=0, σ_{21}=0 and m_1=0 (Fig.4.97a). In the case of a simple shear test, the prescribed horizontal and vertical displacements increased linearly along the both smooth side walls: $u_1=n\Delta u_1(1-x_2/h)$, $u_2=\Delta u_2\,(1-x_2/h)$ and m_1=0 (Δu_2 – vertical displacement increment of the top boundary), Fig.4.97b.

a)

b)

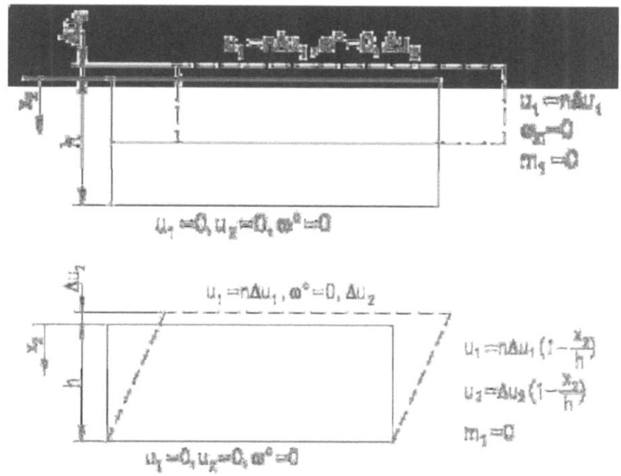

Fig. 4.97. Boundary conditions during direct (a) and simple shearing (b)

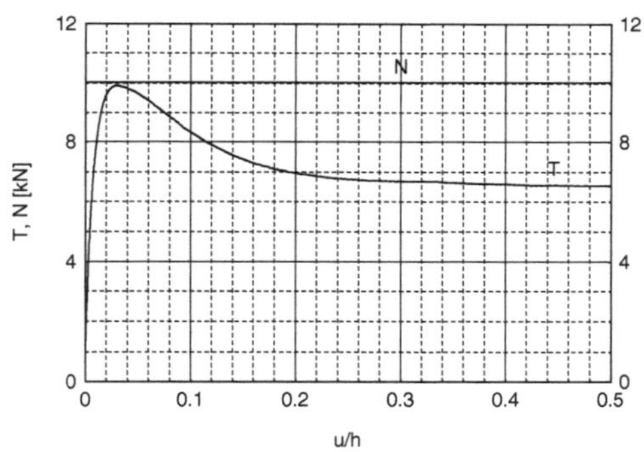

Fig. 4.98. Evolution of the resultant (global) horizontal shear force *T* and vertical normal force on the top *N* versus the normalized horizontal displacement of the top *u/h* during direct shearing of dense sand

The FE analyses of direct and simple shearing for the case of plane strain were carried out with the following material constants for fine Karlsruhe sand: e_{i0}=1.3, e_{d0}=0.51, e_{c0}=0.82, ϕ_c=30°, h_s=190 MPa, α=0.3, n=0.5, β=1, d_{50}=0.5 mm and $a_c=a_1^{-1}$.

For the assumption of an initially dense sand (e_o=0.60), the FE results of direct shearing are demonstrated in Figs.4.98-4.110. In turn, Figs.4.111-4.117 show the FE results for simple shearing.

Direct shear box test

Fig.4.98 presents the evolution of the resultant (global) horizontal shear force *T* and resultant (global) vertical normal force *N* against the normalized horizontal displacement of the top boundary *u/h* (*h* – specimen height). The force *T* was calculated as the sum of all horizontal nodal forces along the top boundary and side boundaries. In

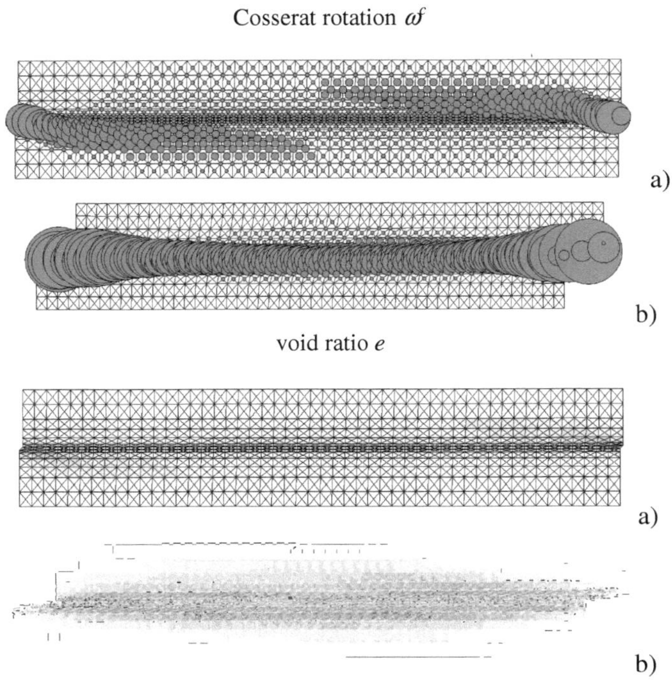

Fig. 4.99. Deformed FE mesh with the distribution of Cosserat rotation ω^c and void ratio e at u/h=0.025 (a) and u/h=0.35 (b) during direct shearing (dense sand)

Fig.4.99, the deformed meshes with the distribution of the Cosserat rotation ω^c and void ratio e at the beginning of shearing and at residual state are presented. The magnitude of Cosserat rotations is marked by circles with a diameter corresponding to the maximum rotation in the entire specimen (ω^c_{max}=0.17 at u/h=0.025 and ω^c_{max}=1.10 at u/h=0.35). The darker region indicates the higher void ratio.

The deformed meshes with the distribution of stresses σ_{ij} at residual state are presented in Fig.4.100. The darker regions indicate lower (negative) normal stresses and higher (positive) shear stresses. The distributions of the Cosserat rotation ω^c along three different vertical cross-sections α, β and γ of Fig.4.96 and its evolution during shearing in the mid-point of the specimen are presented in Fig.4.101. In turn, the distribution of void ratio e (at residual state) along the vertical cross-section in the center of the specimen are presented in Fig.4.102. The evolution of the local internal mobilized friction angle (defined by the Coulomb criterion ϕ=arctan(σ_{12}/σ_{22})) in the center of the specimen and void ratios e along the cross-section in the middle of the specimen versus the normalized horizontal displacement u/h (during initial compression and further shearing) are shown in Figs.4.103 and 4.104. In Figs.4.105 and 4.106, the distribution of normal and shear stresses σ_{ij}, couple stresses m_i along the vertical section in the center of the specimen, and distribution of σ_{12} and σ_{22} along the top boundary of the length l=100 mm are depicted. The numbers in Fig.4.104 denote the location of quadrilateral elements across the layer height (e.g. the element '*1*' is

Fig. 4.100. Deformed FE mesh with the distribution of stresses σ_{11} (a), σ_{22} (b), σ_{12} (c) and σ_{21} (d) during direct shearing at the normalized horizontal displacement of the top u/h=0.35

located at the bottom and the element '*13*' is at the top). The quantities σ_{ij}, m_i and e were taken as the mean values from quadrilateral elements.

The global horizontal force T grows from the beginning of shearing, reaches a maximum for u/h=0.03, shows a pronounced softening and tends to an asymptotic value for u/h=0.30 (Fig.4.98). The overall internal mobilized friction angle, ϕ=arctan(T/N), is equal to 44.7° (at peak, at u/h=0.029) and 33.3° (at residual state). In laboratory experiments with the same sand and specimen size, and under the same initial conditions (Wernick 1978), the values of φ were slightly higher: 45.5° (at peak) and 37° (at residual state), respectively.

The distribution of the void ratio e and stresses σ_{ij} is strongly non-uniform in the entire specimen (Figs.4.99 and 4.100). The horizontal normal stress σ_{11} changes across the specimen height in its centre from -130 kPa up to -210 kPa, the vertical normal stress σ_{22} changes from -110 kPa up to -150 kPa (top), the lateral normal stress σ_{33} changes from -130 kPa up to -150 kPa (top), the shear stress σ_{12} changes from 60 kPa up to 100 kPa, and the shear stress σ_{21} changes from 50 kPa up to 100 kPa (Fig.4.105). The vertical normal stress along the top boundary is strongly non-uniform (high at the left side due to a passive state, and very small at the right side due to an active state), Fig.4.106. The maximum stress σ_{22} occurs nearly at mid-point of the top boundary.

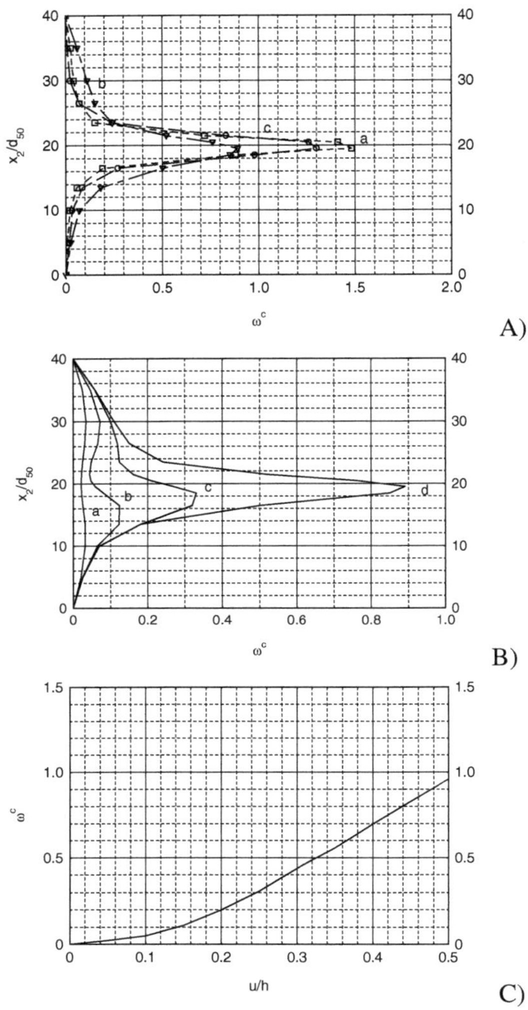

Fig. 4.101. Cosserat rotation ω^c: A) distribution along 3 different vertical planes across the specimen at the normalized horizontal displacement of the top u/h=0.50: a) left side (x_1=14 mm), b) center (x_1=50 mm), c) right side (x_1=86 mm), B) distribution in the center of the specimen at: a) u/h=0.05, b) u/h=0.125, c) u/h=0.25, d) u/h=0.50, and C) evolution in the center of the shear zone during direct shearing

At the beginning of shearing, inclined two shear zones are formed at the mid-point of wall sides (Fig.4.99) (de Borst and Heers 2002, Zhang 2003). They are almost symmetric (small non-symmetry is caused by the presence of the specimen weight). After the peak, the zones rotate to the nearly horizontal plane to create in the middle of the specimen a pronounced shear zone (Figs.4.99 and 4.101). The shear zone is characterized by the occurrence of Cosserat rotations and an increase of the void ratio.

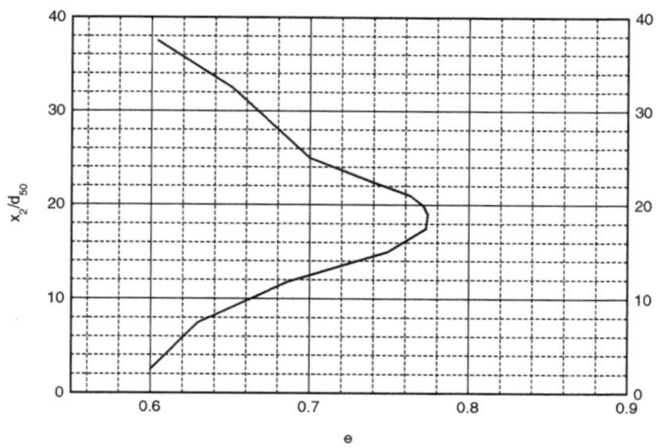

Fig. 4.102. Distribution of void ratio *e* in the vertical central plane across the specimen during direct shearing at the normalized horizontal displacement of the top u/h=0.50

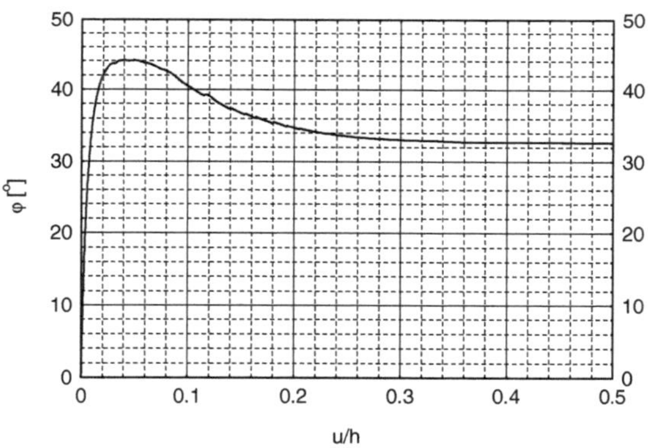

Fig. 4.103. Evolution of the local internal mobilized friction angle ϕ=arctan(σ_{12}/σ_{22}) in the middle of the specimen versus the normalized horizontal displacement of the top u/h (direct shearing)

The direction of the Cosserat rotation is opposite to the direction defined in Fig.3.26. The largest Cosserat rotation is greater at both ends of the specimen than in the middle of the specimen (Fig.4.101a). In contrast to DEM-results (Zhang 2003), the evolution of the Cosserat rotation with shear strain is essentially linear only at residual state (Fig.4.101c).

The thickness of the shear zone varies along the specimen lengths due to the effect of sides (it is the smallest at the left side, slightly larger at the right side and the largest in the middle of the specimen). In the center of the specimen, it is approximately t_s=8 mm (16×d_{50}) on the basis of shear deformation (Fig.4.100), Cosserat rotation (Fig.4.101), and the relative extreme values of the normal stress σ_{11}, shear stress σ_{21} and couple stress m_2 (Fig.4.105).

The evolution of the local internal mobilized friction angle φ=arctan(σ_{12}/σ_{22}) in the center of the shear zone (Fig.4.103) is similar as of the horizontal global force T (Fig.4.98). The local internal mobilized friction angle ϕ is equal to 44° (at peak for u/h=0.04) and 32.5° (at residual state for u/h=0.3). Thus, it is only slightly smaller

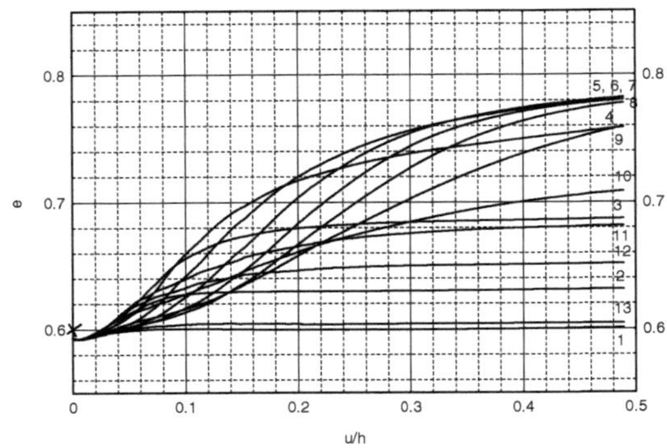

Fig. 4.104. Evolution of void ratios along the central cross-section versus the normalized horizontal displacement of the top *u/h* during direct shearing

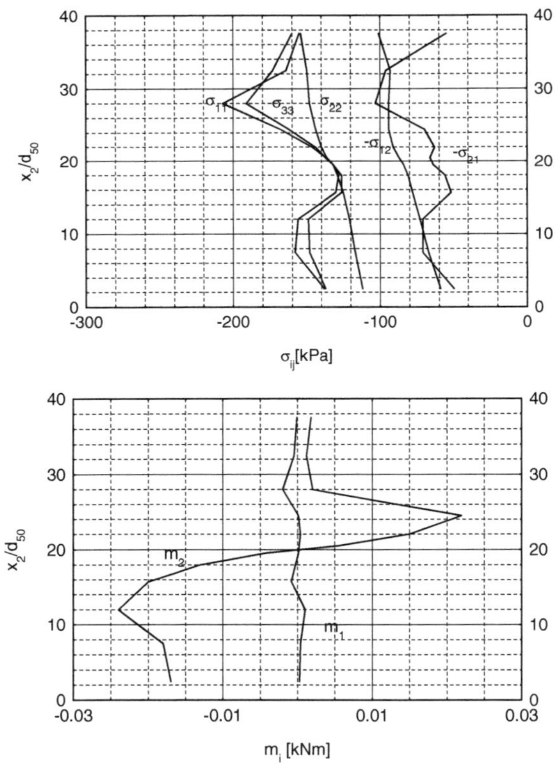

Fig. 4.105. Distribution of stresses σ_{ij} and couple stresses m_i across the central cross-section of the specimen during direct shearing at the normalized horizontal displacement of the top u/h=0.50

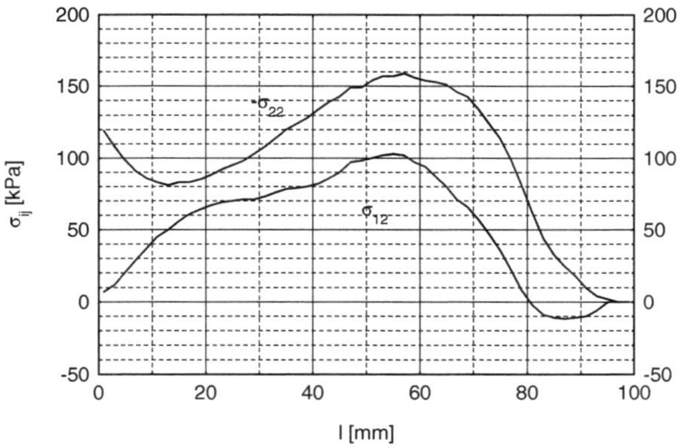

Fig. 4.106. Distribution of stresses σ_{12} and σ_{22} along the top boundary l during direct shearing at the normalized horizontal displacement of the top u/h=0.50

than the overall internal mobilized friction angle of 44.7° (at peak) and of 33.3° (at residual state). This outcome is qualitatively in accordance with DEM-results (Thornton and Zhang 2001, 2003, Zhang 2003).

During initial compression by the load p, the specimen undergoes contractancy (the void ratio changes from 0.600 up to 0.592). At the beginning of direct shearing, the specimen undergoes again slight compression (from 0.592 up to 0.591), and later during progressive shearing only volumetric expansion up to a residual state. The largest dilatancy appears in the middle of the shear zone where the pressure dependent maximum void ratio in the residual state is about e=0.78 (Fig.4.102). The calculated normalized vertical displacement of the top boundary at u/h=0.50, v/h=0.044, is slightly larger than in the experiment (v/h=0.035) (Wernick 1978).

In the middle of the shear zone, the pressure ratios σ_{11}/σ_{22} and σ_{11}/σ_{33} are 1.0 at residual state (Fig.4.105). It is in accordance with results by DEM (Zhang 2003). The stress tensor remains non-symmetric $\sigma_{12}\neq\sigma_{21}$ in the entire granular layer.

Effect of initial void ratio

The calculations were carried out with initially loose sand (e_o=0.90, d_{50}=0.5 mm, p=100 kN/m). In Fig.4.107, the evolution of the resultant friction coefficient T/N versus the normalized horizontal displacement of the top u/h is presented. The resultant friction coefficient T/N continuously increases approaching an asymptote. The overall internal friction angle, ϕ=arctan(T/N), is equal to 31.8° at residual state. In turn, the local internal friction angle φ=arctan(σ_{12}/σ_{22}) in the center of the shear zone is smaller by 1.5° (30.3°). The material is only subject to contractancy. The shear zone spreads over the entire specimen.

Effect of mean grain diameter

The evolution of the resultant friction coefficient T/N versus the normalized horizontal displacement of the top u/h for coarse sand (e_o=0.60, d_{50}=1.0 mm, p=100 kN/m) is presented in Fig.4.108. The maximum and residual overall internal friction angle, and

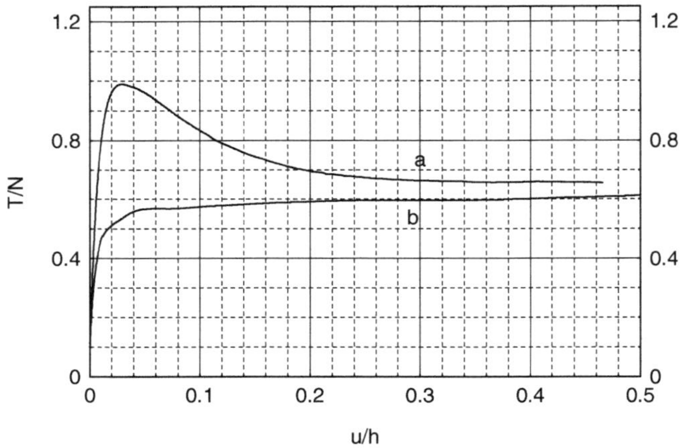

Fig. 4.107. Evolution of the resultant friction coefficient *T/N* versus the normalized horizontal displacement of the top *u/h* (initially dense and loose sand) during direct shearing: a) e_o=0.60, d_{50}=0.5 mm, *p*=100 kN/m, b) e_o=0.90, d_{50}=0.5 mm, *p*=100 kN/m

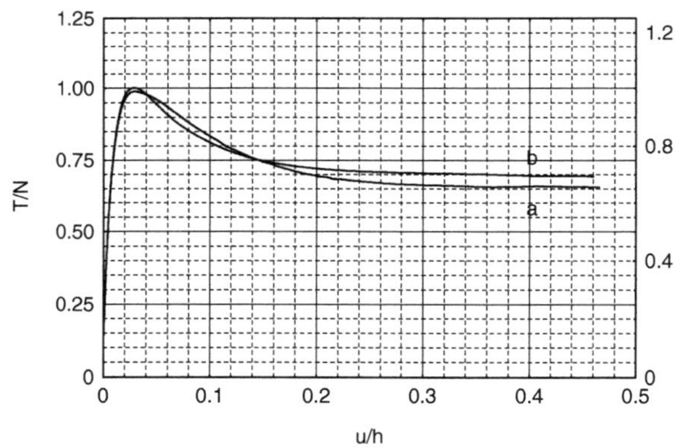

Fig. 4.108. Evolution of the resultant friction coefficient *T/N* versus the normalized horizontal displacement of the top *u/h* (fine and coarse sand) during direct shearing: a) e_o=0.60, d_{50}=0.5 mm, *p*=100 kN/m, b) e_o=0.60, d_{50}=1.0 mm, *p*=100 kN/m

shear zone thickness increase with increasing d_{50}. The overall internal friction angle is equal to 45.1° (at peak) and 34.8° (at residual state), respectively. Thus, it is larger by 0.4-1.5° than for sand with d_{50}=0.5 mm. The local internal friction angle is equal to 44.0° (at peak) and 33.7° (at residual state), respectively. The thickness of the shear zone in the centre of the specimen is approximately t_s=10 mm (10×d_{50}) (i.e. is higher by 25%).

Effect of pressure level

The evolution of the resultant friction coefficient *T/N* versus the normalized horizontal displacement of the top *u/h* for dense sand (e_o=0.60, d_{50}=0.5 mm) under the larger vertical load p=500 kN/m) is shown in Fig.4.109. The maximum and residual overall internal friction angle decrease with increasing p. The overall internal friction angle is equal to 41° (at peak) and 32.5° (at residual state), respectively. The local internal friction angle is equal to 40.1° (at peak) and 32.6° (at residual state), respectively. The thickness of the shear zone is similar as for p=100 kN/m; it is in the centre of the specimen approximately t_s=10 mm (20×d_{50}).

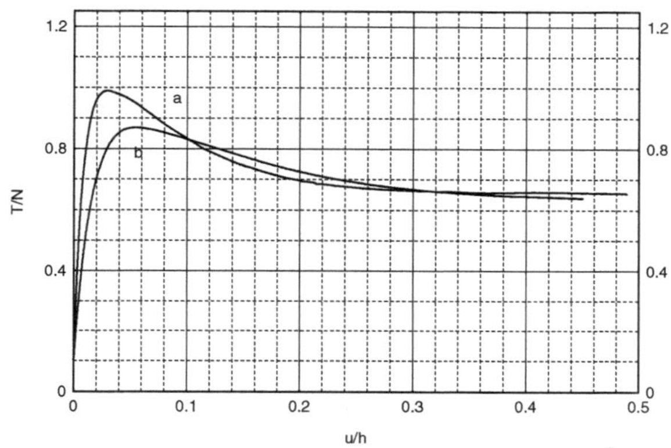

Fig. 4.109. Evolution of the resultant friction coefficient *T/N* versus the normalized horizontal displacement of the top *u/h* (small and large load) during direct shearing: a) e_o=0.60, d_{50}=0.5 mm, p=100 kN/m, b) e_o=0.60, d_{50}=0.5 mm, p=500 kN/m

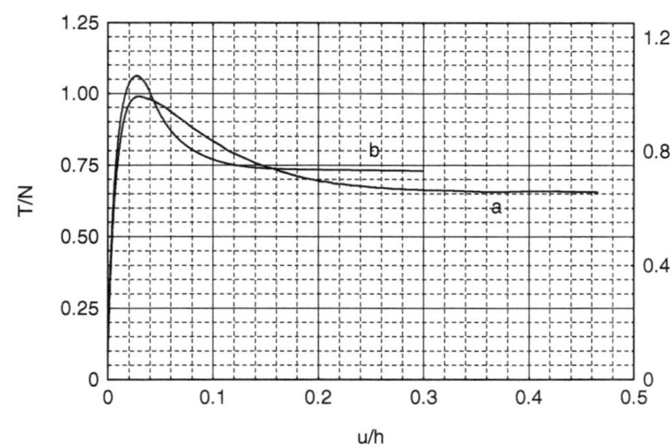

Fig. 4.110. Evolution of the resultant friction coefficient *T/N* versus the normalized horizontal displacement of the top *u/h* and distribution of Cosserat rotation ω^c in the vertical central plane across the specimen at *u/h*=0.30 during direct shearing (shorter and longer specimen with h=20 mm): a) e_o=0.60, d_{50}=0.5 mm, p=100 kN/m, l=100 mm, b) e_o=0.60, d_{50}=0.5 mm, p=100 kN/m, l=60 mm

Effect of specimen length

The calculations were performed with a shorter granular specimen (l=60 mm, h=20 mm), Fig.4.110. The results show that the overall internal friction angle increases both with decreasing specimen length. The overall internal friction angle is equal to 46.7° (at peak at u/h=0.027) and 36.3° (at residual state at u/h=0.30), respectively with a shorter specimen. The thickness of the shear zone in the center of the specimen is smaller, about 6.5 mm.

The FE results concerning the global and local internal friction angle and shear zone thickness for all cases are summarized in Tab.4.1.

Simple shear box test

The results are demonstrated in Figs.4.111-4.117. The behaviour of the local internal mobilized friction angle $\varphi=\arctan(\sigma_{12}/\sigma_{22})$ in the centre of the shear zone (Fig.4.115) is different as in the case of direct shearing (Fig.4.104). The internal mobilized friction angle is smaller by 2° at peak (42.0° for u/h=0.025) and only by 0.4° at residual state (32.1° for u/h=0.8). The shape of both curves is also different. The residual

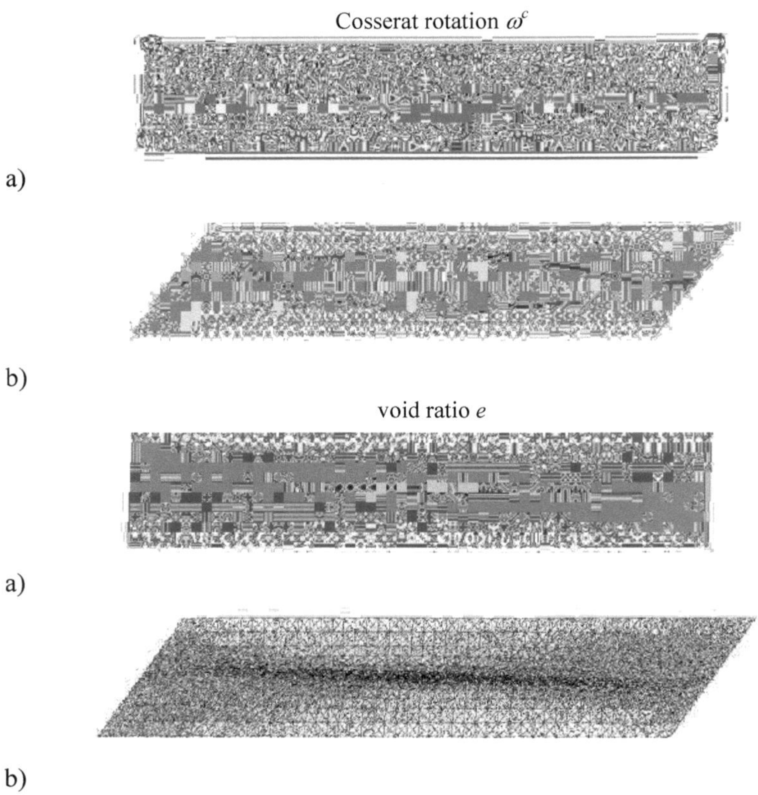

Fig. 4.111. Deformed FE mesh with the distribution of Cosserat rotation ω^c and void ratio e at u/h=0.03 (a) and u/h=0.75 (b) during simple shearing

Table 4.1. The overall and local internal friction angle at peak and residual state and thickness of the shear zone during direct shearing (Tejchman 2005)

Material properties	Overall internal friction angle ϕ [^{o}] at peak	u/h related to the peak	Overall internal friction angle ϕ [^{o}] at residual state	Local internal friction angle φ [^{o}] at peak	u/h related to the peak	Local internal friction angle φ [^{o}] at residual state	Thickness of the shear zone at residual state [mm]
'1' e_o=0.60 d_{50}=0.5 mm p=100 kN/m h=20 mm l=100 mm	44.7	0.029	33.3	44.0	0.040	32.5	8
'2' e=0.90	-	-	31.8	-	-	30.3	20
'3' d_{50}=1.0 mm	45.1	0.029	34.8	44.0	0.035	33.7	10
'4' p=500 kN/m	41.0	0.058	32.5	40.1	0.045	32.6	10
'5' l=60 mm	46.7	0.027	36.3	46.1	0.031	34.3	6.5

friction angle is obtained for a larger shear deformation (u/h=0.80). As compared to results with a simple shearing of an infinite sand layer without side walls of h=20 mm (Tejchman and Gudehus 2001), the internal mobilized friction angle is larger by 1° at peak and by 2° at residual state, respectively.

The distribution of the void ratio e (Figs.4.111 and 4.113) and stresses σ_{ij} (Figs.4.116 and 4.117) is also non-uniform in the entire specimen (however, more uniform than during direct shearing). The magnitude of stresses across the layer height (Fig.4.116) is different than during direct shearing (the stresses in the central part are almost symmetric against the horizontal axis). The stresses σ_{11} and σ_{12} along the top boundary differ strongly from the distribution obtained for direct shearing (they are significantly larger at the right side, Fig.4.117).

The deformation within the specimen is non-uniform from the beginning of shearing. The zone of shear localization is slightly inclined in the mid-region and strongly curved at sides (Fig.4.111). Its thickness varies along the specimen length increasing towards the sides. In the center of the specimen, it is approximately t_s=4.0 mm (8×d_{50}) on the basis of shear deformation (Fig.4.111) and the Cosserat rotation (Fig.4.112b). Thus, it is significantly smaller than during direct shearing. In contrast to direct shearing, the maximum Cosserat rotation occurs in the middle of the specimen. The evolution of the Cosserat rotation with shear strain is linear during the entire process of shearing (Fig.4.112c).

The distribution and evolution of void ratio is similar in the central part (Figs.4.113 and 4.115) as during direct shearing. The maximum residual void ratio in the shear

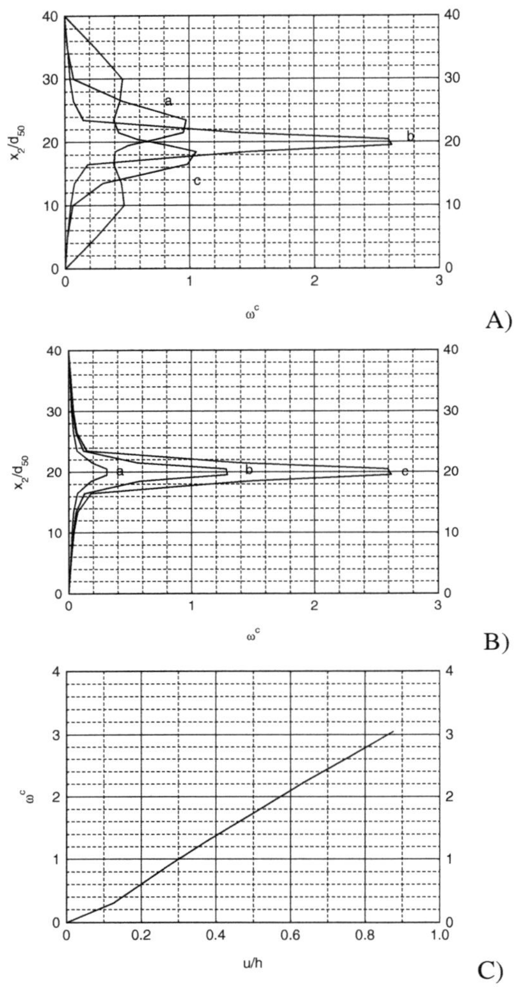

Fig. 4.112. Cosserat rotation ω^c: A) distribution along 3 different vertical planes across the specimen at the normalized horizontal displacement of the top u/h=0.75: a) left side (x_1=14 mm), b) center (x_1=50 mm), c) right side (x_1=86 mm), B) distribution in the center of the specimen at: a) u/h=0.125, b) u/h=0.375, c) u/h=0.75, and C) evolution in the center of the shear zone during simple shearing

zone is slightly larger, 0.81. The calculated normalized vertical displacement of the top boundary at u/h=0.75, v/h=0.067, is also larger as compared to direct shearing.

In the middle of the shear zone, the pressure ratios σ_{11}/σ_{22} and σ_{11}/σ_{33} are 1.0 (similarly as during direct shearing). The stress tensor is non-symmetric $\sigma_{12}\neq\sigma_{21}$ in the entire specimen.

The following conclusions can be drawn on the basis of the performed FE simulations in two different shear testers:

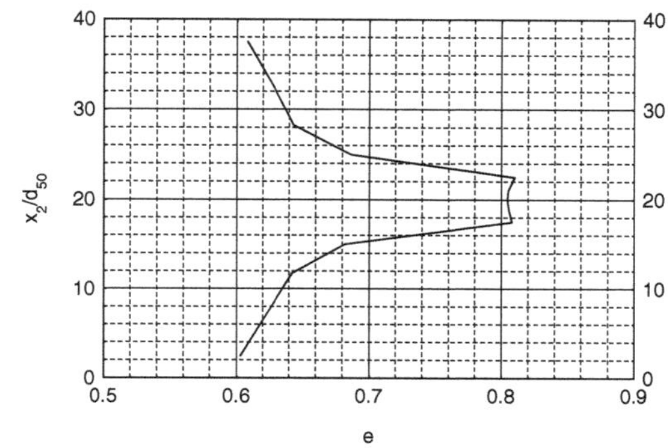

Fig. 4.113. Distribution of void ratio *e* in the vertical central plane across the specimen during simple shearing at the normalized horizontal displacement of the top *u/h*=0.75

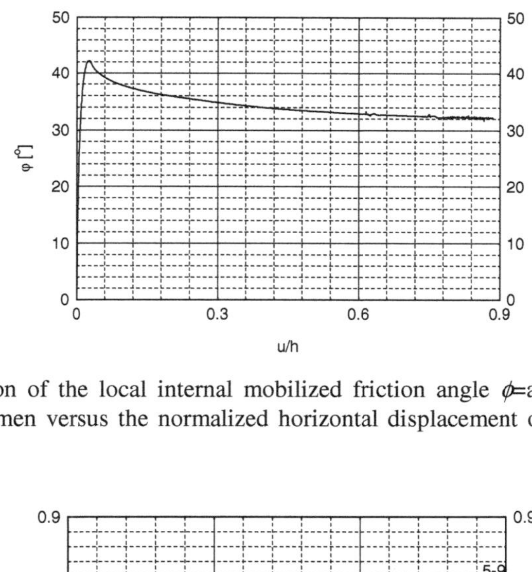

Fig. 4.114. Evolution of the local internal mobilized friction angle ϕ=arctan(σ_{12}/σ_{22}) in the middle of the specimen versus the normalized horizontal displacement of the top *u/h* during simple shearing

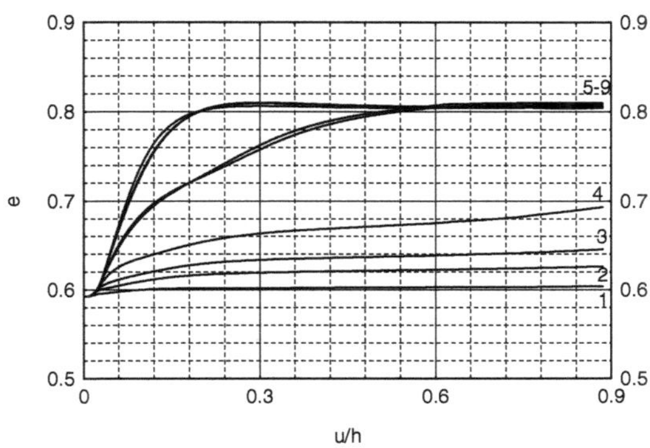

Fig. 4.115. Evolution of void ratios along the central cross-section versus the normalized horizontal displacement of the top *u/h* during simple shearing

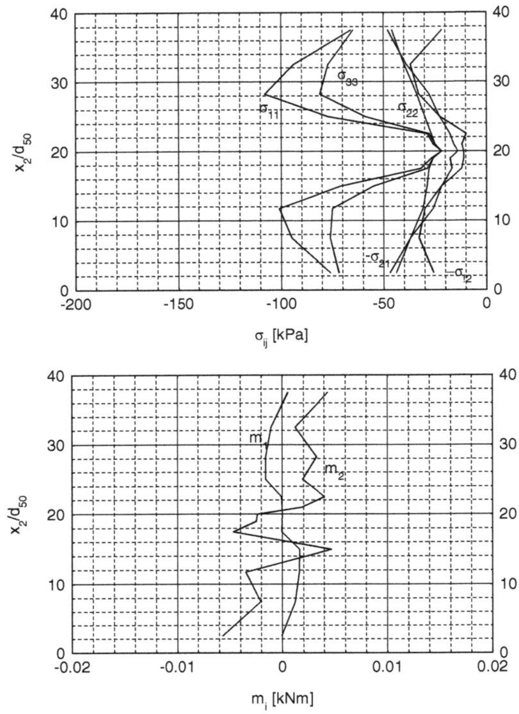

Fig. 4.116. Distribution of stresses σ_{ij} and couple stresses m_i across the central cross-section of the specimen during simple shearing at the normalized horizontal displacement of the top *u/h*=0.75

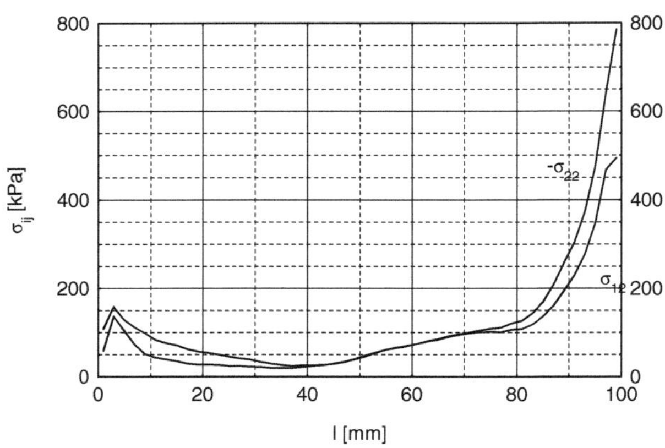

Fig. 4.117. Distribution of stresses σ_{12} and σ_{22} along the top boundary *l* during simple shearing at the normalized horizontal displacement of the top *u/h*=0.75

The method of the shear deformation inducement influences the maximum shear resistance. During direct shearing, the local mobilized friction angle in the mid-point of the shear zone is only slightly smaller than the mean value calculated from the resulting global forces. The local peak friction angle is higher by 2^0 in direct shearing than in simple shearing.

- The deformations and stresses are non-uniform in both shear testers (particularly during a direct shear test).
- Due to the effect of boundary conditions, the thickness and shape of the shear zone appearing along a horizontal mid-section are non-uniform and different in both testers. During direct shearing, the shape of the shear zone is close to the horizontal plane. The shear zone is the widest in the mid-region. During simple shearing, the shear zone is slightly inclined in the mid-region and strongly curved at ends. The shear zone is the widest at ends. The thickness of the shear zone in the mid-region is twice larger during direct shearing than during simple shearing.
- The evolution of the Cosserat rotation with shear strain is linear during the entire process of simple shearing and in the residual state during direct shearing.
- The overall internal friction angle at peak and at residual state increases with decreasing initial void ratio, pressure level and specimen length, and increasing mean grain diameter.
- The thickness of the shear zone increases with increasing initial void ratio, pressure level, mean grain diameter and specimen length.

4.8 Wall Direct Shear Test

The intention was to investigate the deformation and stress field, and wall shear resistance in dry sand with a finite element method during shearing along a very rough wall in a direct shear box (Tejchman 2004b). In the FE analysis, the influence of the layer height, initial void ratio and mean grain diameter on the shear resistance and shear zone thickness was studied. The emphasis was placed on the effect of boundary conditions on the shear zone formation.

Wall shear tests were carried with a parallel-guided direct shear box at Karlsruhe University (Tejchman 1997). This box (Wernick 1977, 1978, Kast 1985) differs from a conventional direct shear box in that the rotation of the top cap is suppressed by horizontal guides, Fig.4.118. Thus, the deformation is expected to be more uniform. The tests were performed to determine a mobilised wall friction angle between dry sand and a very rough steel wall for various layer heights and mean grain diameters of sand. In experiments, mainly fine Karlsruhe sand was used. Some tests were also carried out with coarse Karlsruhe sand (d_{50}=1.4 mm) to investigate the influence of the mean grain diameter on the mobilised wall friction angle. In this case, fine Karlsruhe sand with only particle diameters larger than 1.0 mm was used. The experiments were carried out only with a very rough steel wall. The roughness r_w, expressed by the difference between the highest peak and the lowest valley along a surface profile over a length of few mean grain diameters (Yoshimi and Kishida 1981), was greater than d_{50} for both sands. This was achieved by gluing large sand grains to a smooth steel surface. The tests were carried out always at a constant vertical pressure of p=100.0 kN/m^2 and a constant shear velocity of 0.1mm/min. The gap between

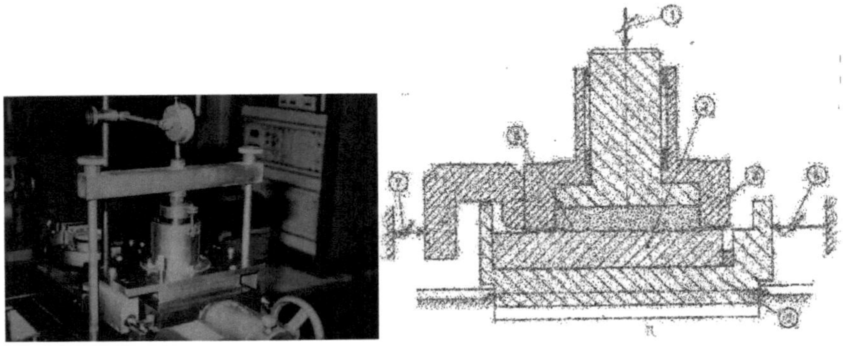

Fig. 4.118. Parallel-guided direct shear apparatus: 1. measurement of vertical displacement, 2. very rough steel wall, 3. upper box, 4. measurement of the horizontal displacement, 5. lower box, 6. sand specimen, 7. measurement of the shear force

the wall and the upper box was chosen to be approximately equal to d_{50}. The shear wall surface, $A=l\times b=0.10\times 0.10=0.01$ m^2, was constant during the entire test. In order to obtain an uniform initial density distribution within the specimen, sand was pluviated by means of a distributing cylinder where the stream of falling grains was homogenized by three sieves (Bauer 1992). The drop height of 0.15 m was kept constant in each test. Thus, the initial packing was expected to be dense. When the final specimen height was obtained, its upper surface was equalised by sucking off all asperities.

Fig.4.119 and Tab.4.2 demonstrate the experimental results of tests with dense fine Karlsruhe sand (d_{50}=0.5 mm) for two different heights of the sand specimen h=20 mm ($40\times d_{50}$) and h=10 mm ($20\times d_{50}$). The initial density was γ_d=17.0±0.05 kN/m^3. The results of the evolution of the global wall friction coefficient $\mu=T/N$ and displacement of the top boundary u_v versus the horizontal box displacement u are shown in Fig.4.119 (T – resultant horizontal shear force, $N=pA=100\times 0.01=1.0$ kN – resultant vertical normal force). Tab.4.2 includes the values of the maximum wall friction angle $(\varphi_w)_p$, wall friction angle in a residual state $(\varphi_w)_{cr}$, and horizontal displacements u_p and u_{cr} corresponding to $(\varphi_w)_p$ and $(\varphi_w)_{cr}$.

Both curves μ=f(u) increase up to the peak and then they experience pronounced softening (Fig.4.119) due to the formation of a wall shear zone along a steel surface. After this, they reach a plateau in a residual state. The dense sand specimen behaves first contractantly, later dilates and reaches a residual state without volume changes. The tests show that the smaller the height of sand mass, the higher $(\varphi_w)_p$ and $(\varphi_w)_{cr}$. The maximum vertical displacement of the specimen top is about u_v=0.38-0.40 mm. For a height layer of h=20 mm, the wall friction angles at peak and residual state are: $(\varphi_w)_p=48.8^{\circ}$ and $(\varphi_w)_{cr}=38^{\circ}$, respectively. They are slightly larger than the angles of internal friction of Karlsruhe sand measured by Wernick (1977) in the same apparatus during experiments with dense specimen (h=20 mm, e_o=0.60): $\phi_p=46.4^{\circ}$ at u_p=2 mm and $\phi_{cr}=36.8^{\circ}$ at u_{cr}=4 mm, respectively.

The results of the evolution of wall friction coefficients μ=f(u) with different heights of the dense sand mass (changing between 2.2 mm and 7.5 mm) are depicted in Fig.4.120 and Tab.4.3. The results demonstrate an inverse effect of the specimen

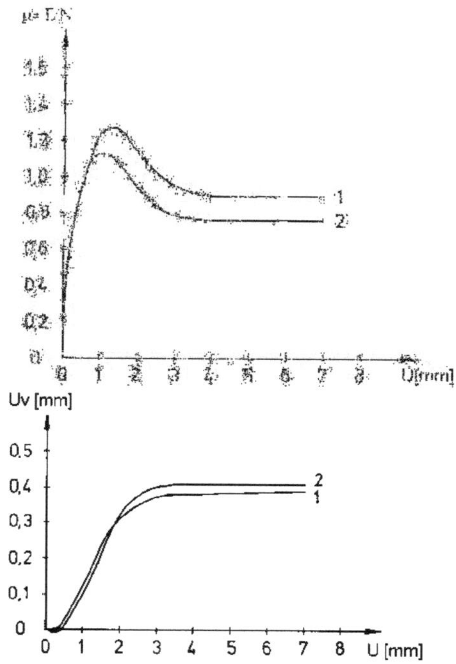

Fig. 4.119. Results of wall shear tests in a parallel-guided direct shear apparatus for different heights h of sand mass (T – wall shear force, N – vertical normal force, u – horizontal displacement, u_v – vertical displacement): 1. h=10 mm, 2. h=20 mm

Table 4.2. Measured wall friction angles $\varphi_w = arctan\mu$ and horizontal displacements u along the wall during wall shear tests with different heights h of sand mass (p – maximum, cr – residual)

h [mm]	$(\varphi_w)_p$ [°]	u_p [mm]	$(\varphi_w)_{cr}$	u_{cr} [mm]
20	48.8	10.0	38.0	40.0
10	51.8	13.3	42.0	40.0

Table 4.3. Measured wall friction angles φ_w= arctanμ and horizontal displacements u along the wall during wall shear tests with different heights h of sand mass ('p' – maximum, 'cr' – residual)

h [mm]	$(\varphi_w)_p$ [°]	u_p [mm]	$(\varphi_w)_{cr}$	u_{cr} [mm]
7.5	43.1	20.9	34.8	46.8
5.3	40.1	35.5	34.8	57.8
3.0	38.4	72.2	38.4	72.2
2.2	35.5	104.8	35.5	-

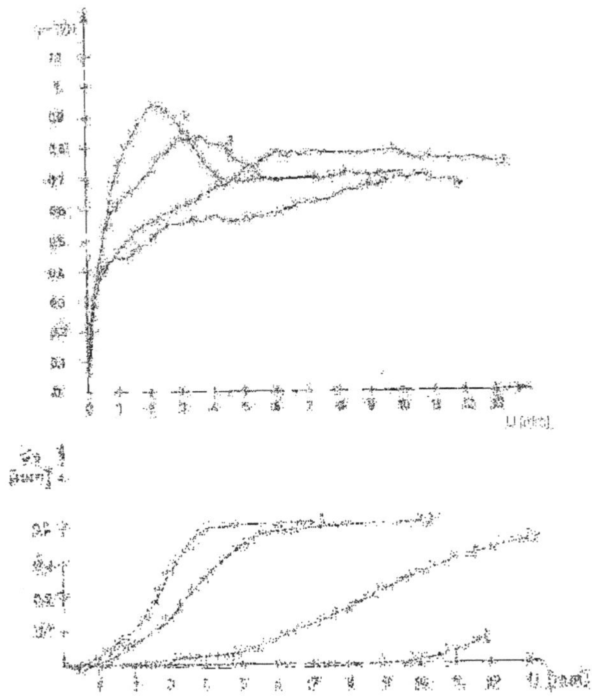

Fig. 4.120. Results of wall shear tests in a parallel-guided direct shear apparatus for different heights *h* of sand mass (*T* – wall shear force, *N* – vertical normal force, *u* – horizontal displacement, u_v – vertical displacement): 1. *h*=7.5 mm, 2. *h*=5.3 mm, 3. *h*=3.3 mm, 4. *h*=2.2 mm

height on the maximum wall friction angle than in the case of *h*=10-20 mm (Fig.4.119), namely the maximum wall friction angle decreases with decreasing layer height. The reason is that the initial density of sand specimens was decreasing with decreasing height (in spite of a constant falling height). An exact determination of the initial density of sand mass for thin layers was not practically possible; so it was estimated on the basis of a dilatancy angle calculated as a maximum inclination of the curve depicting the change of the vertical displacement versus the horizontal displacement in the range of sand loosening (Fig.4.120). The results of Fig.4.120 show evidently a decrease of the dilatancy angle with decreasing sand height. This is caused by material loosening in the contact zones between sand and surrounding boundaries of the box (Denton 1947, Herle 1992). The thinner the specimen in the shear box, the smaller the initial density. In addition, mobilised wall friction and dilatancy angles in thin granular layers are strongly influenced by both grain mixing and grain segregation (Löffelmann 1989). Small particles have a tendency to move both downwards due to gravity and in the direction of a decreasing deformation rate. In experiments by Löffelmann (1989) in a torsional ring shear apparatus, the maximum friction angle rose with the layer height (up to ten grain diameters), afterwards dropped (up to fifteen grain diameters) and reached an asymptote.

Fig.4.121 shows the influence of the mean grain diameter on the wall friction force with h=20 mm. The falling height was the same for fine sand (d_{50}=0.5 mm) and coarse sand (d_{50}=1.4 mm). The initial density of fine sand was about γ_d=17.0 kN/m^3 and coarse sand about 16.3 kN/m^3. The wall friction angle at peak and at residual state is larger by 2°-4° for coarse sand due to their larger rotation resistance and dilatancy (Tejchman and Wu 1995, Tejchman 1997). The maximum wall friction angle is reached at u_p=1 mm (fine sand) and at u_p=2.0 mm (coarse sand). The maximum vertical displacement of sand u_v is almost twice as large for coarse sand (u_v=0.7 mm) than for fine sand (u_v=0.4 mm).

For plane strain simulations of a direct wall shear box test, a sand specimen with a length l=0.1 m, width of b=1.0 m and height of h=20 mm was discretized with 2000 quadrilateral elements composed of four diagonally crossed triangles. The heights of finite elements were equal to (from the bottom up to the top of the sand specimen): 0.5 mm, 1 mm, 1.5 mm, 4×2 mm and 3×3 mm. The width of elements was 2.0 mm. In the calculations with h=10 mm, 1200 quadrilateral elements were used. The heights of

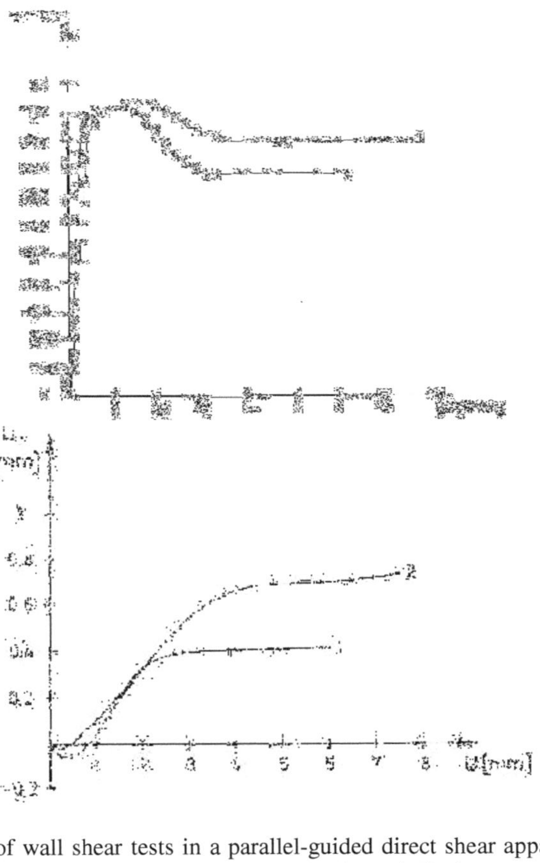

Fig. 4.121. Results of wall shear tests in a parallel-guided direct shear apparatus for different heights h of sand mass (T – wall shear force, u – horizontal displacement, u_v – vertical displacement): 1. d_{50}=0.5 mm, 2. d_{50}=1.4 mm

finite elements were similar as in the case of h=20 mm. They were equal to 1×0.5 mm, 1×1.0 mm, 1×1.5 mm, 2×2 mm and 1×3 mm, respectively.

As the initial stress state in the granular specimen, a K_o-state without polar quantities was assumed ($\sigma_{22}=\gamma_d x_2$, $\sigma_{11}=\sigma_{33}=K_0\gamma_d x_2$, $\sigma_{12}=\sigma_{21}=m_1=m_2=0$); The coefficient was K_0=0.45 (dense specimen) and K_0=0.50 (loose specimen). A constant vertical load p=100 kN/m was prescribed to the entire top boundary.

The bottom of the specimen was very rough (u_1=0, u_2=0 and ω^c=0). Two side boundaries were assumed to be smooth (σ_{21}=0, m_1=0). To simulate a direct wall shear test, the same horizontal displacement increments were prescribed to both side boundaries $u_1=n\Delta u_1$ (Δu_1 – horizontal displacement increment, n – step number). Along a very rough top boundary, the horizontal displacements of all nodes were tied together ($u_1=n\Delta u_1$, ω^c=0). The vertical displacements along the top were constrained to move by the same amount.

The calculations of direct wall shearing were carried out with three different initial void ratios of sand (e_o=0.60, 0.75 and 0.90), two different layer heights (h=10 mm and 20 mm) and two different mean grain diameters (d_{50}=0.5 mm and 1.4 mm) using the same material constants (except of d_{50}).

The calculations were carried out with large deformations and curvatures (Section 4.3.3).

The FE results of a direct wall shear box test with dense sand (initial void ratio e_o=0.60, mean grain diameter d_{50}=0.5 mm), layer height of h=20 mm and very smooth side boundaries are shown in Figs.4.122-4.132.

Fig.4.122 presents the evolution of the global wall friction coefficient $\mu=T/N$ (T - horizontal wall shear force, N=10.0 kN – vertical normal force) and the vertical displacement of the top boundary u_v (after compression by the load p) versus the horizontal displacement u of both side boundaries. The force T was calculated as the sum of all horizontal nodal forces along the top boundary and side boundaries.

In Fig.4.123, the deformed mesh with the distribution of the Cosserat rotation ω^c and void ratio e in the residual state is shown. The deformed meshes with the distribution of normal and shear stresses σ_{ij} in the residual state are presented in Fig.4.124. The darker regions indicate the lower (negative) normal stresses and higher (positive) shear stresses. The distribution of stresses σ_{ij} and couple stresses m_i at residual state along the normalised layer height x_2/d_{50} in the centre of the specimen is shown in Fig.4.125. Figs.4.126 and 4.127 present the distribution of the Cosserat rotation in 5 different cross-sections along the specimen length and evolution of the Cosserat rotation in the centre of the specimen, respectively. The distribution of void ratio at residual state across the specimen centre is depicted in Fig.4.128. In turn, the distribution of the vertical normal stress σ_{22} along the top boundary of l=0.1 m is presented in Fig.4.129. Fig.4.130 demonstrates the evolution of void ratio versus u across the layer height at residual state. The evolution of a mobilised local wall friction angle $\varphi_w=\arctan(\sigma_{12}/\sigma_{22})$ versus u near the bottom at $x_1=l/2$ is presented in Fig.4.131. The quantities σ_{ij}, m_i and e were taken as the mean values from quadrilateral elements.

The calculated global wall friction coefficient reaches a maximum at u=0.3 mm, shows a pronounced softening and tends to an asymptotic value at about u=3 mm (Fig.4.122). The calculated maximum and residual wall friction angles on the basis

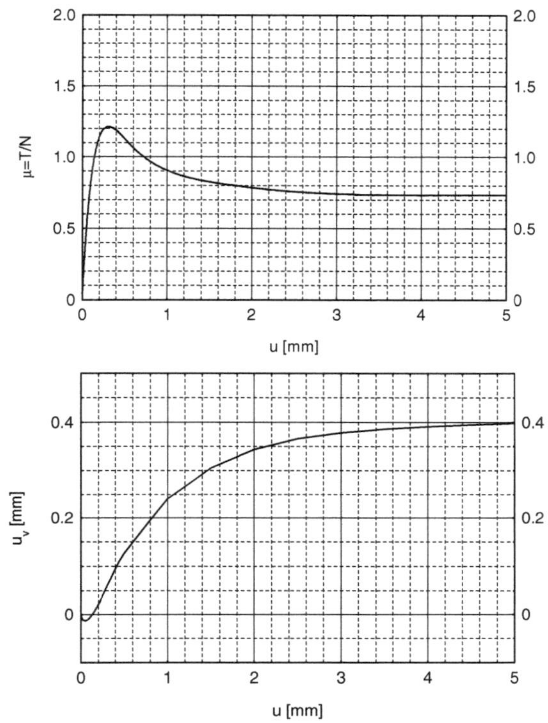

Fig. 4.122. Calculated wall friction coefficient $\mu=T/N$ and vertical displacement of the top boundary u_v versus horizontal displacement u during a direct wall shear test (h=20 mm, d_{50}=0.5 mm, e_o=0.60)

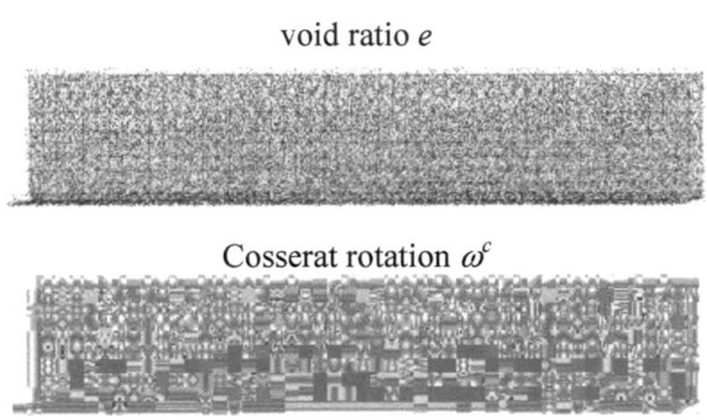

Fig. 4.123. Deformed FE mesh with distribution of Cosserat rotation ω^c and void ratio e during a direct wall shear test at u=3 mm (h=20 mm, d_{50}=0.5 mm, e_o=0.60)

of global forces, φ_w=arctan(T/N), are 50.5° and 36.2°, respectively. The agree well with corresponding experimental quantities (48.8° and 38° in Tab.4.2) and local wall friction angles along the wall (at the mid-point), φ_w=arctan$(\sigma_{12}/\sigma_{22})$, equal to 49.7° and 35.5° (curve '1' in Fig.4.131), respectively. However, the calculated horizontal

Fig. 4.124. Deformed FE mesh with distribution of stresses σ_{11}, σ_{22}, σ_{12} and σ_{21} during a direct wall shear test at u=3 mm (h=20 mm, d_{50}=0.5 mm, e_o=0.60)

displacement corresponding to the maximum force T (u_p=0.32 mm) is significantly smaller than the measured value (u_p=1.0 mm, Fig.4.119). In turn, the calculated horizontal displacement corresponding to the residual force T (u_{cr}=3.0 mm) is close to the measured value (u_{cr}=4.0 mm). First, the material is subject to the contractancy due to the initial compression from the load p (the vertical displacement of the top boundary is u_v=0.68 mm). Next, the material undergoes small contractancy during shearing (u_v=-0.013 mm at u=0.05 mm). Afterwards, material dilatancy occurs. The calculated vertical displacement of the top boundary in the residual state due to dilatancy, u_v=0.4 mm (Fig.4.122), is similar as in the experiment (Fig.4.119).

The distribution of the void ratio e and stresses σ_{ij} is strongly non-uniform in the entire specimen during direct shearing (Figs.4.123, 4.124 and 4.128). Two different stress regions can be distinguished in the specimen; a region with high stresses at the left side boundary (due to compression caused by the side boundary being pushed into a sand specimen) and a region with low stresses at the right side boundary (due to extension caused by the side boundary being pulled out of a sand specimen) where the normal stresses tend asymptotically to zero. The stresses near the wall are significantly higher than in the upper region (Fig.4.132). The horizontal normal stress σ_{11} changes across the specimen height in its centre from 300 kPa (bottom) up to 60 kPa (top), the vertical normal stress σ_{22} changes from 300 kPa (bottom) up to 100 kPa (top), the lateral normal stress σ_{33} changes from 300 kPa (bottom) up to 80 kPa (top), the horizontal shear stress σ_{12} changes from 200 kPa (bottom) up to 0 kPa (top), and the vertical shear stress σ_{21} changes from 60-100 kPa (bottom) up to 0 kPa (top),

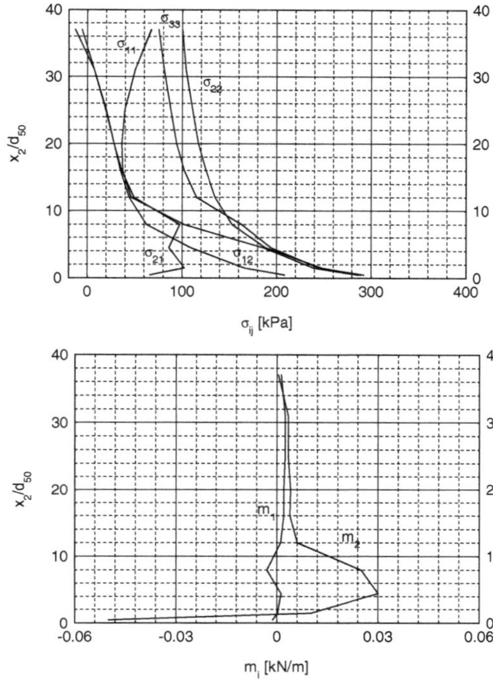

Fig. 4.125. Distribution of stresses σ_{ij} and couple stresses m_2 in the vertical plane across the centre during a direct wall shear test at u=5 mm (h=20 mm, d_{50}=0.5 mm, e_o=0.60)

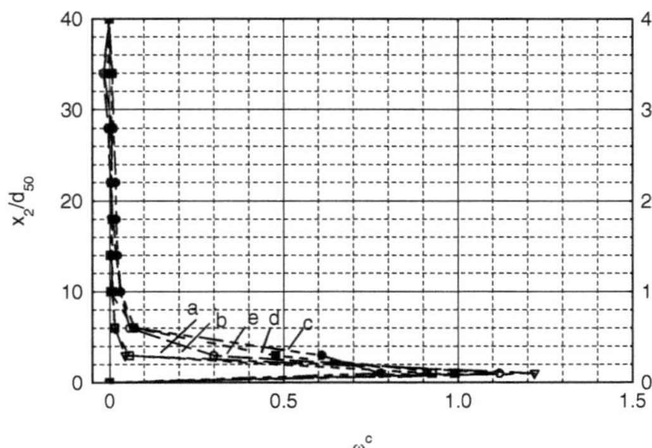

Fig. 4.126. Distribution of Cosserat rotation ω^c in the vertical planes along the specimen length during a direct wall shear test at u=5 mm (h=20 mm, d_{50}=0.5 mm, e_o=0.60): a) x_1=10 mm, b) x_1=25 mm, c) x_1=50 mm, d) x_1=75 mm, e) x_1=90 mm

Fig.4.125. Beyond the region close to the wall, the stress tensor σ_{ij} is symmetric ($\sigma_{12}=\sigma_{21}$) and the couple stress m_2 becomes insignificant. The vertical normal stress along the top boundary is strongly non-uniform (high in the left side, and very small in the right part), Fig.4.129.

Along a very rough bottom, a pronounced shear zone is created characterised by the occurrence of the shear deformation, Cosserat rotation, couple stress, and an increase of void ratio (Figs.4.123 and 4.125). The thickness of the wall shear zone changes along the specimen height (Fig.4.126). It is the largest in the middle of the sand specimen and the smallest at the left side. Thus, it depends on the boundary conditions of the entire system. The shear zone emerges from the beginning of wall shearing (Fig.4.127). In the central part of the specimen, the shear zone thickness is approximately t_s=6×d_{50} (3 mm) on the basis of the shear deformation (Fig.4.124). It is slightly larger on the basis of the Cosserat rotation (Fig.4.127), t_s=(6-10)×d_{50} (3-5 mm), and increasing void ratio (Fig.4.128), t_s=(8-12)×d_{50} (4-6 mm). In turn, on the basis of the couple stress m_2, the shear zone thickness is the largest, namely t_s=12×d_{50} (6 mm), Fig.4.125. This calculated shear zone thickness is slightly larger than the

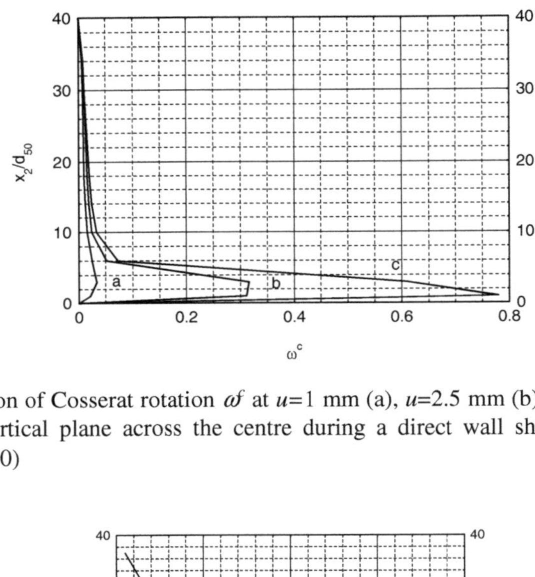

Fig. 4.127. Evolution of Cosserat rotation ω^c at u=1 mm (a), u=2.5 mm (b), and u=5 mm (c) at u=5 mm in the vertical plane across the centre during a direct wall shear test (h=20 mm, d_{50}=0.5 mm, e_o=0.60)

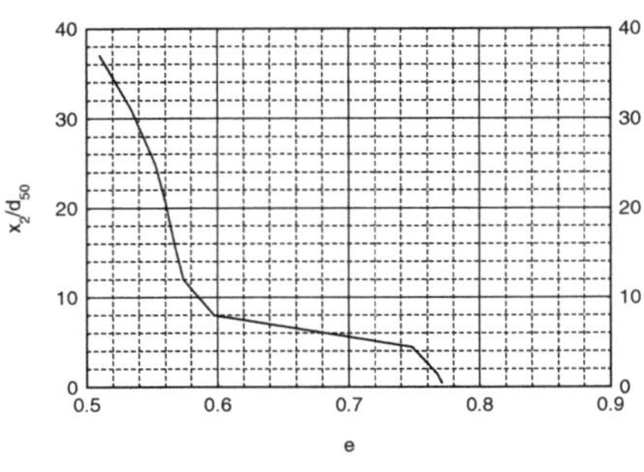

Fig. 4.128. Distribution of void ratio e in the vertical plane across the centre during a direct wall shear test at u=5 mm (h=20 mm, d_{50}=0.5 mm, e_o=0.60)

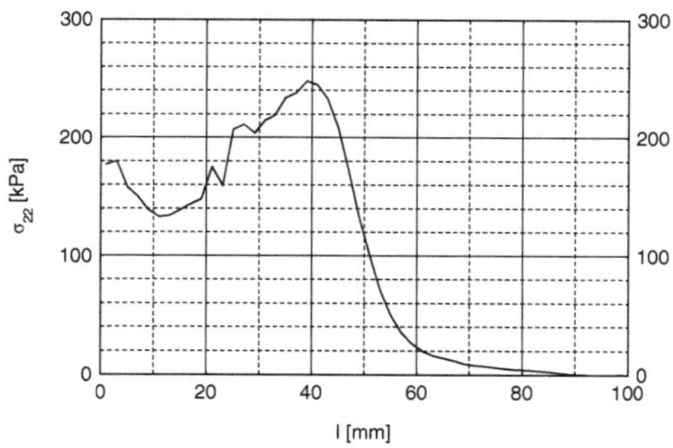

Fig. 4.129. Distribution of vertical normal stress σ_{22} along the top boundary during a direct wall shear test at u=5 mm (h=20 mm, d_{50}=0.5 mm, e_o=0.60)

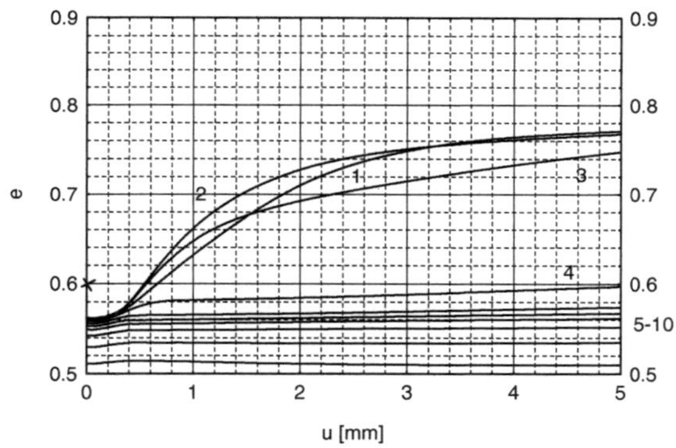

Fig. 4.130. Evolution of void ratio across the layer height versus horizontal displacement u during a direct wall shear test in the middle of the specimen (h=20 mm, d_{50}=0.5 mm, e_o=0.60);'1' – element at the specimen bottom, '10' – element at the specimen top

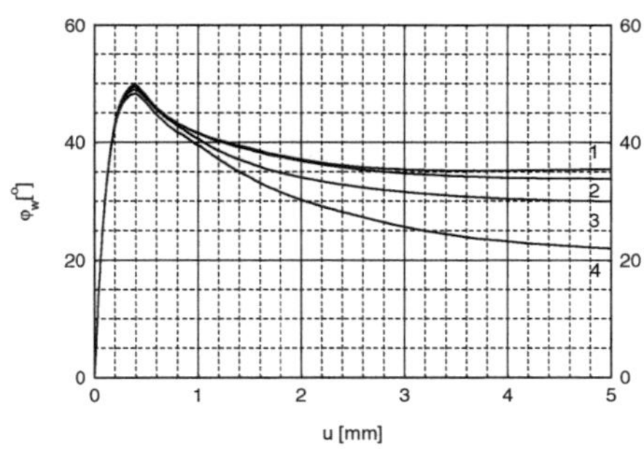

Fig. 4.131. Evolution of a mobilised wall friction angle φ_w=arctan(σ_{12}/σ_{22}) in a wall shear zone versus horizontal displacement u during a direct wall shear test in the middle of the specimen (h=20 mm, d_{50}=0.5 mm, e_o=0.60); curve *1* at x_2=0.25 mm, curve *2* at x_2=1.0 mm, curve *3* at x_2=2.25 mm, curve *4* at x_2=4.0 mm

experimental value, t_s=2.7 mm, estimated approximately on the basis of a simple formula by Wernick (1977):

$$t_s \gamma_d = [t_s + (u_v)_{cr}] \gamma_{cr} \quad , \tag{4.2}$$

wherein γ_d=17.0 kN/m^3, γ_{cr}=15.0 kN/m^3 and $(u_v)_{cr}$=0.41 mm. In the wall shear zone, the calculated void ratio reaches a pressure-dependent critical void ratio e=0.75-0.77 (curves '1-3' in Fig.4.131), and the pressure ratios are: $\sigma_{11}/\sigma_{22}\cong 1$ and $\sigma_{11}/\sigma_{33}\cong 1$ (Fig.4.125).

Influence of initial void ratio

The FE results are depicted in Figs.4.132 and 4.133. The calculated maximum horizontal wall shear force T diminishes with increasing initial void ratio (Fig.4.132). For a medium dense sand specimen (e_0=0.70), the global mobilised wall friction angle is 43.7° (peak value) and 36.3° (residual value), respectively. For a loose sand specimen (e_0=0.90), the global mobilised wall friction angle at residual state is 36.5°. The calculated thickness of the wall shear zone slightly increases with increasing initial void ratio and is about (on the basis of the couple stress m_2): $t_s \approx 16\times d_{50}$ (e_o=0.70) and $t_s \approx 25\times d_{50}$ (e_o=0.90). The calculated maximum vertical displacement of the top during dilatancy of medium dense sand at u=5 mm is u_v=0.23 mm (e_o=0.70).

The calculated maximum wall friction coefficient μ=f(u) increases with decreasing ratio h/d_{50} (Figs.4.134 and 4.135). For a larger mean grain diameter (d_{50}=1.4 mm, e_o=0.60, h=20 mm), the maximum wall shear force T is larger by 8% and the residual one by 20% (similarly as in the experiment, Fig.4.121). The calculated global wall friction angles, φ_w=arctan(T/N), are: 52.5° (maximum) and 42.3° (residual) (the experimental values are: 50^0 and 42°, respectively). The calculated thickness of the wall shear zone in the middle of the specimen increases significantly with increasing mean grain diameter: t_s=15.4 mm=11×d_{50} (d_{50}=1.4 mm) on the basis of the couple stress m_2 (Fig.4.136).

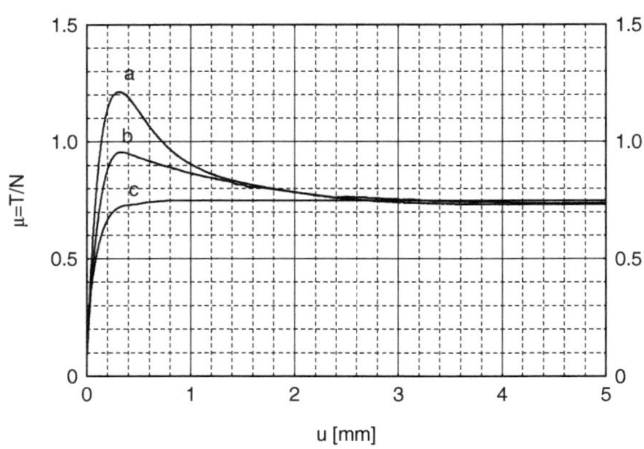

Fig. 4.132. Calculated wall friction coefficient $\mu=T/N$ versus horizontal displacement u during a direct wall shear test (h=20 mm, d_{50}=0.5 mm): a) e_o=0.60, b) e_o=0.70, c) e_o=0.90

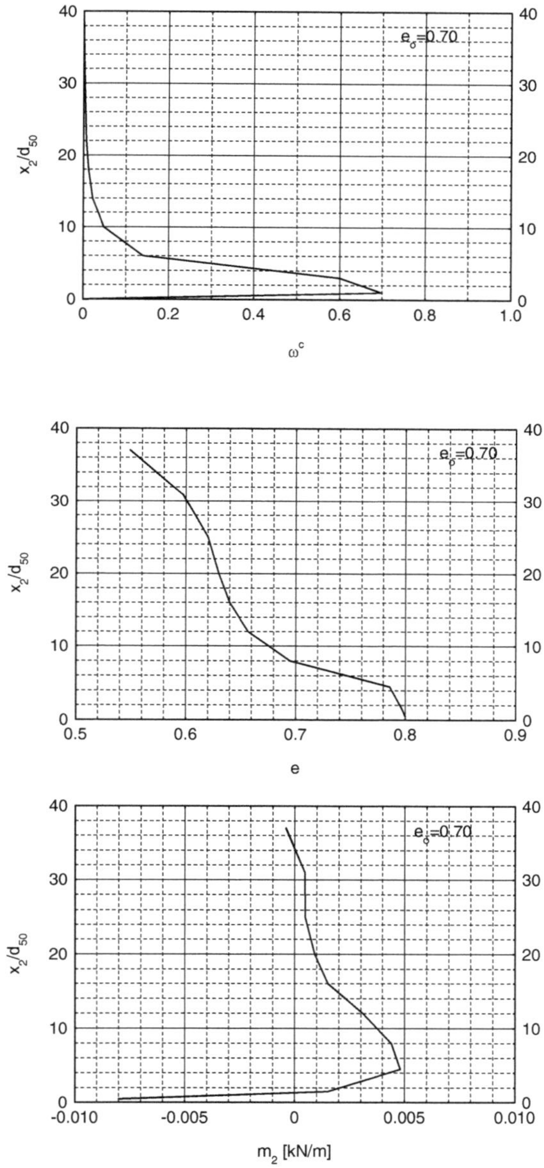

Fig. 4.133. Distribution of Cosserat rotation ω^c, void ratio e and couple stress m_2 in the vertical mid-section of medium dense sand (e_o=0.70) during a direct wall shear test at u=5 mm (h=20 mm, d_{50}=0.5 mm)

Influence of mean grain diameter and layer height

For a smaller layer height (h=10 mm, d_{50}=0.5 mm, e_o=0.60), the calculated maximum wall shear force T is larger by 8% and the residual force is larger by only 15% as

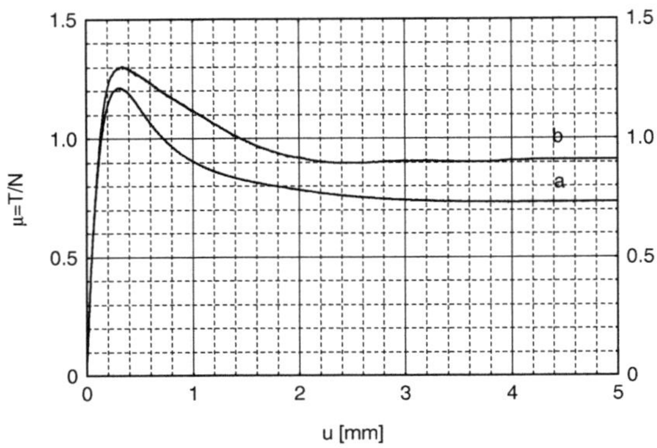

Fig. 4.134. Calculated wall friction coefficient $\mu=T/N$ versus horizontal displacement u during a direct wall shear test (e_o=0.60, h=20 mm): a) d_{50}=0.5 mm, b) d_{50}=1.4 mm

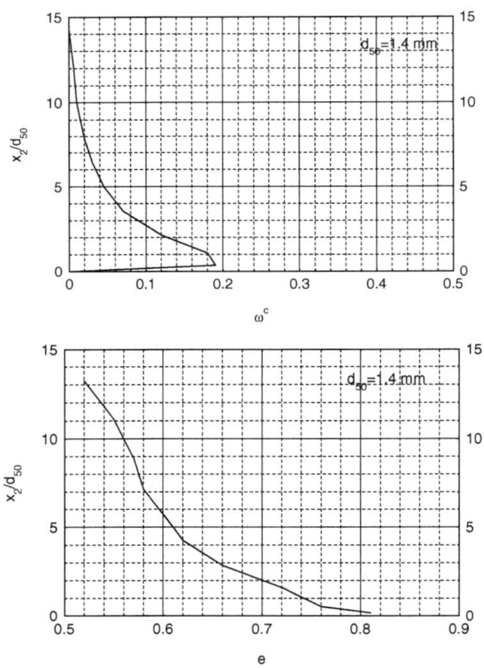

Fig. 4.135. Distribution of Cosserat rotation ω^c, void ratio e and couple stress m_2 in the vertical mid-section of dense sand (e_o=0.60, h=20 mm) during a direct wall shear test at u=5 mm (d_{50}=1.4 mm)

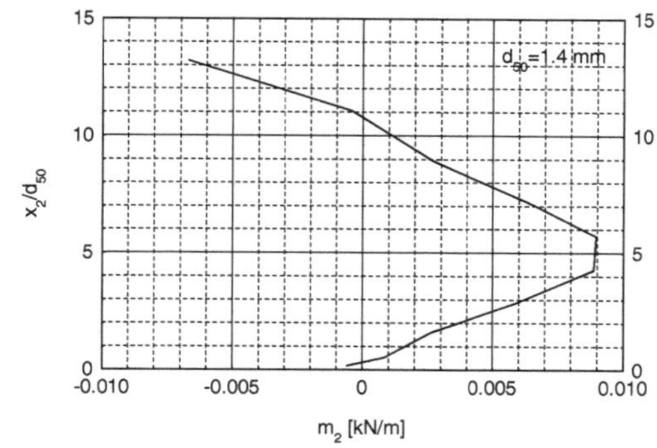

Fig. 4.135. *(continued)*

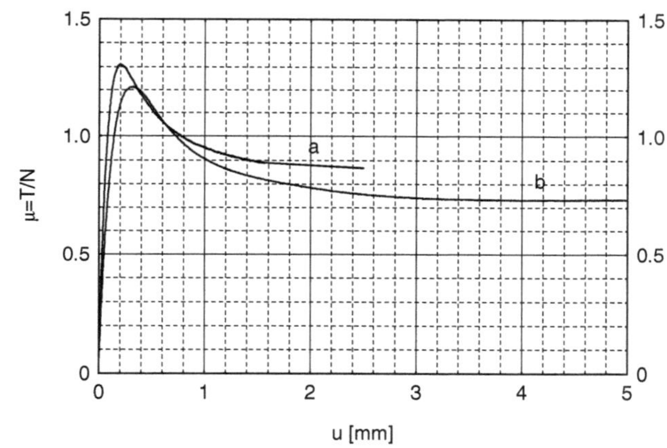

Fig. 4.136. Calculated wall friction coefficient $\mu=T/N$ versus horizontal displacement u during a direct wall shear test (e_o=0.60, d_{50}=0.5 mm): a) h=10 mm, b) h=20 mm

compared to the results with h=20 mm. The calculated global wall friction angles, $\varphi_w=\arctan(T/N)$, are: 52.4° (maximum) and 40.7° (residual value) (Fig.4.136) are in a good accordance with the experimental values: 51.8° and 42°, respectively. However, the evolution of the curve after the peak differs from the experiment (Fig.4.119). The calculated thickness of the wall shear zone in the middle of the specimen does not change with decreasing h and is $t_s \cong 10\times d_{50}$ on the basis of the Cosserat rotation.

The calculated displacement of the top boundary in the residual state at u=5 mm is u_v=0.60 mm (d_{50}=1.4 mm) and u_v=0.45 mm (h=10 mm). These values are close to the experimental ones (Figs.4.119 and 4.121).

- The maximum horizontal wall shear force increases with increasing mean grain diameter and decreasing layer height and initial void ratio.
- The mobilized wall friction angle calculated on the basis of global forces is similar as the local mobilized Coulomb's wall friction angle at the interface centre.
- The material dilatancy increases with increasing mean grain diameter, layer height and decreasing initial void ratio.

- The deformations, stresses and void ratios are strongly non-uniform in the entire sand specimen,
- The thickness of the induced wall shear zone appearing along the very rough bottom increases with increasing mean grain diameter and initial void ratio of the granulate.
- The thickness of the wall shear zone changes along the specimen's bottom due to the presence of side walls.

The performed FE simulations on a direct wall shear box test show a satisfactory agreement with experiments:

4.9 Contractant Shear Zones

The majority of experiments were performed with initially dense and medium dense granular specimens under shearing, i.e. where the initial void ratio was smaller than the pressure-dependent critical void ratio e_c (Eq.3.82) The reason was the fact that an initially dense granular specimen usually promotes shear localization due to the presence of dilatancy connected to material softening. Only a few experiments were carried out with initially loose dry cohesionless specimens. However, they evidently showed that shear localization could be also created inside of initially loose granular specimens, e.g. during plane strain compression (Desrues and Viggiani 2004), passive earth pressure on a retaining wall (James 1965, Lord 1969) and along very rough walls during confined flow in silos (Tejchman 1989).

The plane strain calculations were performed for 3 different rate boundary value problems in initially loose sand specimens where $e_o>e_c$.(Tejchman 2007). The fluctuating initial void ratio in with a mean value of 0.90 was assumed to be $e_0=0.85+0.1r$ ($0.85<e_0<0.95$).

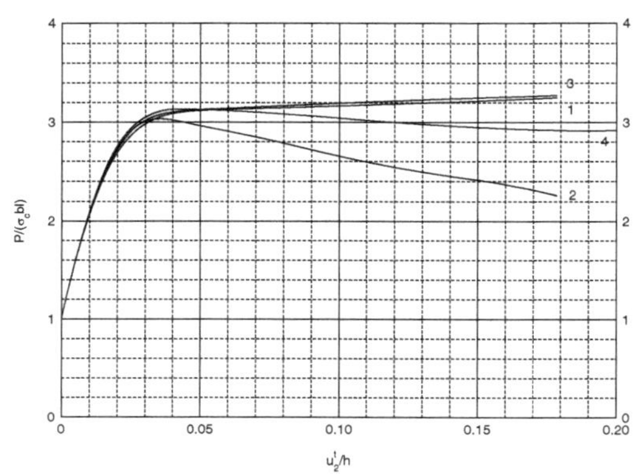

Fig. 4.137. Normalized load-displacement curve during compression with a loose specimen ($e_o=0.85+0.1r$): 1) smooth boundaries (small deformations and curvatures), 2) smooth boundaries (large deformations and curvatures), 3) very rough boundaries (small deformations and curvatures), 4) very rough boundaries (large deformations and curvatures)

Plane strain compression

FE calculations of plane strain compression tests were performed with a sand specimen which was h=14 cm high and b=4 cm wide. In total, 896 quadrilateral elements (0.25×0.25 cm^2) divided into 3584 triangular elements were used. In addition, the calculations were carried out also with small deformations and curvatures (i.e. without Jaumann terms and a change of the varying configuration and volume of the body) which are often applied in simulations of plane strain compression for the sake of simplicity.

A quasi-static deformation in sand was imposed through a constant vertical displacement increment Δu prescribed at nodes along the upper edge of the specimen. Smooth and very rough boundaries were assumed. Smooth boundary conditions of the sand specimen implied no shear stress at the smooth top and smooth bottom. To preserve the stability of the specimen against horizontal sliding along the top boundary (smooth edges), the node in the middle of the top edge was kept fixed. In the case of very rough boundaries, the horizontal displacements and Cosserat rotations were constrained along the top and bottom (u_1=0 and ω^c=0). The vertical displacement increments were chosen as $\Delta u/h$=0.0001.

As the initial stress state, a K_0-state with $\sigma_{22}=\gamma_d x_2$ and $\sigma_{11}=K_0\gamma_d x_2$ was assumed in the specimen; γ_d=14.5 kN/m^3 and K_0=0.50 is the earth pressure coefficient at rest. Next, the confining pressure of σ_c=200 kPa was prescribed.

The normalized load-displacement curves for smooth and very rough boundaries (small and large deformations and curvatures) are depicted in Fig.4.138 (P – resultant vertical force on the top and u_2^t – vertical displacement of the top). The specimen length l (size in the direction perpendicular to the deformation plane) was 1.0 m due to two-dimensional calculations. In turn, Fig.4.138 shows the deformed FE mesh with the distribution of the void ratio e at residual state. In the case of very rough edges (using a formulation with large deformations and curvatures), the results of two simulations with a random distribution of e_o were shown (Fig.4.138d).

Fig.4.139 demonstrates the evolution of void ratio in 3 points inside of the lower shear zone and in 2 points far beyond of the lower shear zone for the sand specimen with very rough boundaries of Fig.4.138d. For points beyond the shear zone, the point '1' is located below, and the point '2' above the shear zone. In the case of the points inside of the shear zone, the point '3' is located in the middle of the shear zone, and the points '4' and '5' lie near the edges of the shear zone.

The effect of the type of the formulation (large or small deformations and curvatures) is significant in an initially loose specimen with respect to the load-displacement diagram after the peak and shear localization. In the case of small deformations and curvatures, the resultant vertical force P continuously increases and approaches its asymptote (Section 4.1). The shape of the curve is not influenced by the edge roughness. The multiple interior shear zones are created before the peak on the load-displacement curve. However, they do not develop later and become inactive.

In the case of large deformations and curvatures (where the change of the body configuration and body volume is considered), the resultant vertical force increases first, shows a peak and drops later (it does not reach a residual state). For very rough boundaries, it reaches a clear residual state at u_2^t/h=0.20. For smooth boundaries, it decreases and the residual state has not been obtained yet at u_2^t/h=0.175. The mobilized overall residual internal friction angles at peak are about ϕ_p=30 (Eq.4.1). The internal

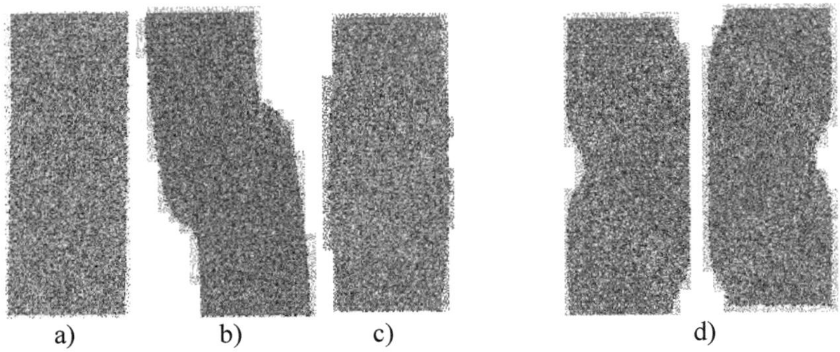

Fig. 4.138. Deformed FE mesh with the distribution of void ratio e during plane strain compression (e_o=0.85+0.1r): a) smooth boundaries (small deformations and curvatures), b) smooth boundaries (large deformations and curvatures), c) very rough boundaries (small deformations and curvatures), d) very rough boundaries (large deformations and curvatures) (light and dark grey correspond to small and large void ratio, respectively)

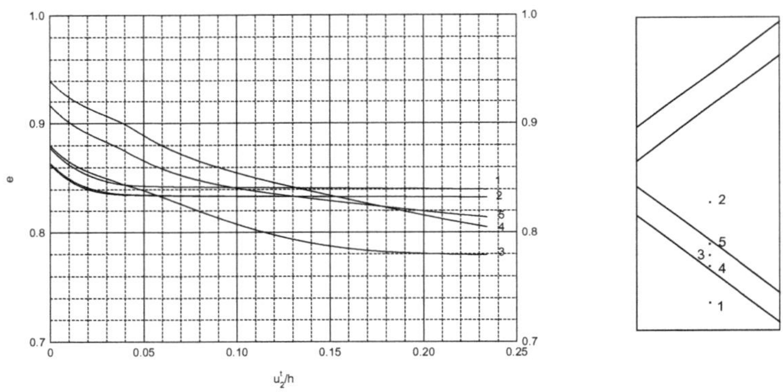

Fig. 4.139. Evolution of void ratio in loose specimen during plane strain compression (e_o=0.85+0.1r): inside of the shear zone (points '3-5') and beyond the shear zone (points '1-2') (very rough boundaries and large deformations and curvatures)

friction angles at residual state are smaller. Thus, material softening occurs. One shear zone occurs for smooth boundaries similarly as in dense specimen (Section 4.1). In turn, two zones appear for very rough boundaries which develop from the mid-point at the left or right side depending on the distribution of the initial void ratio (Fig.4.138d). The shear zone thickness on the basis of shear deformation is about 18 mm (36×d_{50}) and is larger than in a dense specimen. The void ratio decreases in the entire specimen at the beginning of loading (Fig.4.139). Next, it stops to change in the material beyond the shear zone (e≈0.0.83-84 in the points '1' and '2'); the material in this region behaves later as a rigid body. However, the void ratio decreases continuously in the shear zone approaching the pressure-dependent critical void ratio. The critical void ratio has been already reached in the middle of the shear zone (e=0.78 in the point '3') at

u_2^t/h=0.17. It has not been reached at the shear zone edges yet ($e \approx$0.80-0.82 in the points '4' and '5' at u_2^t/h=0.23).

Biaxial compression

Three different sets of conditions along boundaries of the sand specimen were assumed (Section 4.4): a) with three non-deforming rigid boundaries and one free boundary, and b) and c) with four non-deforming rigid boundaries (Fig.4.140).

In the first case, the granular specimen was placed on the smooth fixed bottom, its smooth top was subject to the uniform vertical pressure p=100 kPa, and its vertical moving smooth sides were subjected to the equal horizontal displacement increments directed to the specimen inside. The boundary conditions were along the bottom: $u_2^b=0$, $\sigma_{12}=0$ and $m_1=0$, along the top: $\sigma_{22}=p$, $\sigma_{12}=0$ and $m_1=0$, along the left side: $u_1^l=n\Delta u$, $\sigma_{21}=0$ and $m_2=0$, and along the right side: $u_1^r=-n\Delta u$, $\sigma_{21}=0$ and $m_2=0$ (n denotes the number of the time steps and Δu is the constant displacement increment in one step, 'r' – right, 'l' - left). As the initial stress state in the granular specimen, a K_o-state without polar quantities ($\sigma_{22}=\gamma_d x_2$, $\sigma_{11}=\sigma_{33}=K_o\gamma_d x_2$, $\sigma_{12}=\sigma_{21}=m_1=m_2=0$) was assumed ($\gamma_d$=14.5 kN/m^3). In the second case, the granular specimen was also placed on the smooth fixed bottom. The moving vertical smooth sides were subject to equal horizontal displacement increments directed to the specimen inside. In turn, the vertical displacements along the top boundary were constrained to move by the same amount Δu_2. The boundary conditions were along the bottom: $u_2^b=0$, $\sigma_{12}=0$ and $m_1=0$, along the top: $u_2^t=\Delta u_2$, $\sigma_{12}=0$ and $m_1=0$, along the left side: $u_1^l=n\Delta u$, $\sigma_{21}=0$ and $m_2=0$, and along the right side: $u_1^r=-n\Delta u$, $\sigma_{21}=0$ and $m_2=0$. As the initial stress state in the granular specimen, a K_o-state without polar quantities ($\sigma_{22}=p+\gamma_d x_2$, $\sigma_{11}=\sigma_{33}=p+K_o\gamma_d x_2$, $\sigma_{12}=\sigma_{21}=m_1=m_2=0$) was assumed ($p$=100 kPa - pressure). In the third case, the granular specimen was also placed on the smooth fixed bottom. The moving vertical smooth sides were subject to equal horizontal displacement increments directed to the specimen inside and the smooth top was subject to uniform vertical displacement increments directed to the specimen outside. The vertical displacement increments were slightly smaller than the horizontal ones ($u_1^t=0.8u_1^l=0.8u_1^r$). The boundary conditions were along the bottom: $u_2^b=0$, $\sigma_{12}=0$ and $m_1=0$, along the top: $u_2^t=n0.8\Delta u$, $\sigma_{12}=0$ and $m_1=0$, along the left side: $u_1^l=n\Delta u$,

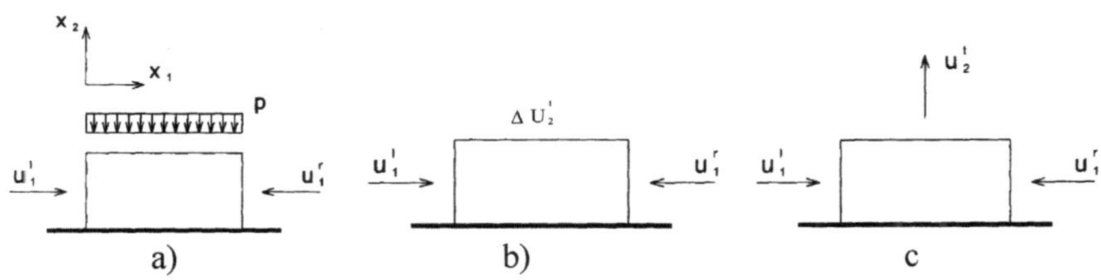

Fig. 4.140. Biaxial compression: a) deformation produced by horizontal displacements of both sides ($u_1^l=u_1^r$) and vertical pressure on the top p, b) deformation produced by horizontal displacements of both sides ($u_1^l=u_1^r$), and vertical displacement of the top ($u_1^t=u_1^l=u_1^r$); l –left, r – right, t -top

$\sigma_{21}=0$ and $m_2=0$, and along the right side: $u_1^r=-n\Delta u$, $\sigma_{21}=0$ and $m_2=0$. As the initial stress state in the granular specimen, a K_o-state without polar quantities ($\sigma_{22}=\gamma_d x_2$, $\sigma_{11}=\sigma_{33}=K_o\gamma_d x_2$, $\sigma_{12}=\sigma_{21}=m_1=m_2=0$) was assumed.

The normalized load-displacement curves for an initially loose specimen are shown in Figs.4.141, 4.143 and 4.147. In turn, Figs.4.142, 4.144, 4.145, 4.146, 4.148 and 4.149 show the deformed FE mesh with the distribution of the Cosserat rotation and void ratio at residual state for different boundary conditions. The Cosserat rotation is

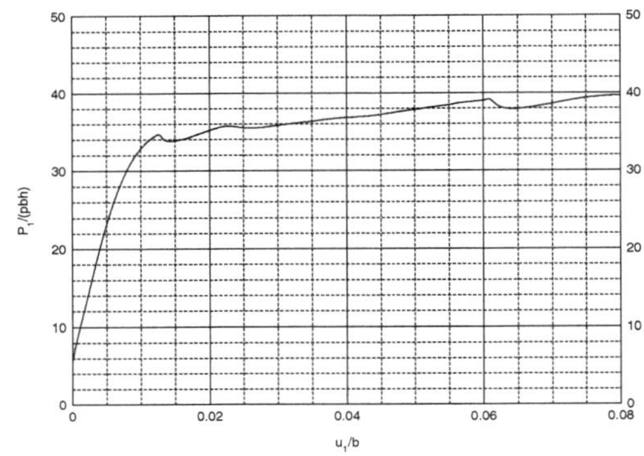

Fig. 4.141. Normalized load-displacement curve during a biaxial test with a loose specimen (e_o=0.85+0.1r, boundary conditions of Fig.4.141a)

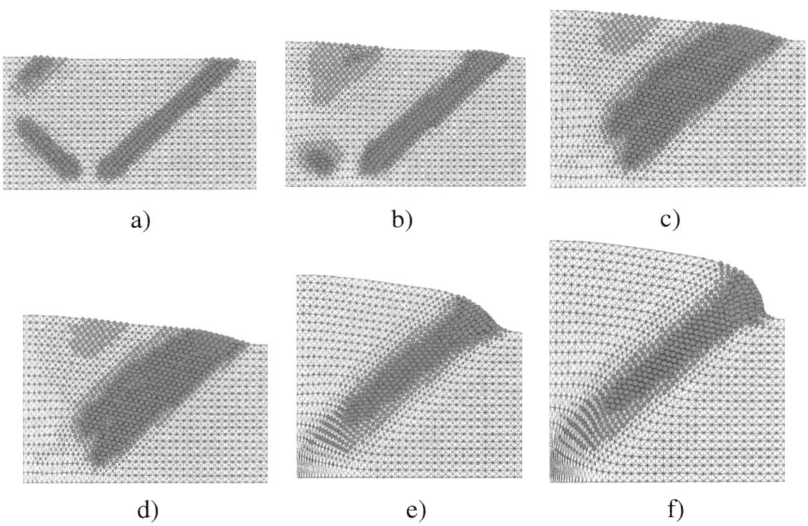

Fig. 4.142. Deformed FE mesh with the distribution of Cosserat rotation ω^c during biaxial test (e_o=0.85+0.1r, boundary conditions of Fig.4.141a): $_1/b$=0.025, b) u_1/b=0.05, c) u_1/b=0.075, d) u_1/b=0.10, e) u_1/b=0.125, f) u_1/b=0.15

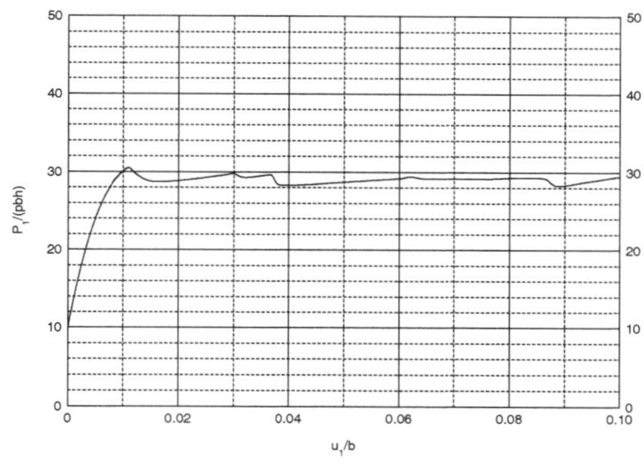

Fig. 4.143. Normalized load-displacement curve during a biaxial test with a loose specimen (e_o=0.85+0.1r, boundary conditions of Fig.4.141b)

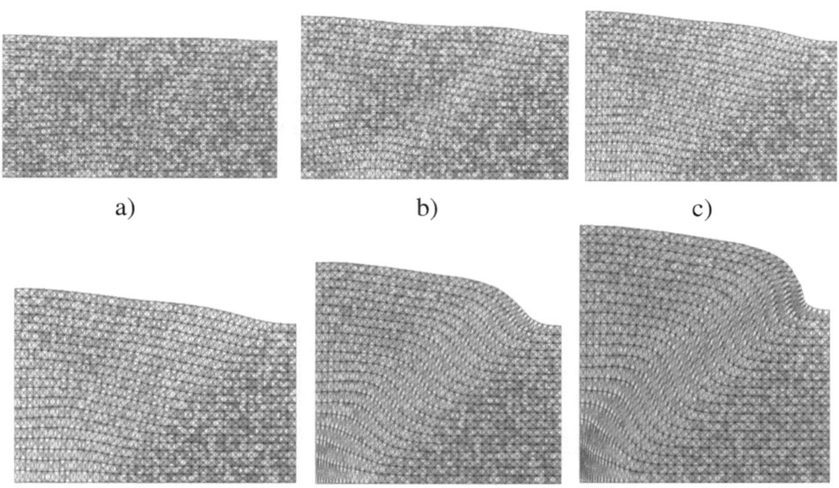

Fig. 4.144. Deformed FE mesh with the distribution of void ratio e during a biaxial test (e_o=0.85+0.1r, boundary conditions of Fig.4.141a): a) u_1/b=0.025, b) u_1/b=0.05, c) u_1/b=0.075, d) u_1/b=0.10, e) u_1/b=0.125, f) u_1/b=0.15 (light and dark grey correspond to small and large void ratio, respectively)

marked by circles with a diameter corresponding to the magnitude of the rotation in the given step.

In the case of three rigid and one deformable wall (Fig.4.140a), at the beginning three shear zones are created (on the basis of the Cosserat rotation) and later only one dominant shear zone appears (Figs.4.142 and 4.144). The rigid fixed and moving boundaries can reflect and free top boundary cannot reflect a shear zone (Section 4.4).

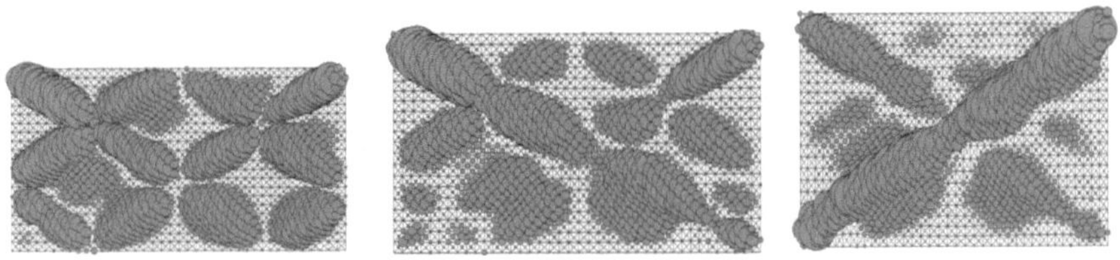

a) b) c)

Fig. 4.145. Deformed FE mesh with the distribution of Cosserat rotation ω^c during a biaxial test (e_o=0.85+0.1r, boundary conditions of Fig.4.141b: a) u_1/b=0.025, b) u_1/b=0.05, c) u_1/b=0.10

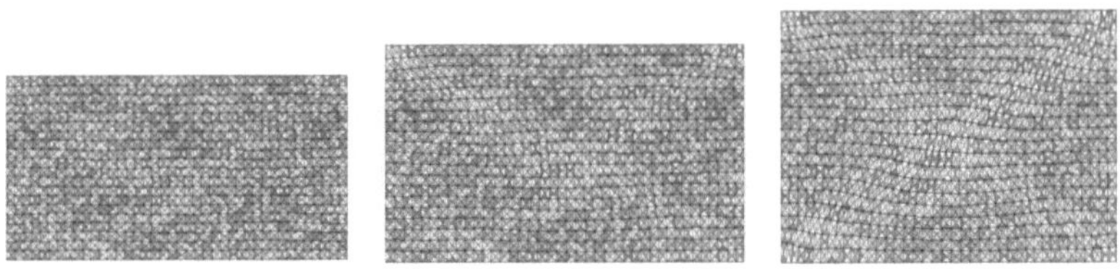

a) b) c)

Fig. 4.146. Deformed FE mesh with the distribution of void ratio e during a biaxial test (e_o=0.85+0.1r, boundary conditions of Fig.4.141b: a) u_1/b=0.025, b) u_1/b=0.05, c) u_1/b=0.10 (light and dark grey correspond to small and large void ratio)

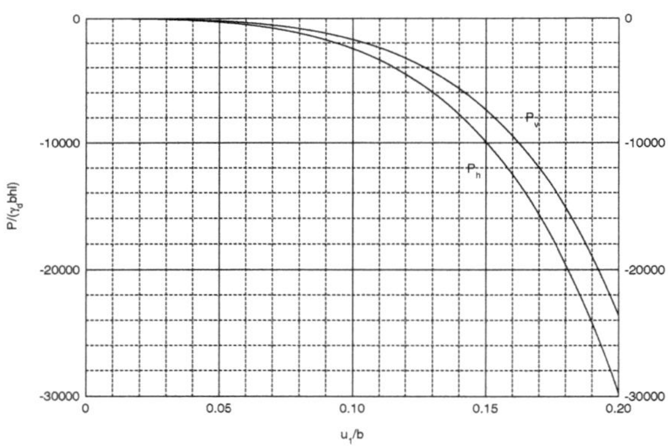

Fig. 4.147. Normalized load-displacement curve during a biaxial test with a loose specimen (e_o=0.85+0.1r, boundary conditions of Fig.4.141c)

The resultant horizontal force on sides continuously increases showing small drops during the loading (Fig.4.141). The thickness of the shear zone on the basis of shear strain is about 22 mm ($44\times d_{50}$) and is larger than in a dense specimen (Section 4.4).

For the case of four rigid boundaries with boundary conditions of Fig.4.140b, a pattern of shear zones occurs at the beginning of loading (since shear zones can be reflected from all 4 boundaries), and later two dominant intersecting shear zones form which spread between the specimen corners (Figs.4.145 and 4.146). The resultant horizontal force on sides increases and approaches its asymptote showing small oscillations (Fig.4.143).

For the case of four rigid boundaries with boundary conditions of Fig.4.141c, a pattern of shear zones occurs as usually at the beginning of deformation (Figs.4.148 and 4.149). However after it, they become inactive (in contrast to a dense specimen where they remained active during the entire loading process, Section 4.4). The void ratio becomes entirely uniform at residual state (Fig.4.149). The resultant vertical and horizontal forces continuously increase (Fig.4.147).

Passive earth pressure on retaining walls

The calculations were carried out with retaining wall moving horizontally against the backfill (passive case) (Section 6.9). The sand body was h=200 mm high and b=400 mm wide (Section 4.6). Totally, about 3400 triangular elements were used. The mesh was refined along the wall (twice) and at the wall foot (4 times). The height of the retaining wall located at the right side of the sand body was assumed to be h_w=170 mm (h_w/h=0.85).The mean grain diameter was increased up to 2.0 mm to preserve the ratio of the element size and d_{50} to be not larger than 5 (Tejchman and Bauer 1996).

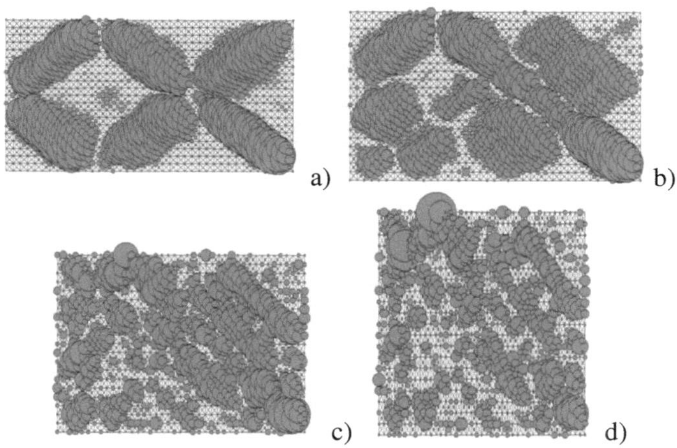

Fig. 4.148. Deformed FE mesh with the distribution of Cosserat rotation ω^c during a biaxial test (e_o=0.85+0.1r, boundary conditions of Fig.4.141c): a) u_1/b=0.025, b) u_1/b=0.05, b) u_1/b=0.10, d) u_1/b=0.2

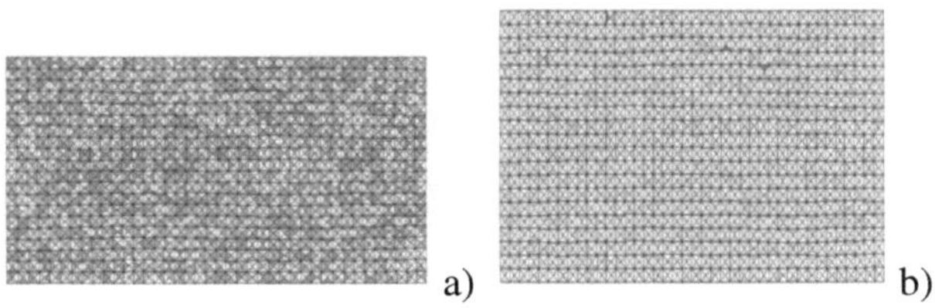

Fig. 4.149. Deformed FE mesh with the distribution of void ratio e during a biaxial test (e_o=0.85+0.1r, boundary conditions of Fig.4.141c): a) u_1/b=0.025, b) u_1/b=0.10 (light and dark grey correspond to small and large void ratio, respectively)

Two sides and the bottom of the sand specimen were assumed to be very rough: u_1=0, u_2=0 and ω^c=0. The top of the sand specimen was traction and moment free. The retaining wall was assumed to be stiff and very rough ($u_1=n\Delta u$, u_2=0 and ω^c=0). The maximum horizontal displacement increments were chosen as $\Delta u/h$=0.00001. The initial stresses were generated using a K_o-state without polar quantities: $\sigma_{22}=\gamma_d\, x_2$, $\sigma_{11}=\sigma_{33}=K_0\gamma_d x_2$, $\sigma_{12}=\sigma_{21}=m_1=m_2=0$ (γ_d=14.5 kN/m^3).

Fig.4.150 presents the evolution of the normalized horizontal earth pressure force $2E_h/(\gamma h^2)$ versus the normalized horizontal wall displacement u/h for a passive wall translation. The force E_h was calculated as the integral of mean horizontal normal stresses σ_{11} from quadrilateral elements along the retaining wall. In Fig.4.151, the deformed meshes with the distribution of the void ratio and Cosserat rotation in the residual state are shown.

The horizontal force on the wall continuously increases. The maximum normalized horizontal earth pressure force is about $2E_h/\gamma h^2$=7 (at u_1/h_w=5%). It is high due to the assumption of the very rough wall and large ratio of d_{50}/h_w.

Three clear shear zones are obtained (Fig.4.151): one shear zone projecting horizontally from the wall base, one inclined (slightly curved) zone spreading between the

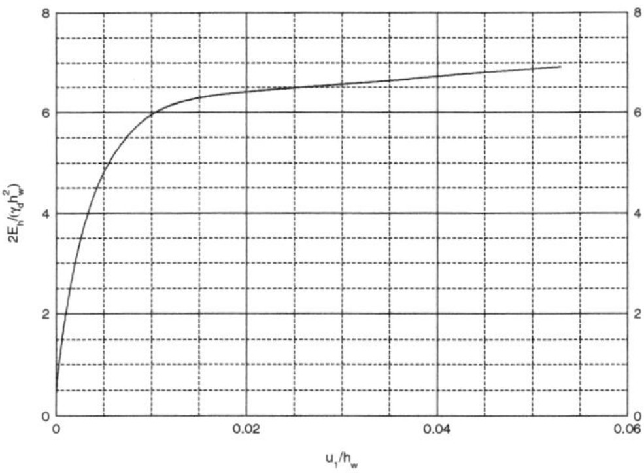

Fig. 4.150. Evolution of the normalized horizontal force on the wall $2E_h/(\gamma_d h_w^2)$ during a passive wall translation u_2/h_w in a loose specimen (e_o=0.85+0.1r)

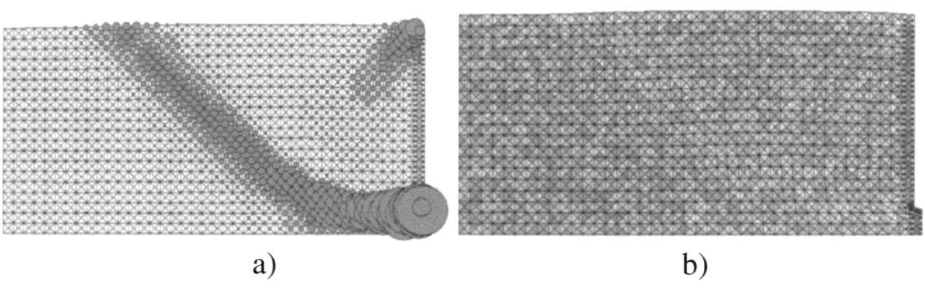

a) b)

Fig. 4.151. Deformed FE mesh with the distribution of Cosserat rotation ω^c (a) and void ratio e (b) during a passive wall translation (e_o=0.85+0.1r) at u_1/h_w=0.05 (light and dark grey correspond to small and large void ratio, respectively) wall bottom and free boundary, and one radial oriented shear zone starting at the wall top (which has not been fully developed yet at u_l/h_w=5%). The inclined shear zone becomes dominant in the course of deformation. The horizontal shear zone develops only at the beginning of the wall translation. The thickness of the dominant shear zone is about 25 mm (12.5×d_{50}) and its inclination from the bottom is about θ=40°. The void ratios decrease during the wall translation reaching the pressure-dependent critical values e_c. The calculated geometry of shear zones inside of the loose specimen is similar as this in a dense one (Section 4.6).

The following conclusions can be drawn:

- Shear localization with a continuous densification can be also created in initially loose granular materials with a random distribution of the initial void ratio. The global material softening is not necessary to obtain shear localization whose formation mainly depends on the boundary conditions of the entire system.
- Depending upon boundary conditions of the entire system, a single shear zone or several shear zones can be created in contractant granular materials.
- The thickness of shear zones in initially loose granulates is larger than in dense ones.
- The consideration of large deformations and curvatures in the calculations promotes shear localization during plane strain compression. In this case, a contractant internal shear zone inside of the material is created.

References

Bauer, E.: Zum mechanischen Verhalten granularer Stoffe unter vorwiegend ödometrischer Beanspruchung. Publication Series of the Institute for Rock and Soil Mechanics at the University Karlsruhe, 130 (1992)

Bauer, E.: Calibration of a comprehensive hypoplastic model for granular materials. Soils Found. 36(1), 13–26 (1996)

Bauer, E.: Conditions of embedding Casagrande's critical states into hypoplasticity. Mech. of Coh. -Frict. Mater. 5, 125–148 (2000)

Bauer, E., Huang, W.: Evolution of polar quantities in a granular Cosserat material under shearing. In: Mühlhaus, H.B., Dyskin, A., Pasternak, E. (eds.) Bifurcation and Localization Theory in Geomechanics, Balkema, pp. 227–239 (2001)

Bauer, E., Huang, W., Wu, W.: Investigations of shear banding in an anisotropic hypoplastic material. Int. J. Solids Structures 41, 5903–5919 (2004)

Bätcke, W.: Tragfähigkeit gedrungener Körper im geneigten Halbraum. Dissertation at the University Braunschweig (1982)

Behringer, R.P., Miller, B.: Stress fluctuations for sheared 3D granular materials. In: Behringer, R.P., Jenkins, R. (eds.) Proc. Powder and Grains, Balkema, pp. 333–336 (1997)

Bilgili, E., Yepes, J., Stephenson, L., Johanson, K., Scarlett, B.: Stress homogeneity in powder specimens tested in the Jenike shear cell: myth or fact? In: Kalman, H., Gyenis, J. (eds.) Proc. 4th Intern. Conf. For Conveying and Handling of Particulate Solids, 5.25–5.30(2003)

Bjeruum, L., Landva, A.: Direct simple shear tests on a Norwegian quick clay. Geotechnique 16(1), 1–20 (1966)

de Borst, R., Heeres, O.M.: A unified approach to the implicit integration of standard, non-standard and viscous plasticity models. Int. J. Num. Anal. Meth. Geomech. 26(11), 1059–1070 (2002)
Bransby, P.L.: Stress and strain in sand caused by rotation of a model wall. PhD Thesis at the University of Cambridge (1968)
Bransby, P.L., Milligan, G.W.E.: Soil deformations near cantilever sheet pile walls. Geotechnique 25(2), 175–195 (1975)
Budhu, M.: Nonuniformities imposed by simple shear apparatus. Can. Geotech. J. 20, 125–137 (1984)
Budhu, M., Britto, A.: Numerical analysis of soils in simple shear devices. Soils Found. 27(2), 31–41 (1987)
Caquot, A., Kerisel, J.: Tables for the calculation of passive pressures, active pressures and bearing capacity of foundations. Gauthier-Villars, Paris (1948)
Chandler, R.J., Hamilton, P.S.: On the measurement of the undrained strength of discontinuities in the direct shear box. Geotechnique 49(5), 615–620 (1999)
Chen, J., Yuan, H., Kalkhof, D.: A nonlocal damage model for elastoplastic materials based on gradient plasticity theory. Report Nr.01-13, Paul Scherrer Institute, pp. 1–130 (2001)
Christian, J.T., Hagmann, A.J., Marr, W.A.: Incremental plasticity analysis of frictional soils. Int. J. Num. Anal. Meth. Geomech. 1, 343–375 (1977)
Cole, E.R.L.: The behaviour of soils in simple shear apparatus. PhD Thesis, Cambridge University (1967)
Coulomb, C.A.: Essai sur une application. Science et Industrie, Paris (1775)
Darwin, G.H.: On the horizontal thrust of a mass of sand. Proc. Inst. Civ. Eng. 71, 350–378 (1883)
Dembicki, E.: Active, passive earth pressure and bearing capacity (in Polish), Warszawa (1979)
Denton, W.H.: The changing of porosity with wall distance. Chem. Engng. Progr. 43 (1947)
Desai, C.S., Drumm, E.C., Zaman, M.M.: Cyclic testing and modeling of interfaces. J. Geot. Engng. ASCE 111(6) (1985)
Desrues, J., Hammad, W.: Shear banding dependency on mean pressure level in sand. In: Dembicki, E., et al. (eds.) Proc. Int. Workshop on Numerical Methods for Localization and Bifurcation of Granular Bodies, Gdansk, Poland, pp. 57–67 (1989)
Desrues, J., Chambon, R., Mokni, M., Mazerolle, F.: Void ratio evolution inside shear bands in triaxial sand specimens studied by computed tomography. Géotechnique 46(3), 529–546 (1996)
Fourie, A.B., Potts, D.M.: Comparison of finite element and limiting equilibrium analyses for an embedded cantilever retaining wall. Geotechnique 39(2), 175–188 (1989)
Garga, V.K., Infante Sedano, J.A.: Steady State Strength of Sands in a Constant Volume Ring Shear Apparatus. Geotech. Test. J. 25, 414–421 (2002)
Geng, J., Hartley, R.R., Howell, D., Behringer, R.P., Reydellet, G., Clement, E.: Fluctuations and Instabilities in granular materials. In: Labuz, J., Drescher, A. (eds.) Bifurcations and Instabilities in Geomechanics, Swets and Zeitlinger, pp. 79–108 (2003)
Gudehus, G.: Engineering approximations for some stability problems in geomechanics. Advances in Analysis of Geotechnical Instabilities 13, 1–24 (1978)
Gudehus, G.: Einige Beiträge der Bodenmechanik zur Entstehung und Auswirkung von Diskontinuitäten. Felsbau 4, 190–195 (1986)
Gudehus, G., Schwing, E.: Standsicherheit historischer Stützwände. Internal Report of the Institute of Soil and Rock Mechanics, University Karlsruhe (1986)
Gudehus, G.: Attractors, percolation thresholds and phase limits of granular soils. In: Behringer, Jenkins (eds.) Powder and Grains, pp. 169–183 (1997)

Gudehus, G.: Shear localization in simple grain skeletons with polar effects. In: Adachi, T., Oka, F., Yashima, A. (eds.) Localization and Bifurcation Theory for Soils and Rocks, Balkema, pp. 3–11 (1998)

Gudehus, G., Loukachev, I., Pralle, N.: Inelastic behaviour of grain skeletons with propagation of plane shear waves. In: Kishino (ed.) Powders and Grains, pp. 125–128. Swets, Zeitlinger, Lisse (2001)

Gudehus, G., Nübel, K.: Evolution of shear bands in sand. Geotechnique 54(3), 187–201 (2004)

Hartl, J., Ooi, J.Y.: Experiments and simulations of direct shear tests: porosity, contact friction and bulk friction. Granular Matter (2008), doi:10.2007/s10035-008-0085-3

Herle, I.: Influence of the height of sand mass on its initial density. Internal Report of the Institute for Soil Mechanics, Karlsruhe University (1992)

Herle, I.: Hypoplastizität und Granulometrie einfacher Korngerüste. PhD thesis, University of Karlsruhe, 142 (1997)

Hicks, M.A., Yap, T.Y., Bakar, A.A.: Adaptive and fixed mesh study of localisation in a strain-softening soil. In: Mühlhaus, H.B., Dyskin, A., Pasternak, E. (eds.) Bifurcation and Localisation Theory in Geomechanics, Balkema, pp. 147–155 (2001)

Houlsby, G.T., Wroth, C.P.: Direct solution of plasticity problems in soils by the method of characteristics. In: Proc. 4th Int. Conf. on Num. Meth. in Geomech., Edmonton, Canada, pp. 1059–1071 (1982)

Huang, W., Bauer, E.: Numerical investigations of shear localization in a micro-polar hypoplastic material. J. Num. Anal. Meth. Geomech. 27, 325–352 (2003)

Iwashita, K., Oda, M.: Shear band Development in modified DEM: importance of couple stress. TASK Quarterly 2, 443–461 (1998)

James, R.G.: Stress and strain fields in sand. PhD Thesis, University of Cambridge (1965)

James, R.G., Bransby, P.L.: A velocity field for some passive earth pressure problems. Geotechnique 21(1), 61–83 (1971)

Jarzombek, G.: Experimentelle und numerische Untersuchungen zum Spannungs Verformungsverhalten von trockenem Sand am Beispiel einer Streifengründung. PhD thesis, Universität Weimer, Germany (1989)

Jenike, A.W.: Storage and Flow of Solids. Bulletin, 123. Engineering Experiment Station, University of Utah (1964)

Jenike, A.W., Johanson, J.R., Carson, J.W.: Bin loads – part 2, 3 and 4. J. Engng. Ind. ASME, 1–6 (1973)

Jewell, R.A., Wroth, C.P.: Direct shear tests on reinforced sand. Geotechnique 37(1), 53–68 (1987)

Johanson, K., Hamey, R., Pinto, E.: Measurements of local stress and strain in Schulze Testers using pressure sensitivity pads and X-ray tomography. In: Kalman, H., Gyenis, J. (eds.) Proc. 4th Intern. Conf. For Conveying and Handling of Particulate Solids, 15.14–15.19 (2003)

Kanatani, K.I.: Distribution of directional data and fabric tensor. Int. J. Engng. Science 22, 149–160 (1984)

Kast, K.: Geführtes Rahmenschergerät für direkte Scherversuche. Geotechnik 3, 109–110 (1985)

Kishida, H., Uesugi, M.: Tests of the interface between sand and steel in the simple shear apparatus. Geotechnique 37, 45–52 (1987)

Kitsabunnarat, A., Alsaleh, M., Helwany, S.: Capturing strain localization in reinforced soils. Acta Geotechnica (2008) (under review)

Lai, T.Y., Borja, R.I., Duvernay, B.G., Meehan, R.L.: Capturing strain localization behind a geosynthetic-reinforced soil wall. I. J. Num. Anal. Meths. Geomech. 27, 425–451 (2003)
Leśniewska, D., Mróz, Z.: Limit equilibrium approach to study the evolution of shear band systems in soil. Geotechnique 50, 389–403 (2000)
Leśniewska, D., Mróz, Z.: Study of evolution of shear band systems in sand retained by flexible wall. Int. J. Numer. Anal. Meth. Geomech. 25, 909–932 (2001)
Leśniewska, D., Mróz, Z.: Shear bands in soil deformation processes. In: Labuz, J., Drescher, A. (eds.) Bifurcations and Instabilities in Geomechanics, Swets, Zeitlinger, pp. 109–119 (2003)
Lord, J.A.: Stress and strains in an earth pressure problem. PhD Thesis, University of Cambridge (1969)
Maier, T.: Numerische Modellierung der Entfestigung im Rahmen der Hypoplastizität. PhD Thesis, Dortmund University, Germany (2002)
Milligan, G.W.E.: The behaviour of rigid and flexible retaining walls in sand. Geotechique 26(3), 473–494 (1974)
Milligan, G.W.E.: Soil deformations near anchored sheet-pile walls. Geotechnique 33(1), 41–55 (1983)
Morgenstern, N.R., Tchalenko, J.S.: Microscopic structures in kaolin subjected to direct shear. Geotechnique 17(4), 309–328 (1967)
Nakai, T.: Analysis of earth pressure problems considering the influence of wall friction and wall deflection. In: Proc. Int. Conf. Num. Meth. Geomech., Nagoya, pp. 765–772 (1985)
Nakase, H., Motegi, H.: Couple stress in DEM analysis. In: Adachi, T., Oka, F., Yashima, A. (eds.) Localisation and Bifurcation Theory for Soils and Rocks, Balkema, Rotterdam, pp. 289–295 (1998)
Negre, R.: Sur une methode approchee de la repatition des equilibre limite des massifs plans a faible frottement internet, pp. 3118–3120. C. R. A. S., Paris (1959)
Nübel, K.: Experimental and numerical investigation of shear localisation in granular materials. Publication Series of the Institute of Soil and Rock Mechanics, University Karlsruhe (2002)
Oda, M.: Initial fabrics and their relations to mechanical properties of granular material. Soils and Foundations 12(1), 17–36 (1972)
Oda, M., Kazama, H.: Micro-structure of shear band and its relation to the mechanism of dilatancy and failure of granular soils. Géotechnique 48(4), 465–481 (1998)
Oda, M., Iwashita, K.: Study on couple stresses and shear band development in granular media based on numerical simulation analyses. Int. J. of Engineering Science 38, 1713–1740 (2000)
Osinov, V.A.: The role of dilatancy in the plastodynamics of granular solids. In: Kishino (ed.) Powders and Grains, pp. 135–138. Swets, Zeitlinger, Lisse (2001)
Potts, D.M., Fourie, A.B.: The behaviour of a propped retaining wall: results of a numerical experiment. Geotechnique 34(3), 383–404 (1984)
Potts, D.M., Dounias, G.T., Vaughan, P.R.: Finite element analysis of the direct shear box test. Geotechnique 37(1), 11–23 (1987)
Potyondy, J.G.: Skin friction between various soils and construction materials. Geotechnique 4, 339–353 (1961)
Pradhan, T.B.S., Tatsuoka, F.: Experimental stress-dilatancy relations of sand subject to cyclic loading. Soils and Foundations 29, 45–64 (1989)
Pradhan, T.: Characteristics of shear band in plane strain compression test of sands. In: Asaoka, A., Adachi, T., Oka, F. (eds.) Deformation and Progressive Failure in Geomechanics, pp. 241–246. Pergamon, Oxford (1997)

Roscoe, K.H.: An apparatus for the application of simple shear to soil samples. In: Proc. 3rd Intern. Conf. Soil Mechanics and Foundation Eng., Zurich, vol. 58, pp. 1009–1012 (1953)

Roscoe, K.H., Bassett, R.H., Cole, E.R.I.: Principal axes observed during simple shear of sand. In: Proceedings of Geotech. Conf., Oslo, pp. 231–237 (1967)

Roscoe, K.H.: The influence of strains in soil mechanics. Geotechnique 20(2), 129–170 (1970)

Saada, A.S., Townsend, F.C.: State of the art: laboratory strength testing of soils. Laboratory shear strength of soil. ASTM Spec. Tech. Publicat. 740, 7–77 (1981)

Scarpelli, G., Wood, D.M.: Experimental observations of shear band patterns in direct shear test. In: IUTAM Conference on Deformation and Failure of Granular Materials, Delft, pp. 473–484 (1982)

Shibuya, S., Mitachi, T., Tamate, S.: Interpretation of direct shear box testing of sands as quasi-simple shear. Geotechnique 47, 769–790 (1997)

Simpson, B.: The use of finite element technique in soil mechanics with particular reference to deformation in plane strain. PhD Thesis, University of Cambridge (1972)

Simpson, B., Wroth, C.P.: Finite element computations for a model retaining wall in sand. In: Eisenstein, Z. (ed.) Proc. Int. Conf. on Num. Meth. in Geomech., Edmonton, pp. 85–93 (1974)

Skempton, A.W.: Aleksander Collin: a note on his pioneer work in soil mechanics. Geotechnique 1(4), 216–221 (1949)

Smith, I.: Stress and strain in a sand mass adjacent to a model wall. PhD thesis, University of Cambridge (1972)

Sokolovski, VV.: Statics of granular media. Pergamon Press, Oxford (1965)

Steenfelt, J.S.: Scale effect on bearing capacity factor. Report of the Danish Geotechnical Institute, Copenhagen (1979)

Szczepiński, W.: Limit states and kinematics of granular materials (in Polish). PWN Warszawa (1974)

Tatsuoka, F., Okahara, M., Tanaka, T., Tani, K., Morimoto, T., Siddiquee, M.S.A.: Progressive failure and particle size effect in bearing capacity of footing on sand. In: Proc. of the ASCE Geotechnical Engineering Congress, vol. 27(2), pp. 788–802 (1991)

Tatsuoka, F., Siddiquee, M.S.A., Yoshida, T., Park, C.S., Kamegai, Y., Goto, S., Kohata, Y.: Testing methods and results of element tests and testing conditions of plane strain model bearing capacity tests using air-dried dense Silver Leighton Buzzard sand. Internal Report of the Tokyo University, pp. 1-120 (1994)

Tatsuoka, F., Goto, S., Tanaka, T., Tani, K., Kimura, Y.: Particle size effects on bearing capacity of footing on granular material. In: Asaoka, A., Adachi, T., Oka, F. (eds.) Deformation and Progressive Failure in Geomechanics, pp. 133–138. Pergamon, Oxford (1997)

Taylor, D.W.: Fundamentals of Soil Mechanics. John Wiley, Chichester (1948)

Tejchman, J.: Scherzonenbildung und Verspannungseffekte in Granulaten unter Berücksichtigung von Korndrehungen. Publication Series of the Institute of Soil and Rock Mechanics, University Karlsruhe, vol. 117, pp. 1–236 (1989)

Tejchman, J., Wu, W.: Experimental and numerical study of sand-steel interfaces. Int. J. Num. Anal. Meth. Geomech. 19(8), 513–537 (1995)

Tejchman, J., Bauer, E.: Numerical simulation of shear band formation with a polar hypoplastic model. Comp. Geotech. 19(3), 221–244 (1996)

Tejchman, J.: Modelling of shear localisation and autogeneous dynamic effects in granular bodies. Publication Series of the Institute of Soil and Rock Mechanics, University Karlsruhe, vol. 140, pp. 1–353 (1997)

Tejchman, J., Herle, I., Wehr, J.: FE studies on the influence of initial void ratio, pressure level and mean grain diameter on shear localisation. Int. J. Num. Anal. Meth. Geomech. 23, 2045–2074 (1999)

Tejchman, J., Herle, I.: A class "A" prediction of the bearing capacity of plane strain footings on granular material. Soils and Foundations 39(5), 47–60 (1999)

Tejchman, J.: Behaviour of granular bodies in induced shear zones. Granular Matter 2(2), 77–96 (2000)

Tejchman, J., Gudehus, G.: Shearing of a narrow granular strip with polar quantities. Int. J. Num. Anal. Meth. Geomech. 25, 1–28 (2001)

Tejchman, J., Dembicki, E.: Numerical analysis of active and passive earth pressure of cohesionless soils on retaining walls (in Polish). Inżynieria Morska i Geotechnika 6, 365–368 (2001)

Tejchman, J.: Patterns of shear zones in granular materials within a polar hypoplastic continuum. Acta Mech. 155(1-2), 71–95 (2002a)

Tejchman, J.: Evolution of shear localisation at earth pressure problems of a retaining wall. Task Quarterly 6(3), 387–410 (2002b)

Tejchman, J., Bauer, E.: Effect of cyclic shearing on shear localisation in granular bodies. Granular Matter 5, 201–212 (2004)

Tejchman, J.: FE analysis of patterning of shear zones in granular bodies for earth pressure problems of a retaining wall. Arch. Hydro-Enging and Environ. Mech. 51(4), 317–348 (2004a)

Tejchman, J.: FE simulations of a direct wall shear box test. Soils Found.s 44(4), 67–81 (2004b)

Tejchman, J.: Effect of heterogeneity on shear zone formation during plane strain compression. Arch. Hydro-Engng. Environ. Mech. 51(2), 149–183 (2004c)

Tejchman, J.: Influence of a characteristic length on shear zone formation in hypoplasticity with different enhancements. Comp. Geotech. 31(8), 595–611 (2004d)

Tejchman, J., Bauer, E.: FE simulations of a direct and a true simple shear test within a polar hypoplasticity. Comp. Geotech. 21(1), 1–16 (2005)

Tejchman, J.: FE Analysis of Shearing of Granular Bodies in a Direct Shear Box. Int. J. Partic. Scien. Tech. 23(3), 229–248 (2005)

Tejchman, J.: Effect of fluctuation of current void ratio on the shear zone formation in granular bodies within micro-polar hypoplasticity. Comp. Geotech. 33(1), 29–46 (2006)

Tejchman, J.: FE analysis of contractant shear zones in loose granular materials. Granular Matter 9(1-2), 49–67 (2007)

Tejchman, J., Bauer, E., Tantono, S.F.: Influence of initial density of cohesionless soil on evolution of passive earth pressure. Acta Geotech. 2(1), 53–63 (2007)

Terzaghi, K.: Mecanique theorique des sols. Dunod, Paris (1951)

Terzaghi, K., Peck, R.B.: Soil mechanics in engineering practice, New York. Wiley, Chichester (1948)

Thornton, C., Zhang, L.: A DEM comparison of different shear testing devices. In: Kishino (ed.) Powders and Grains, Swets, Zeitlinger, Lisse, pp. 183–190 (2001)

Thornton, C., Zhang, L.: Numerical simulations of the direct shear test. Chem. Eng. Technol. 26(2), 1–4 (2003)

Tykhoniuk, R., Luding, S., Tomas, J.: Simulation der Scherdynamik kohäsiver Pulver. Chem.-Ing.-Technik 76, 59–62 (2004)

Uesugi, M., Kishida, H.: Influential factors of friction between steel and dry sand. Soils Found. 26(2), 33–46 (1986)

Uesugi, M.: Friction between dry sand and construction. PhD thesis, Tokyo Institute of Technology (1987)
Wang, Y.Z.: Distribution of earth pressure on a retaining wall. Geotechnique 50(1), 83–88 (2000)
Wernick, E.: Zusammenhänge zwischen der Mantelreibung von zylindrischen Ankern und Ergebnissen von direkten Scherversuchen. Bautechnik 8, 263–267 (1977)
Wernick, E.: Tragfähigkeit zylindrischer Anker in Sand unter besonderer Berücksichtigung des Dilatanzverhaltens. Publication Series of the Institute for Rock and Soil Mechanics, University Karlsruhe, 75 (1978)
von Wolffersdorff, P.A.: A hypoplastic relation for granular materials with a predefined limit state surface. Mechanics Cohesive-Frictional Materials 1, 251–271 (1996)
Wood, D.M., Budhu, M.: The behaviour of Leighton Buzzard sand in cyclic simple shear tests. In: Int. Symp. on Soils and Cyclic and Transient Loading, Swansea, pp. 9–21 (1980)
Vardoulakis, I.: Scherfugenbildung in Sandkörpern als Verzweigungsproblem. Dissertation, Institute for Soil and Rock Mechanics, University of Karlsruhe, 70 (1977)
Vardoulakis, I.: Shear band inclination and shear modulus in biaxial tests. Int. J. Num. Anal. Meth. Geomech. 4, 103–119 (1980)
Vesic, A.S.: Bearing capacity theory from experiments. Journal of Soil Mechanics and Foundation Engineering ASCE, Discussion 99, 575–577 (1973)
Yamada, Y., Ishihara, K.: Anisotropic deformation characteristics of sand under three dimensional stress conditions. Soils and Foundations 19(2), 79–94 (1979)
Yoshida, T., Tatsuoka, F., Siddiquee, M.S.A.: Shear banding in sands observed in plane strain compression. In: Chambon, R., Desrues, J., Vardoulakis, I. (eds.) Localisation and Bifurcation Theory for Soils and Rocks, Balkema, Rotterdam, pp. 165–181 (1994)
Youd, T.L.: Maximum density of sand by repeated straining in simple shear. Highway Research Record 374, 1–6 (1971)
Zhang, L.: The behavior of granular material in pure shear, direct shear and simple shear. PhD Thesis, Aston University (2003)
Ziegler, M.: Berechnung des verschiebungsabhängigen Erddruckes in Sand. PhD thesis, Institute for Soil and Rock Mechanics. Karlsruhe University (1986)

5 Finite Element Calculations: Advanced Results

Abstract. This chapter presents advanced numerical results of different quasi-static boundary value problems in granular bodies including shear localization. Calculations were carried out with the finite element method on the basis of a micro-polar hypoplastic model. The following problems were considered: sandpiles, direct symmetric cyclic shearing under constant normal stiffness condition, wall boundary conditions, deterministic and statistical size effects, non-coaxiality and stress-dilatancy rule and textural anisotropy. Attention was paid to the thickness and spacing of shear zones. Numerical solutions were compared with corresponding laboratory tests.

5.1 Sandpiles

The static and dynamic behaviour of sandpiles has attracted much attention. Simple experiments with prismatic and conical piles of granular materials indicate, contrary to intuition, that the maximum vertical normal stress does not always appear directly beneath the pile vertex but at a certain distance from the apex (Hummel and Finnan 1920, Smid and Novosad 1981, Liffman et al. 1994, 2001, Watson 1996, Wittmer et al. 1996, Brockbank et al. 1997, Savage 1997, 1998, Huntley 1999, Vanel et al. 1999). It was found later that the occurrence of the stress dip at the heap centre strongly depends on the method of pile construction (Vanel et al. 1999). In the case of the raining procedure by means of a sieve located above the heap, the pressure maximum occurs at the centre of the sandpile. However, when a funnel procedure (centric flow out of a hopper) is used, a pressure peak is obtained away from the centre, where a significant pressure dip appears. The pressure dip is usually more pronounced in conical heaps than in prismatic ones (Hummel and Finnan 1920).

There exist a number of theories explaining this phenomenon. The two newest continuum theories were recently reported by Michalowski and Park (2004, 2007) and Al Hattamleh et al. (2005). According to Michalowski and Park (2004, 2007), the occurrence of the stress dip at the heap centre is caused by arching understood as an ability of the structure to adopt itself during the loading process. For the case of plane strain, they performed a limit analysis based on the Mohr-Coloumb yield condition. The active limit state exhibited a maximum stress at the centre of the base, whereas the passive state encountered a discontinuity along a radius at some distance from the centre

J. Tejchman: Shear Localiz. in Gran. Bodies, SSGG, pp. 213–311, 2008.
springerlink.com

and a substantial depression at the pile centre. Al Hattamleh et al. (2005) reported a FE analysis with an elasto-plastic multi-slip model to simulate the construction methods of uniform raining and centric flow. For the raining procedure, the maximum vertical normal stress always reached its peak at the apex. For sandpiles constructed from a localized source, the maximum vertical stress occurred beyond the apex under the condition that the orientation of the initial active slip lines were different from $45°+\phi/2$ and $-45°-\phi/2$, respectively (ϕ – the angle of repose).

Below the stress distribution under sandpiles is studied (Tejchman and Wu 2008a) without imposing additional condition (e.g. the orientation of the initial slip lines in the elasto-plastic multi-slip model, Al Hattamleh et al. 2005). The analyses were performed with a micro-polar hypoplastic model which is suitable to investigate the phenomenon of the granular heap construction since it takes into account the effect of the direction of deformation rate (Eqs.3.68 and 3.69). The calculations were carried out with prismatic and conical heaps composed mainly of an initially dense cohesionless sand. The effect of the following parameters was investigated: a) construction method, b) mean grain diameter, c) base roughness, d) heap inclination and d) initial void ratio of sand.

The analyses were carried out for a plane strain case and an axisymmetric case. In the calculations, the symmetry axis was assumed. The pile was discretized with 200 triangular elements. The heap inclination to the bottom was assumed to be $\alpha=30°$, which was equal to the critical internal friction angle of sand (Eq.3.90). The height of the pile was $H=28\times d_{50}$ (H=70 mm) and the width B (or diameter D) was $B=D=100\times d_{50}$ ($B=D$=250 mm) with d_{50}=2.5 mm (Fig.5.1). Thus, the size of elements was not larger than $5\times d_{50}$, which was sufficient to obtain mesh-independent numerical results (Tejchman and Bauer 1996).The construction of the heap was simulated in 10 stages using two different methods (Fig.5.2), viz. the raining procedure (Fig.5.2a) and the funnel procedure (Fig.5.2b). The sandpile was subject only to gravitational load in the vertical direction. The following boundary conditions were prescribed: fully constrained Cosserat rotation and displacements along the very rough base ($u_1=0$, $u_2=0$, $\omega^c=0$) and fully constrained Cosserat rotation and horizontal displacement along the symmetry axis ($u_1=0$, $\omega^c=0$, $\sigma_{21}=0$).

Prismatic heaps

FE calculations were performed for the case of plane strain. Fig.5.3 shows the evolution of the normalized resultant vertical force on the base using two different construction methods shown in Fig.5.2 (for dense sand with the initial void ratio of e_o=0.60, initial volume weight γ=16.8 kN/m^3, d_{50}=2.5 mm). The vertical force was calculated as the sum of all nodal vertical forces at the base. The distribution of the normalized vertical normal stress under the sandpile for two different methods of the heap construction is presented in Fig.5.4 (after 5 and 10 loading stages, respectively). The evolution of the normalized vertical normal stresses in ten vertical sections along the sandpile bottom is demonstrated in Fig.5.5. In addition, the sand settlement along the symmetry axis (displacement of sand from the time of deposition to the end of construction) is given in Fig.5.6.

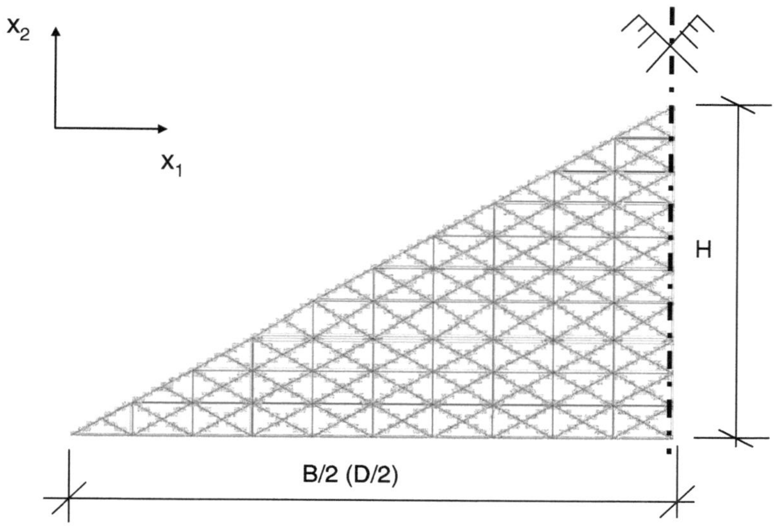

Fig. 5.1. FE mesh for granular heap

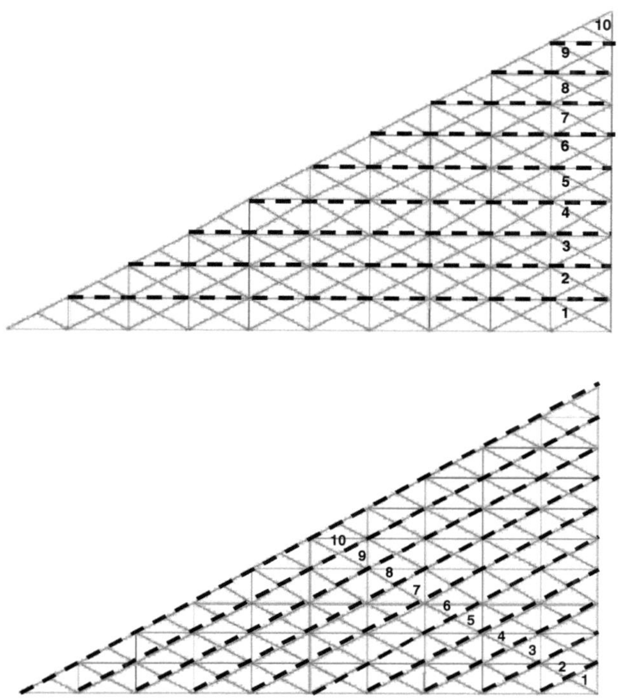

Fig. 5.2. Sequential staged loading for granular heap: a) raining procedure, b) funnel procedure

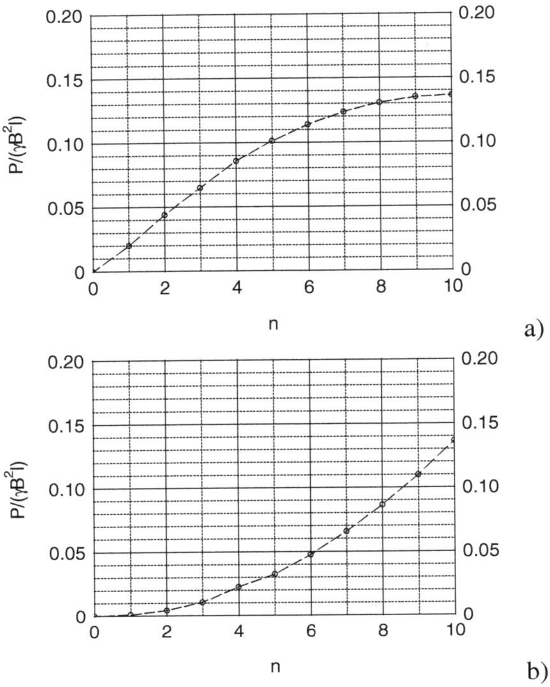

Fig. 5.3. Evolution of the normalized resultant total vertical force at the base $P/(\gamma B^2 l)$ of the prismatic granular heap against the loading stages n (dense sand with the initial void ratio of e_o=0.60, initial volume weight γ=16.7 kN/m^3, d_{50}=2.5 mm, heap inclination α=30°): a) raining procedure of Fig.5.154a, b) funnel procedure of Fig.5.154b (B=250 mm, l=1.0 m)

The FE results show that the construction method, similarly as in the experiments (Vanel et al. 1999), is of primary importance for the stress distribution beneath the sandpile. The stress profile increases monotonically with the stress maximum beneath the apex using the raining procedure (Fig.5.4a). For the funnel procedure, however, the stress distribution shows a significant stress dip of about 20% beneath the apex (Fig.5.4b). The pile settlement for the raining procedure was about 20% larger than for the funnel procedure (Fig.5.6).

Next, the analysis was performed with a smaller mean grain diameter (d_{50}=1.0 mm), initially loose sand (e_o=0.90), very smooth base (assuming u_2=0, ω^c=0, σ_{12}=0 and one fixed bottom point in the symmetry axis) and higher heap inclination (α=33°). The results have demonstrated that the initial void ratio and mean grain diameter of sand, heap inclination and wall roughness of the base have only minor influence on the FE results.

Conical heaps

FE calculations were carried out for the case of axisymmetry. Fig.5.7 presents the distribution of the normalized vertical normal stress under the pile using two

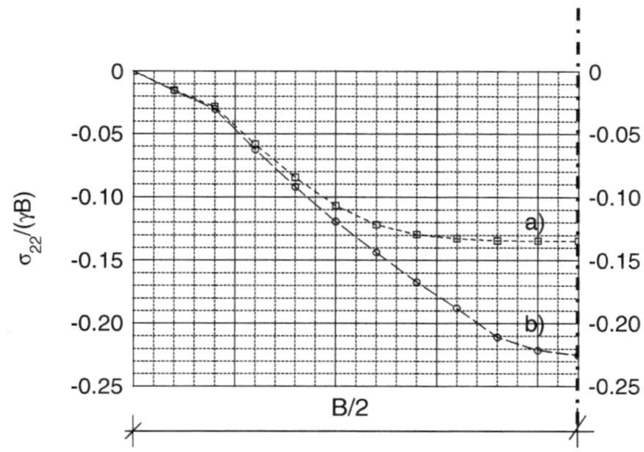

a)

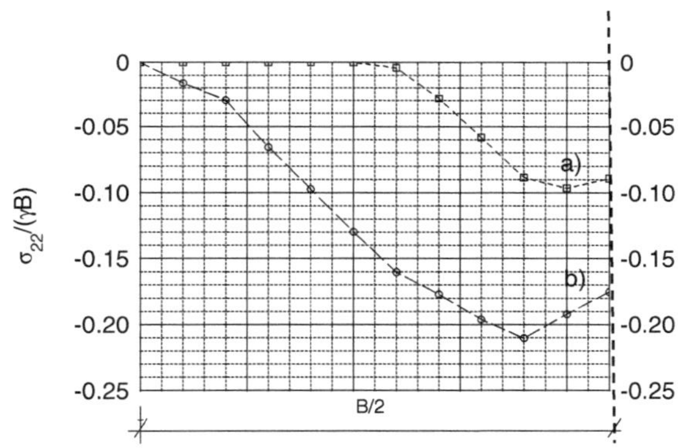

b)

Fig. 5.4. Distribution of the normalized vertical normal stress at the base $\sigma_{22}/(\gamma B)$ of the prismatic granular heap for A) raining procedure of Fig.5.154a and B) funnel procedure of Fig.5.154b (dense sand with the initial void ratio of e_o=0.60, initial volume weight γ=16.7 kN/m^3, d_{50}=2.5 mm, heap inclination α=30°): (a) after 5 loading stages, b) after 10 loading stages)

construction methods after 10 loading stages (dense sand with the initial void ratio of e_o=0.60, initial volume weight γ=16.8 kN/m^3, d_{50}=2.5 mm).

The results are qualitatively similar to those for the case of plane strain. The stress distribution depends similarly as for prismatic piles on the method of the pile construction. In the case of the funnel procedure, the stress depression was observed under the heap centre.

The following conclusions can be drawn:

- The vertical stress distribution is dependent on the method of the heap construction.
- The stress increases monotonically up to the apex of the sandpile for the raining procedure.

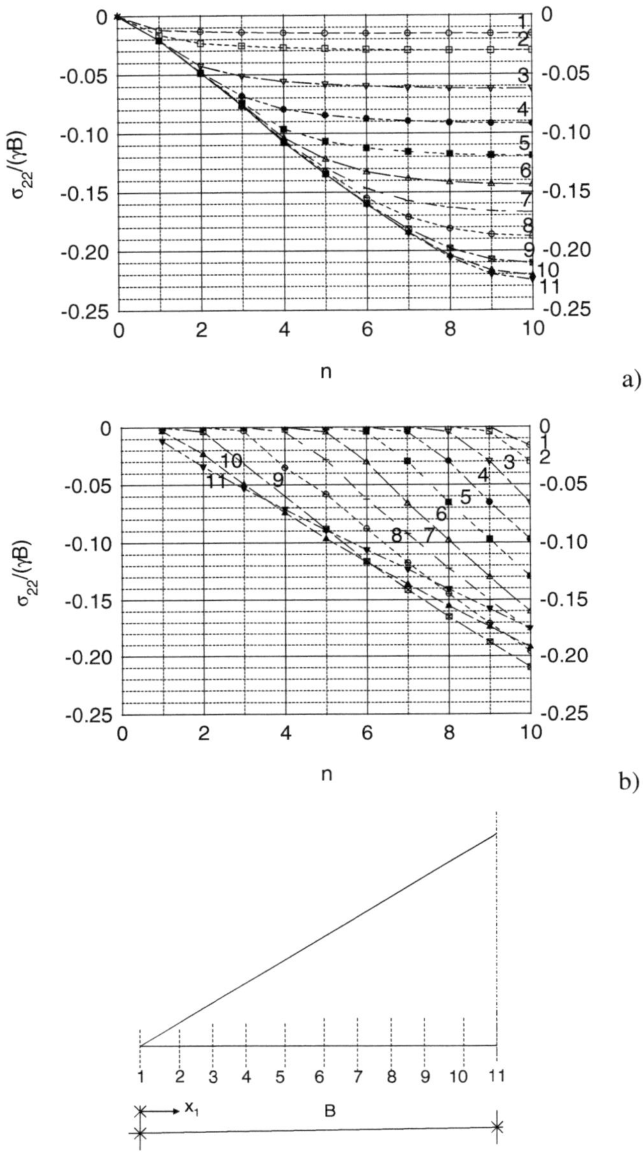

Fig. 5.5. Evolution of the vertical normal stress at the base σ_{22} of the plane strain granular heap in vertical sections 1-11 (dense sand with the initial void ratio of e_o=0.60, initial volume weight γ=16.7 kN/m^3, d_{50}=2.5 mm, heap inclination α=30°): a) raining procedure of Fig.2a, b) funnel procedure of Fig.2b (1) x_1/B=0, 2) x_1/B=0.1, 3) x_1/B=0.2, 4) x_1/B=0.3, 5) x_1/B=0.4, 6) x_1/B=0.5, 7) x_1/B=0.6, 8) x_1/B=0.7, 9) x_1/B=0.8, 10) x_1/B=0.9, 11) x_1/B=1.0, n – loading stages)

- In turn the stress shows a maximum beyond the heap mid-point for the funnel procedure.
- The stress distribution does not depend upon the initial void ratio, mean grain diameter, heap inclination and base roughness.
- The results are similar for prismatic and conical sandpiles.

The FE results confirm the experimental results by Vanel et al. (1999) and numerical results by Al Hattamleh et al. (2005). However, in contrast to the numerical results by Al Hattamleh et al. (2005), no additional condition (as orientation of initial slip lines) was imposed. The non-uniform distribution of the vertical normal stress beneath the sandpile during a funnel procedure was a natural numerical outcome.

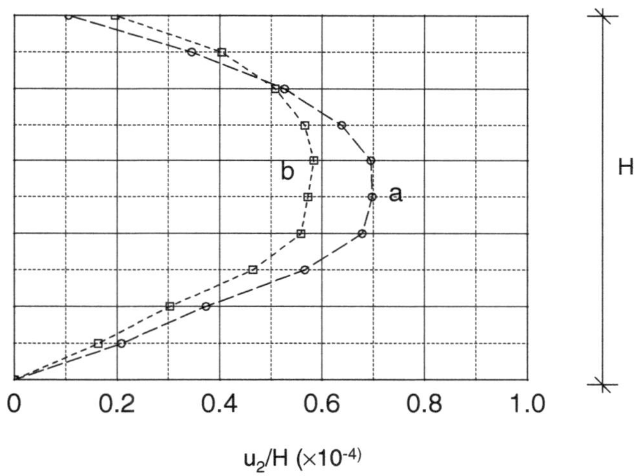

Fig. 5.6. Normalized settlement u_2/H of the prismatic granular heap along the symmetry axis (dense sand with the initial void ratio of e_o=0.60, initial volume weight γ=16.7 kN/m^3, d_{50}=2.5 mm, heap inclination α=30^0): a) raining procedure of Fig.5.154a, b) funnel procedure of Fig.5.154b (H=70 mm)

5.2 Direct Cyclic Shearing under Constant Normal Stiffness (CNS) Condition

The cyclic shearing behaviour of a soil-structure interface is responsible for the response of many geotechnical systems (e.g. piles, foundations). Of a major importance is the estimation of a decrease in maximum mobilized shear resistance with the number of cycles primarily due to the reduction of normal stress caused by cumulative contraction of the material in the neighborhood of the soil-structure interface (Poulos 1989). From experiments with cohesionless granular materials like sand it is well known that under cyclic shearing the material can exhibit contractancy or dilatancy depending on the initial density, stress state, magnitude of the shear amplitude, interface friction mechanism and far field boundary conditions surrounding a granular

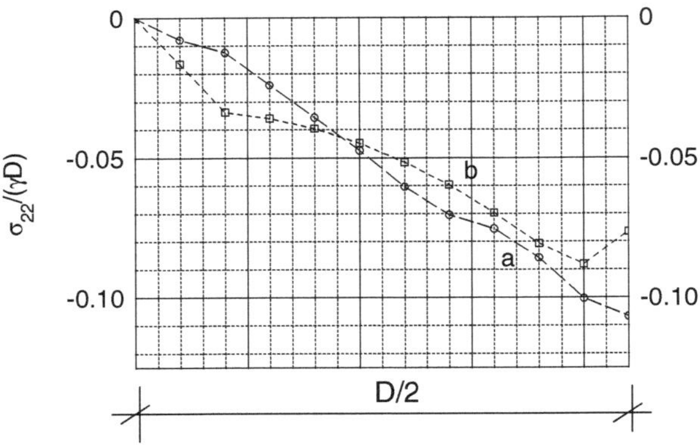

Fig. 5.7. Distribution of the normalized vertical normal stress at the base $\sigma_{22}/(\gamma D)$ of conical granular heap after 10 loading stages: a) raining procedure of Fig.5.154a and b) funnel procedure of Fig.5.154b (dense sand with the initial void ratio of e_o=0.60, initial volume weight γ=16.8 kN/m^3, d_{50}=2.5 mm, heap inclination α=30°)

body (Wood and Budhu 1980, Pradhan and Tatsuoka 1989, Paikowsky et al. 1995, DeJong et al. 2003, 2006).

A cyclic interface direct shear test under a constant normal stiffness (CNS) condition was modeled (Tejchman and Bauer 2008) with boundary conditions similar to that in corresponding experiments carried out by DeJong et al. (2003, 2006). This CNS confinement condition approximately simulates the far-field stiffness around the soil-structure interface. The calculations were performed for small cyclic shear amplitudes with +/-1 mm (similarly as in the experiments). For a very rough interface surface, the assumption was made that no relative displacements between the particles in contact with the interface surface took place.

In order to investigate the influence of the distance of the lateral boundaries of the direct shear test, the FE results were compared with those with lateral unrestricted boundaries which were simulated on the basis of boundary conditions for a lateral infinite narrow granular layer (Sections 4.2 and 4.3).

There are few experimental studies in the literature on cyclic shearing of granular materials. Tests with sand specimens were performed with the so-called simple shear devices (Poulos 1989, Airey et al. 1992) where global quantities were measured. They indicate that the evolution of an overall mean void ratio within a sand specimen strongly depends on the magnitude of ~~a~~ the cyclic shear amplitude. For medium dense sand and a small cyclic shear amplitude, the mean value of the void ratio becomes lower while a global dilatant behaviour can be observed for a larger shear amplitude. Recently, the degradation of shear resistance during cyclic direct shear and under a constant normal stiffness was experimentally investigated by DeJong et al. (2003, 2006). The top plate of the direct shear device was coupled with a spring to simulate elastic far-field stiffness. To prevent a rotation of the top platen the plate was fixed to the loading piston. The experiments were performed with an uniform subrounded silica sand (with a mean grain diameter of d_{50}=0.73 mm and the limit void ratios e_{max}=0.679 and e_{min}=0.5). The specimen was prepared at a medium dense state

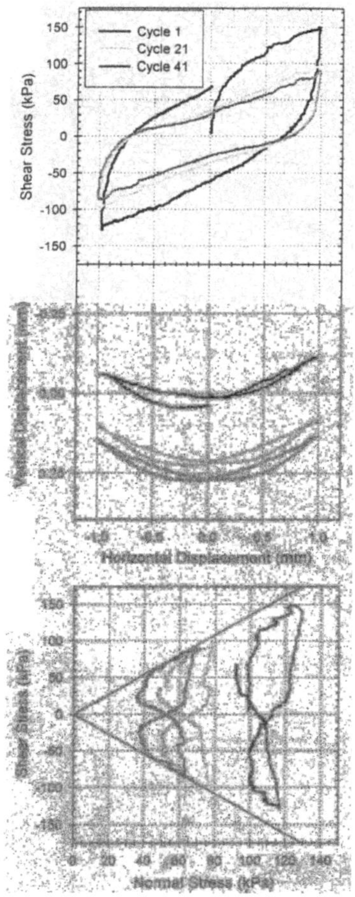

Fig. 5.8. Experimental results by DeJong et al. (2006) for uncemented silica: evolution of shear stress τ and vertical displacement δ_v versus horizontal displacement δ_h, and evolution of shear stress τ versus normal stress σ

(e_o=0.567). The initial vertical normal stress was 100 kPa and the vertical stiffness of the top was k=250 kPa/mm. The horizontal displacement rate was 0.5 mm/min and the cyclic limits were ±1.0 mm. The shearing had place along a very rough interface surface. Totally, 45 shear cycles were performed. A selection of the experimental results is shown in Fig.5.8. It is obvious that mobilized global shear stress decreases with number of cycles, where the degradation is more pronounced within the first cycles. The displacement of the top surface in each cycle begins with a small contractive phase before pronounced dilation start. During reversal shearing the forgoing dilation disappears and a net contraction remains. Thus, with a number of cycles, a significant cumulative contraction can be detected. Accompanied with a global contraction of the sand specimen is a reduction of the normal stress caused by

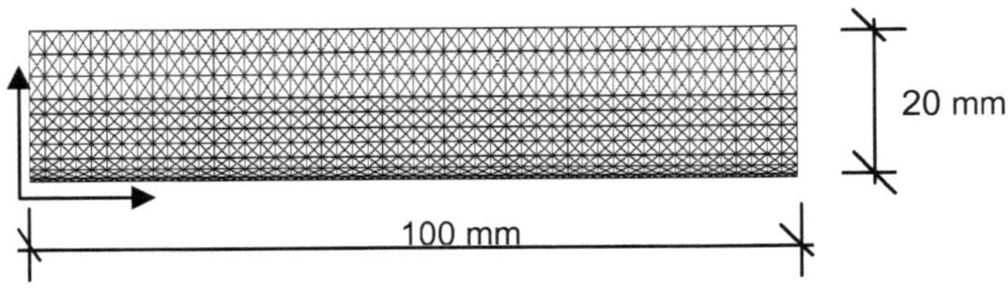

Fig. 5.9. FE mesh used for calculations of direct cyclic wall shearing (h=20 mm, b=100 mm)

the relaxation of the vertical spring located at the top surface. Fig.5.8 indicates that the maximum mobilized shear stress decreases almost proportional to the normal stress although a slight non-linearity of the failure envelope is evident. The non-linearity of the failure envelope is caused by a higher peak friction angle at lower pressure levels. With the help of particle image velocimetry, a localization of shear deformation close the interface was detected. The mean thickness of the localized zone is approximately 6 mm ($8\times d_{50}$). From the cyclic direct shear tests under CNS confinement conditions it can be concluded that the degradation of maximum mobilized shear resistance with the number of cycles is primarily due to the reduction of normal stress caused by global cumulative contraction of the sand specimen.

FE calculations of plane strain direct wall shearing under a constant normal stiffness (CNS) were performed for a granular specimen with a width of b=100 mm, a depth of l=1.0 m and an initial height of h_o=20 mm. 2000 quadrilateral finite elements composed of four diagonally crossed triangles were applied to avoid volumetric locking. In expectation of shear localization close to the bottom smaller elements were generated. The heights of finite elements were equal to (from the bottom up to the top): 0.5 mm, 1.0 mm, 1.5 mm, 4×2.0 mm and 3×3.0 mm (Fig.5.9). The width of elements was 2 mm. Thus, the size of elements in the shear zone along the interface (width and height) was not greater than $5\times d_{50}$ (with d_{50}=0.5 mm) to obtain mesh-independent results. The calculations were carried out with large deformations and curvatures (Section 3.5).

As the initial stress state in the granular specimen, a K_0-state without polar quantities was assumed ($\sigma_{22}=p+\gamma_d x_2$, $\sigma_{11}=\sigma_{33}=K_0(p+\gamma_d x_2)$, $\sigma_{12}=\sigma_{21}=m_1=m_2=0$), where: γ_d=15.5 kN/m^3 - initial volume weight of sand, K_0=0.50 – pressure coefficient at rest and p=100 kPa - an initial constant vertical pressure. The initial void ratio (e_o=0.70) of medium dense sand (before the vertical compression) was assumed to be homogeneous in the specimen.

The bottom of the sand specimen was assumed to be very rough: u_1=0, u_2=0 and ω^c=0 i.e. sliding and rotation of particles against the bounding surface were excluded (Fig.5.10). The horizontal displacements of all nodes along the very rough top boundary were tied together: $u_1=n\Delta u_1$ and ω^c=0. The index *1* denotes the co-ordinate in the direction of shearing, and the index *2* denotes the co-ordinate perpendicular to the direction of shearing, n is the number of time steps, Δu_1 is the constant displacement increment in one step. In order to simulate a constant normal stiffness elastic springs were vertically arranged along the top note of the specimen. Therefore the movement of the top surface was restricted by the constant normal stiffness and the requirement that all nodes of the top surface have the same vertical displacement u_2^t. With respect

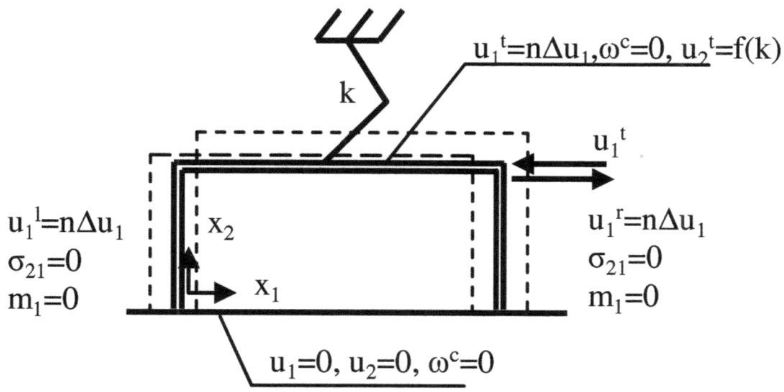

Fig. 5.10. Boundary conditions during direct wall shearing ('*r*' – right, '*l*' – left, 't' – top, *k* – spring stiffness)

to the initial normal pressure *p* and the spring stiffness *k* the resultant vertical force *N* is determined by $N=(p+u_2^t k)bl$. In accordance with the experiments by DeJong et al. (2006), the spring stiffness was takes as 250 kPa/mm. To simulate a direct shear test, the same horizontal displacement increments were prescribed to two smooth side walls: $u_1=n\Delta u_1$, $\sigma_{21}=0$ and $m_1=0$. The specimen was subject to shearing in one direction up to $u_1^t=1$ mm (u_1^t denotes the horizontal displacement of the top). Afterwards, the direction of shearing was repeatedly changed by applying a symmetric cyclic shear amplitude of $u_1^t=\pm1$ mm, which was the same as in the experiments. The shear amplitude was large enough to initiate shear localization within each cycle. Totally, 5 full shear cycles were performed.

FE calculations of plane strain shearing under a constant normal stiffness (CNS) were also performed for an infinite granular layer with a height of $h_o=20$. The calculations were performed with a section of an infinite shear layer with a width of $b=h_o$, discretised by 20 quadrilateral elements composed of four diagonally crossed triangles. The height of the elements was always $\leq 5\times d_{50}$ (with $d_{50}=0.5$ mm). The initial stress state was similar as in a direct shear tester.

A quasi-static shear deformation was initiated through constant horizontal displacement increments prescribed at the nodes along the top of the layer. Both bottom and top surfaces were assumed to be very rough. The boundary conditions were along the bottom: $u_1=0$, $u_2=0$ and $\omega^c=0$, and along the top: $u_1=n\Delta u$, $\omega^c=0$ with the vertical displacement hindered by the springs $u_2=f(k)$. Totally, 21 full shear cycles were performed.

Cyclic direct wall shearing

Figs.5.11-5.20 show the numerical results obtained for a constant normal spring stiffness of *k*=250 kPa/mm at the top of the granular specimen, an initial void ratio of $e_o=0.70$ and for a prescribed horizontal cyclic displacement amplitude of $u_1^t=\pm1$ mm subjected to the lateral boundaries of the specimen. For quasi-static cyclic direct wall shearing, the evolution of a global shear force and global normal force and also the evolution of state values at certain cross-sections and local points were outlined and

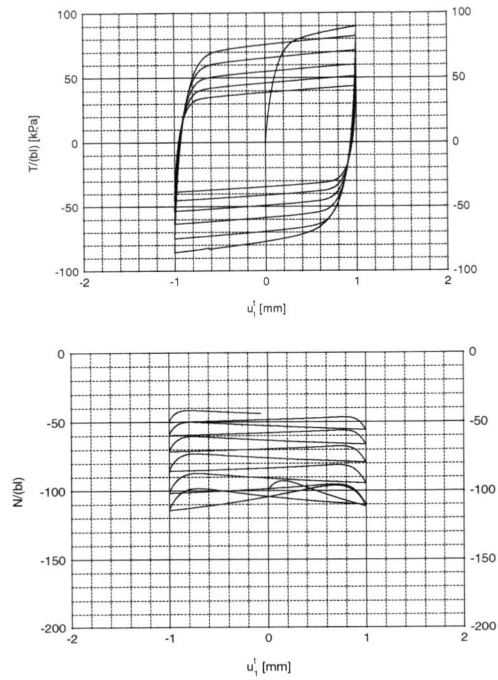

Fig. 5.11. Evolution of: average horizontal shear stress *T/(bl)* and average vertical normal stress N/*(bl)* versus u_1^t during direct wall shearing

discussed. The resultant horizontal shear force *T* was calculated as the integral of the shear stresses along the interface between the granular specimen and the very rough bottom surface and the resultant vertical force *N* as the integral of the normal stresses along the interface. With respect to smooth lateral boundaries, the force *N* at the interface is the same as the force acting at the top surface. The local state quantities σ_{ij}, m_i and *e* were taken at the mid-point of the interface as the mean values from quadrilateral elements composed of four triangle elements. The evolution of the average horizontal shear stress *T/(bl)* and average vertical normal stress *N*/*(bl)* versus the prescribed horizontal displacement u_1^t is shown in Fig.5.11. In turn, the evolution of the local horizontal shear stress σ_{12} and vertical normal stress σ_{22} at the mid-point of the bottom versus u_1^t is depicted in Fig.5.12. Figs.5.13 and 5.14 present the evolution of the average mobilized friction angle ϕ=arctan*(T/N)* and evolution of local mobilized interface friction angle ϕ=arctan$(\sigma_{12}/\sigma_{22})$ at the mid-point of the bottom, respectively. The evolution of the average horizontal shear stress *T/(bl)* versus vertical normal stress *N/(bl)* and evolution of horizontal shear stress σ_{12} versus vertical normal stress σ_{22} at the mid-point of the bottom are demonstrated in Figs.5.15 and 5.16, respectively. Fig.5.17 shows the evolution of vertical displacement of the top boundary u_2^t/h_o and Fig.5.18 the evolution of void ratio *e* at mid-point of the bottom (x_2/d_{50}=1) and at the top boundary (x_2/d_{50}=37). In Fig.5.19 the deformed meshes are demonstrated together with the distribution of Cosserat rotation and void ratio after the

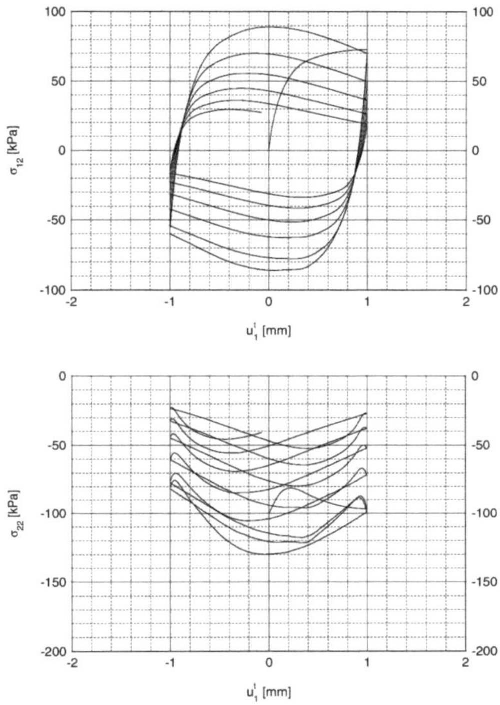

Fig. 5.12. Evolution of local horizontal shear stress σ_{12} and vertical normal stress σ_{22} at the mid-point of the bottom versus u_1^t during direct wall shearing

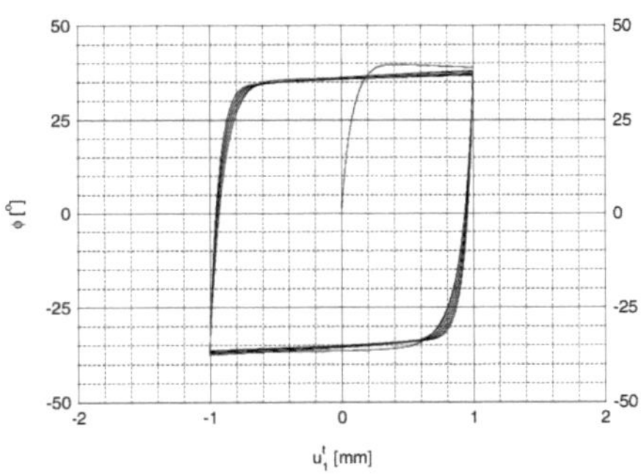

Fig. 5.13. Evolution of average mobilized friction angle ϕ=arctan(T/N) versus u_1^t during direct wall shearing

initial shearing, and after five full shear cycles. The magnitude of Cosserat rotations is marked by circles with a diameter corresponding to the rotation magnitude in the entire specimen. Finally, Fig.5.20 presents the distribution of Cosserat rotation ω^c, void ratio e, stresses σ_{ij} and couple stress m_2 across the normalized height x_2/d_{50} in the middle of the specimen after the initial shearing, and after five full shear cycles.

All state variables (stresses and void ratio) show a tendency to reach asymptotically stationary values in each cycle. The stationary state, however, is not fully reached because of the too small shear deformation. As a consequence of the constant normal stiffness assumed along the top surface and the accumulation of contractancy and dilatancy during cyclic shearing, the maximum values of state variables significantly change with the number of shear cycles. This general tendency is in agreement with the experiments (Fig.5.8) and can be detected for arbitrary cohesionless granular materials. In particular, the maximum values of the average shear stress (Fig.5.11), the average normal stress (Fig.5.11) and the maximum local values (Fig.5.12) significantly decrease during cyclic shearing. Similarly as in the experiment (Fig.6.8), the calculated shear stress has not reached its asymptote yet. A decrease of the magnitude of the maximum normal stress (Figs.5.11 and 5.12) is caused by the global densification of the specimen. While for the stress variables, the magnitude of degradation within each cycle is significant, the degradation of the stress ratio and the mobilized friction angle (Figs.5.13 and 5.14) is after the first two cycles less pronounced.

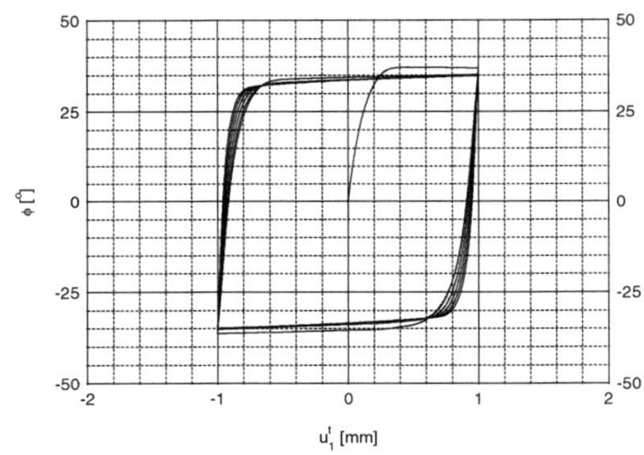

Fig. 5.14. Evolution of local mobilized friction angle ϕ=arctan(σ_{12}/σ_{22}) at the mid-point of the bottom versus u_1^t during direct wall shearing

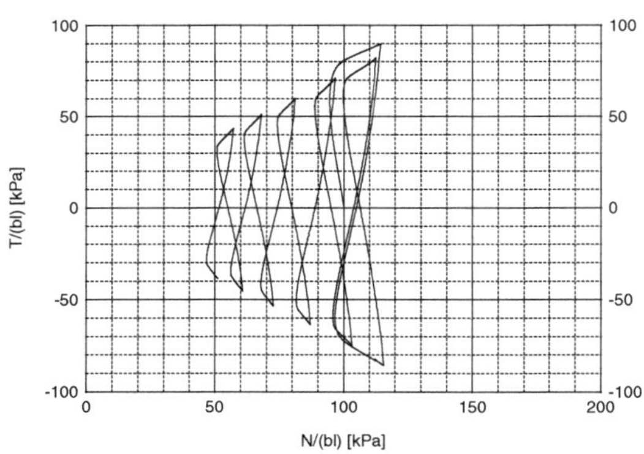

Fig. 5.15. Evolution of average horizontal shear stress *T/(bl)* versus resultant vertical normal stress *N/(bl)* during direct shearing

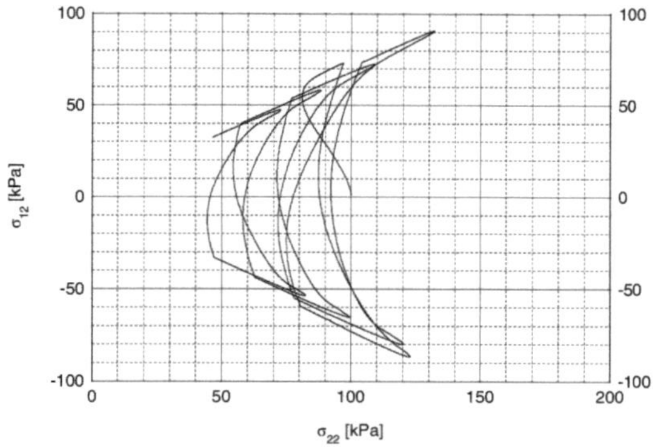

Fig. 5.16. Evolution of local horizontal shear stress σ_{12} versus local vertical normal stress σ_{22} at the mid-point of the bottom during direct wall shearing

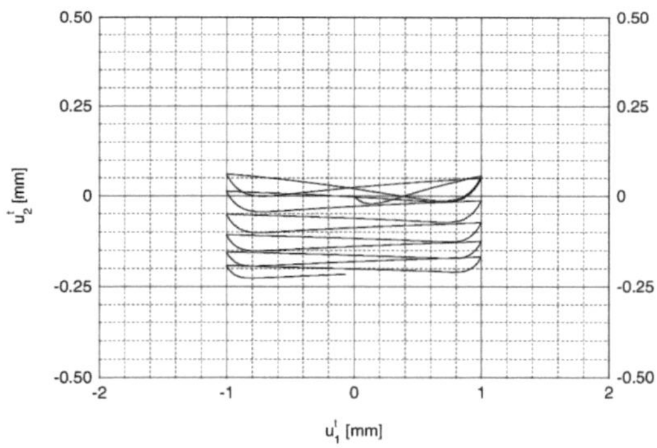

Fig. 5.17. Evolution of vertical displacement of the top boundary u_2^t during direct wall shearing

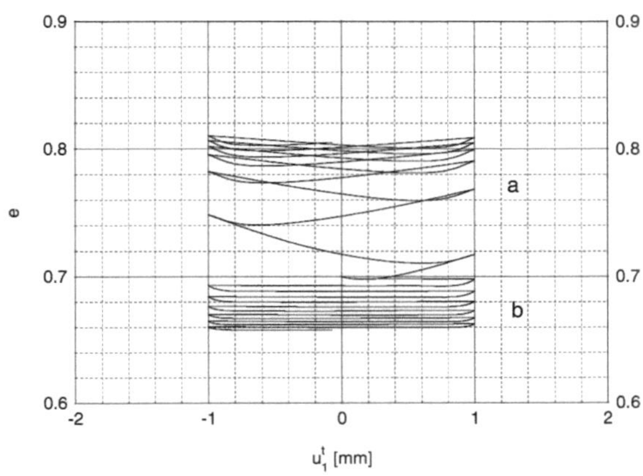

Fig. 5.18. Evolution of void ratio e versus u_1^t during direct wall shearing at the middle section of the specimen: a) x_2/d_{50}=1 (bottom), b) x_2/d_{50}=37 (top)

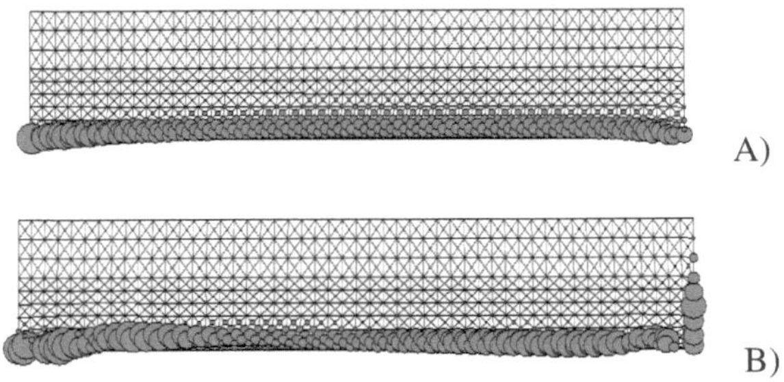

Fig. 5.19. Deformed mesh with the distribution of Cosserat rotation ω^c during wall direct shearing: (A) after the initial shearing, (B) after five full shear cycles

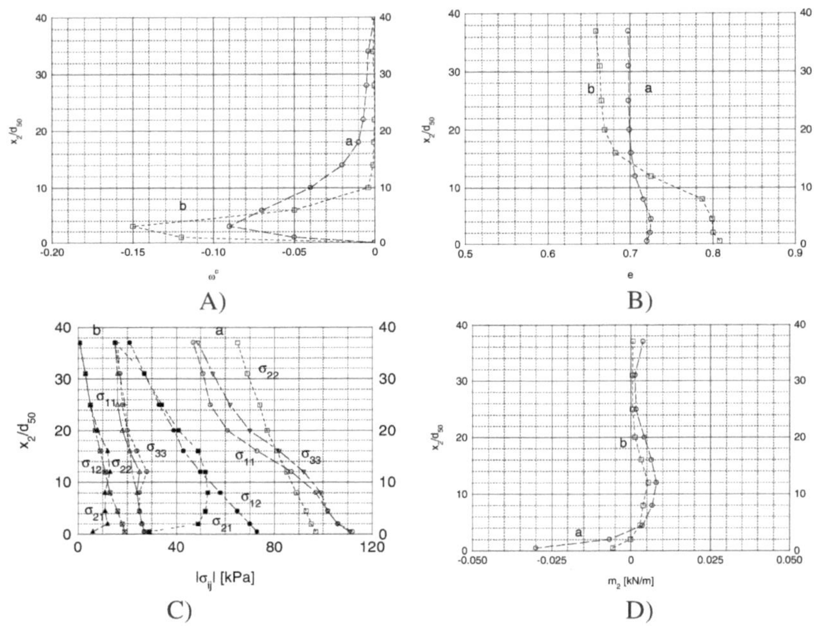

Fig. 5.20. Distribution of Cosserat rotation ω^c (A), void ratio e (B), stresses σ_{ij} (C) and couple stress m_2 (D) across the normalised height x_2/d_{50} in the middle of the specimen during direct wall shearing (a - after the initial shearing, b - after five full shear cycles)

The comparison of Figs.5.11 and 5.13 with Figs.5.12 and 5.14 respectively shows that the average values can strongly deviate from the local values. In this context it is important to note that the evolution of local values is influenced by the width of the specimen. The maximum mobilized average friction angle (obtained during initial shearing) is 42° and after 5 cycles becomes 38°. In turn, the maximum mobilized local friction angle (obtained during initial shearing) is 38° and after 5 cycles the value reduces to 35°. It is obvious that the differences between the average values and local values show no tendency to become smaller with number of cycles. Since the maximum shear stress decreases almost proportionally to the normal stress, the stress states at the end of each cycle are close to the envelope of the critical stress states and the envelope of the critical stress states in the (σ_{12}-σ_{22})-plane is linear (Figs.5.15 and 5.16). The comparison with experiment (Fig.5.8) shows that the evolution of the stress path is qualitatively similar.

The evolution of the calculated volume changes indicates the same tendency as in the experiment (Fig.5.8). Fig.5.17 shows that during initial shearing, the top boundary moves first slightly down (the specimen is subject to contractancy) and afterwards it moves significantly up (the specimen is subject to dilatancy). After the first two cycles, the top surface moves continuously down, which indicates that the specimen is globally subjected to cumulative contractancy. During each change of the shearing direction, the boundary moves first down (small contractancy occurs) and later moves up (dilatancy takes place). However, a local inspection of the evolution of the void ratio (Fig.5.18) shows that only a part of the entire specimen is subjected to contractancy. While the void ratio in the upper part (Fig.5.18b) of the specimen decreases the void ratio in a zone close to the interface (Fig.5.18a) increases. It is obvious that the increase of the void ratio is more pronounced within the first two cycles and then the change of the void ratio stagnates. As the global behaviour show contractancy and a local decrease of the void ratio (Fig.5.18b) is smaller compared to a local increase of the void ratio (Fig.5.20A), the zone of dilatancy must be much smaller than the zone of contractancy (this is also clearly visible in Fig.5.20B).

An increase of the void ratio and the presence of the micro-rotation (Cosserat rotation) are a consequence of localization of shear deformation along a zone close to the bottom (Figs.5.19, 5.20A and 5.20B). The distribution of the micro-rotations in the specimen (Fig.5.20) shows that the shear zone becomes non-uniform in the course of shearing. In the middle of the specimen (Fig.5.20A), the thickness of the shear zone slightly decreases. It is about (16-18)$\times d_{50}$ after the initial shearing and $10\times d_{50}$ after 5 full shear cycles. In the shear zone, the stress tensor is non-symmetric ($\sigma_{12} \neq \sigma_{21}$) and the shear stress σ_{12} is slightly larger than σ_{21} (Fig.5.20C). The magnitude of the normal stresses decreases with the number of cycles and after 5 cycles, the stress ratios σ_{11}/σ_{22} and σ_{11}/σ_{33} become almost equal to 1, which is in accordance with the conditions for critical states. The distribution of the couple stress m_2 does not significantly change during cyclic shearing with the exception at the interface where the magnitude of m_2 increases (Fig.5.20D). The couple stress m_1 is negligible.

Cyclic shearing of an infinite granular layer

In order to investigate the influence of boundary conditions during wall shearing, the results of the forgoing section for the direct shear test with a certain distance between the lateral vertical boundaries are compared with those with lateral unrestricted boundaries (Tejchman and Bauer 2008). Such a behaviour can be simulated using lateral boundary conditions for an infinite granular layer (Section 4.3).

Figs.5.21-5.25 show the numerical results obtained for cyclic shearing of an lateral infinite granular layer with the initial void ratio of e_o=0.70, a constant normal stiffness of k=250 kPa/mm, and for a prescribed horizontal cyclic displacement amplitude of u_1^t=±1 mm subjected to the top surface of the specimen. Fig.5.21 presents the

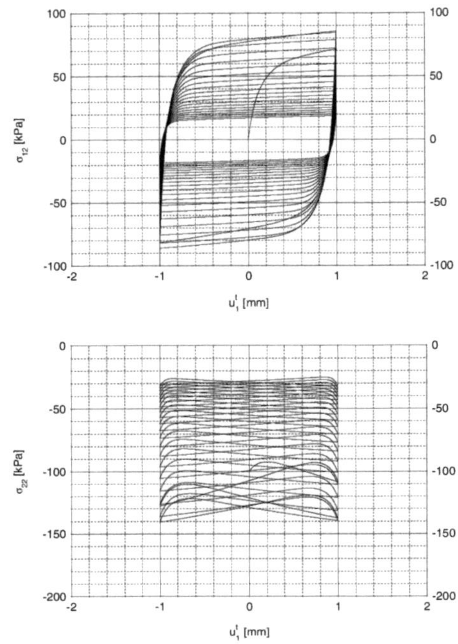

Fig. 5.21. Evolution of horizontal shear stress σ_{12} and vertical normal stress σ_{22} versus u_1^t in an infinite granular layer

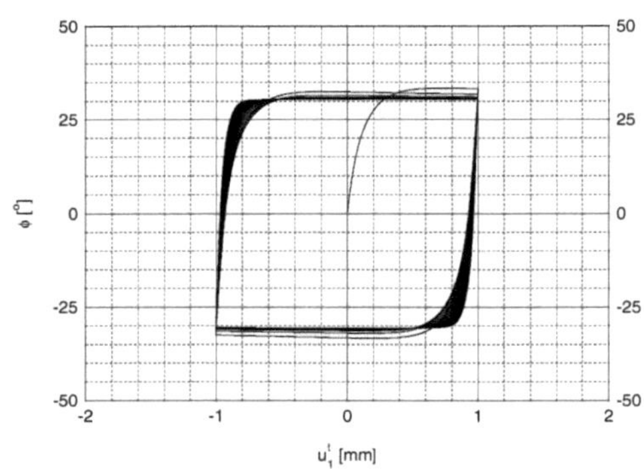

Fig. 5.22. Evolution of mobilized friction angle ϕ=arctan$(\sigma_{12}/\sigma_{22})$ versus u_1^t in an infinite granular layer

evolution of shear the stress σ_{12} in the direction of shearing against the horizontal displacement u_1^t of the top. In Fig.5.22, the mobilized friction angle ϕ=arctan$(\sigma_{12}/\sigma_{22})$ versus u_1^t is shown. The mobilised friction angle ϕ is related to the entire granular layer since the stresses σ_{12} and σ_{22} are independent of both the height and length of the layer. In turn, the evolution of σ_{12} versus σ_{22} with number of shear cycles is demonstrated in Fig.5.23. The vertical displacement u_2^t of the top boundary (Fig.5.24) is

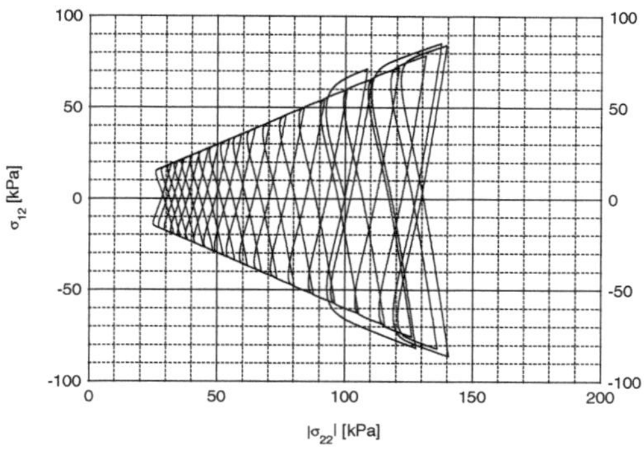

Fig. 5.23. Evolution of horizontal shear stress σ_{12} versus vertical normal stress σ_{22} in an infinite granular layer

positive for a global volume increase and negative for a global compaction of the specimen. In turn, the local evolution of the volume change expressed by the void ratio e is demonstrated in Fig.5.24b at the bottom (curve 'a') and in the middle of the layer (curve 'b'). Finally, Fig.5.25 presents the distribution of the Cosserat rotation ω^c, the void ratio e, the horizontal σ_{12} and vertical shear stress σ_{21} across the normalised height x_2/d_{50} after the initial shearing and after 21 shear cycles.

Although the behaviour of the granular material is qualitatively similar as during direct shearing, there are some considerable differences in the local evolution of state variables in the infinite narrow layer. While in the infinite shear layer the maximum shear stress σ_{12} (Fig.5.21) increases during the first two cycles, the average maximum shear stress under direct shearing is already reached at the end of the first shearing. An increase of the normal stress σ_{22} (Fig.5.21) within the first two cycles is much higher than under direct shearing (Fig.5.12). With the number of shearing, the maximum shear stress and maximum normal stress significantly decrease. The maximum values of normal stress σ_{12} decrease almost proportionally to the corresponding normal stresses σ_{22} (Fig.5.23) which is in a good agreement with the experiment (Fig.5.8). Across the height of the layer, the values of σ_{12} and σ_{22} are independent of the co-ordinates x_1 and x_2 as it is required for the equilibrium of an infinite shear layer. In the middle of the layer, the values of σ_{12} and σ_{21} are almost equal and the stress ratios σ_{11}/σ_{22} and σ_{11}/σ_{33} become equal to 1 in the course of cyclic shearing (Fig.5.25C).

The maximum mobilized peak friction angle during initial shearing is ϕ_p=34° (Fig.5.22) and in the course of cyclic shearing it decreases down to 31° which is close to the internal friction angle of the granular material in the critical state. A comparison with the higher average friction angles of 38° obtained under direct shearing indicates that the results are considerably influenced by the lateral vertical boundaries of the direct shear device. As a result of dilation in the front and contraction in the rear, the distribution of stress and density become inhomogeneous within the specimen also in the direction of shearing. Consequently the average stress ratio along the interface is higher than the one that develops in the infinite granular layer with unaffected boundaries. It can be concluded from the vertical movement of the top boundary that

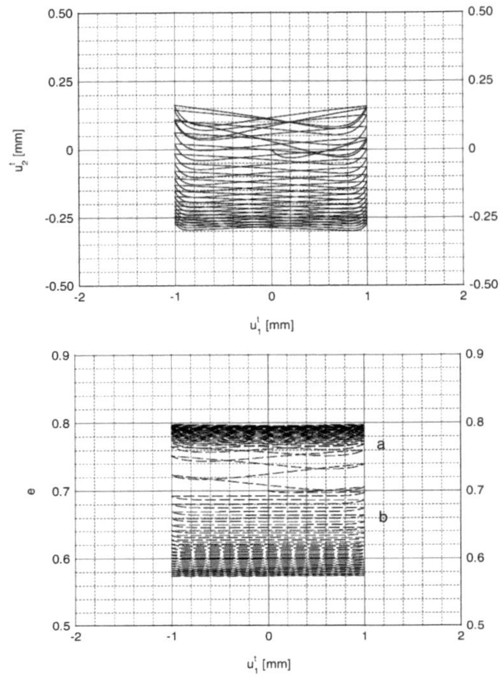

Fig. 5.24. Evolution of vertical displacement of the top boundary u_2^t and void ratio e versus u_1^t at: (a) x_2/d_{50}=1 (bottom), (b) x_2/d_{50}=19 (mid-point) in an infinite granular layer

during the first two cycles the global dilatancy (the mean value of the void ratio increases) of the specimen is significantly higher in the infinite granular layer (Fig.5.24). With advanced cyclic shearing the specimen is subject to cumulative contractancy (the mean value of the void ratio decreases). In this context it can be noted that further investigations indicate that the question as to whether the mean value of the void ratio across the shear layer increases or decreases strongly depends on the magnitude of the shear amplitude (Section 4.3). For very small cyclic shear amplitudes the material becomes generally denser, while for larger shear amplitudes and an initially dense state the void ratio may increase. The evolution of the void ratio close to the bottom and top surface and in the middle of the shear zone (Fig.5.24 and 5.25) is different to that under direct shearing (Fig.5.18). The void ratio close to the boundaries continuously decreases while the void ratio in the middle of the layer increases during each cycle and it tends towards the pressure-dependent critical value. A comparison of the distribution of the Cosserat rotation in Fig.5.20A with Fig.5.25A indicate that in contrast to direct shearing, the zone of intense shear localization occurs in the middle of the infinite granular layer and it is slightly thicker. The thickness of the shear zone significantly decreases during cyclic shearing and it is about $37\times d_{50}$ after the initial shearing and $17\times d_{50}$ after 21 full shear cycles. In this context it can be noted that the location and thickness of shear localization in an infinite granular shear layer

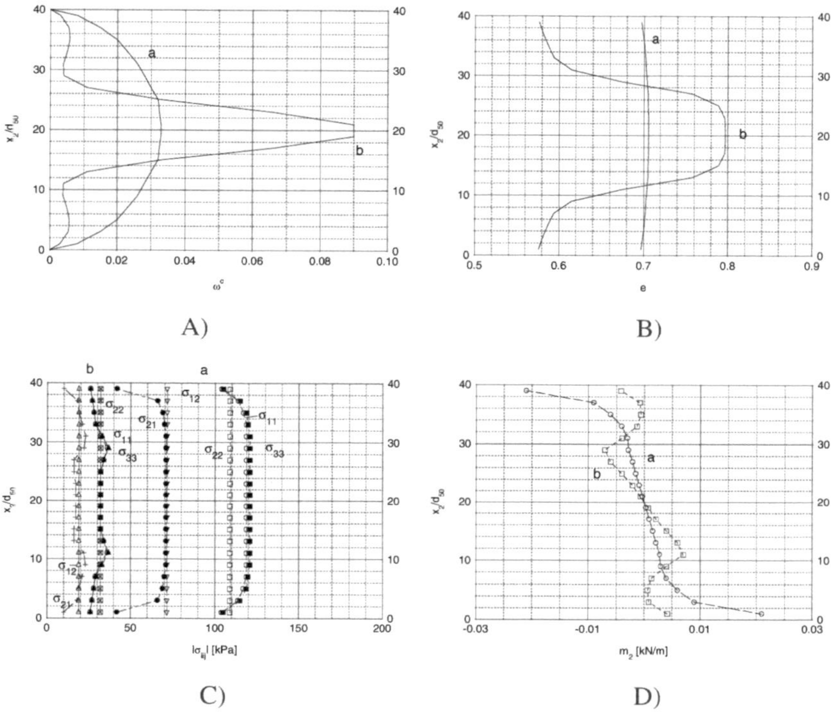

Fig. 5.25. Distribution of: (A) Cosserat rotation ω^c, (B) void ratio *e*, (C) stresses σ_{ij}, *(D)* couple stress m_2–across the normalised height x_2/d_{50} in an infinite granular layer ('a' - after initial shearing, 'b' - after twenty full shear cycles)

mainly depend both on micro-polar boundary conditions assumed at the top and bottom boundary and a mean grain diameter.

In addition, some calculations were carried out during direct wall shearing with varying initial void ratio, mean grain diameter of sand and vertical normal stiffness (Tejchman 2008). Figs. 5.26-5.30 show the effect of the initial void ratio e_o, vertical normal stiffness *k* and mean grain diameter d_{50} on the degradation of shear resistance, volume changes and thickness changes of the wall shear zone in a sand specimen subject to strain-controlled cyclic direct shearing with a small cyclic amplitude of $u_1^t=\pm1$ mm. The additional calculations were performed with two different initial void ratios of sand: e_o=0.60 (initially dense sand) and e_o=0.90 (initially loose sand), two different magnitudes of the vertical normal stiffness: *k*=25 kPa/mm and *k*=2500 kPa/mm and a larger mean grain diameter of sand: d_{50}=2.0 mm.

The degradation of shear resistance after 5 shear cycles increases with increasing initial void ratio and vertical normal stiffness (Fig.5.26). The effect of the mean grain diameter is rather negligible. The shear resistance degradation is about 30% (e_o=0.60), 90% (e_o=0.90), 15% (*k*=25 kPa/mm), 80% (*k*=2500 kPa/mm) and 45% (d_{50}=2 mm), respectively. The material is always subject to global contractancy during cyclic shearing which increases mainly with increasing initial void ratio and

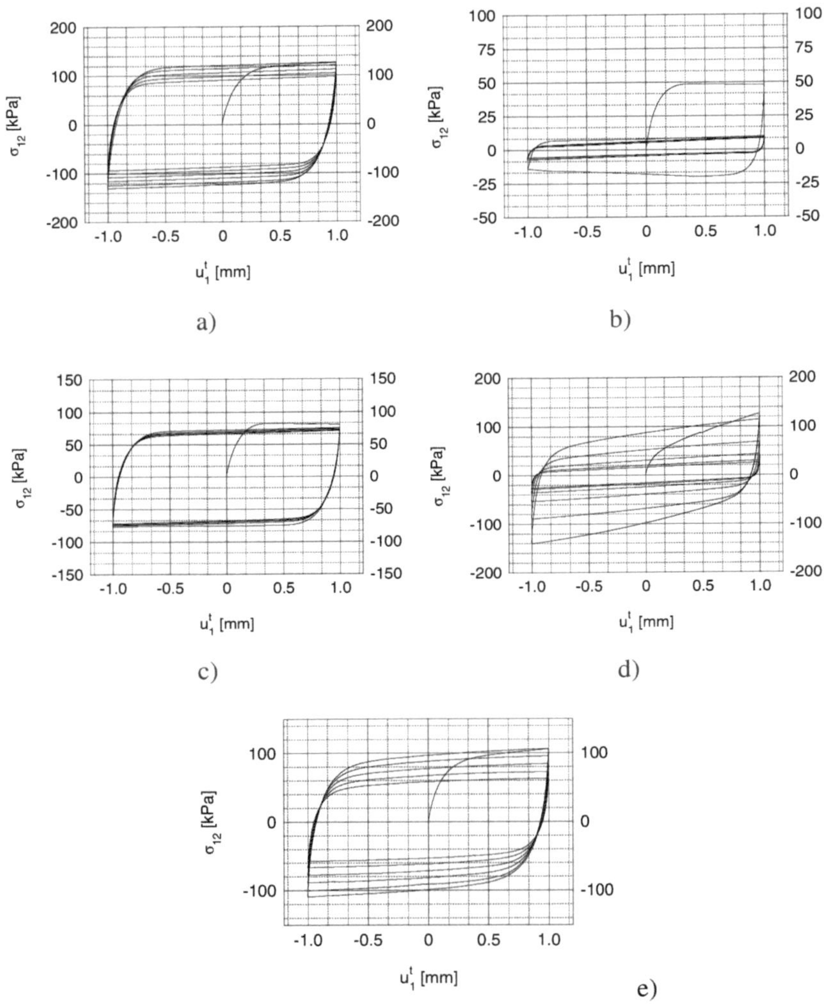

Fig. 5.26. Evolution of mean horizontal shear stress $T/(bl)$ versus u_1^t during direct wall shearing: a) e_o=0.60, d_{50}=0.5 mm, k=250 kPa/mm, b) e_o=0.90, d_{50}=0.5 mm, k=250 kPa/mm, c) e_o=0.70, d_{50}=0.5 mm, k=25 kPa/mm, d) e_o=0.70, d_{50}=0.5 mm, k=2500 kPa/mm, e) e_o=0.70, d_{50}=2.0 mm, k=250 kPa/mm

decreasing vertical normal stiffness (Fig.5.27). The effect of the mean grain diameter is again insignificant. The global contractancy u_v/h_o after 5 shear cycles is about: 1% (e_o=0.60), 1.25% (e_o=0.70), 1.9% (e_o=0.90), 1.5% (k=25 kPa/mm), 0.25% (k=2500 kPa/mm) and 1.25% (d_{50}=2 mm), respectively.

The thickness of the wall shear zone after initial shearing (when the residual state has not been reached yet) is similar for d_{50}=0.5 mm independently of e_o and k. It is about 20 mm ($10\times d_{50}$) for d_{50}=2.0 mm (Fig.5.28) on the basis of the Cosserat rotation.

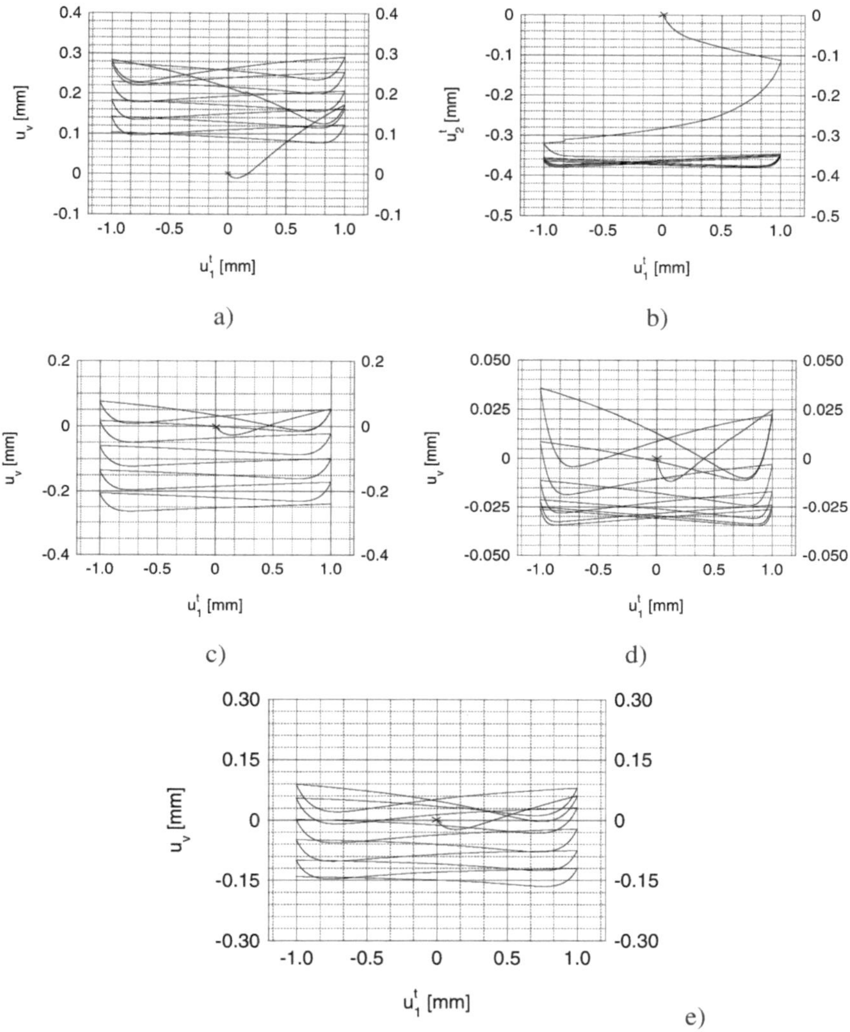

Fig. 5.27. Evolution of vertical displacement of the top boundary u_v during direct wall shearing: a) e_o=0.60, d_{50}=0.5 mm, k=250 kPa/mm, b) e_o=0.90, d_{50}=0.5 mm, k=250 kPa/mm, c) e_o=0.70, d_{50}=0.5 mm, k=25 kPa/mm, d) e_o=0.70, d_{50}=0.5 mm, k=2500 kPa/mm, e) e_o=0.70, d_{50}=2.0 mm, k=250 kPa/mm

The thickness decrease in the course of cyclic shearing mm cannot be noticed for a larger diameter. A decreases of the thickness of the wall shear zone after 5 shear cycles mainly increases with decreasing initial void ratio and increasing vertical normal stiffness. The thickness of the wall shear zone is 3 mm (6×d_{50}) for e_o=0.60, 4.5 mm (9×d_{50}) for e_o=0.70 mm and 5 mm (10×d_{50}) for e_o=0.90 mm, 5 mm (10×d_{50}) for k=25 kPa/mm, 4 mm (8×d_{50}) for k=2500 kPa/mm and 20 mm (10×d_{50}) for d_{50}=2.0 mm (Fig.5.28).

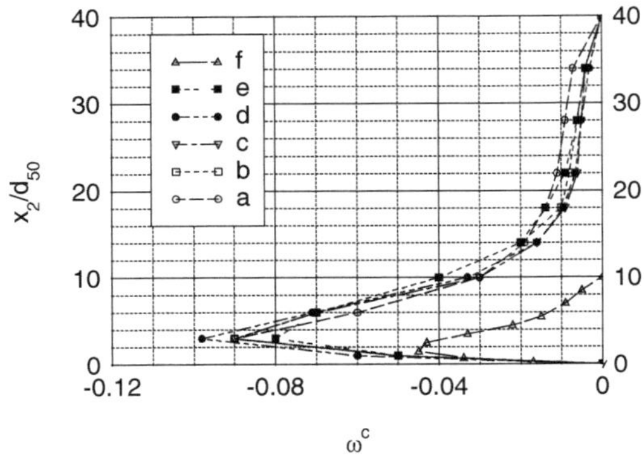

Fig. 5.28. Distribution of Cosserat rotation ω^c across the normalised height x_2/d_{50} in the middle of the specimen after the initial shearing: a) e_o=0.60, d_{50}=0.5 mm, k=250 kPa/mm, b) e_o=0.70, d_{50}=0.5 mm, k=250 kPa/mm, c) e_o=0.90, d_{50}=0.5 mm, k=250 kPa/mm, d) e_o=0.70, d_{50}=0.5 mm, k=25 kPa/mm, e) e_o=0.70, d_{50}=0.5 mm, k=2500 kPa/mm, f) e_o=0.70, d_{50}=2.0 mm, k=250 kPa/mm

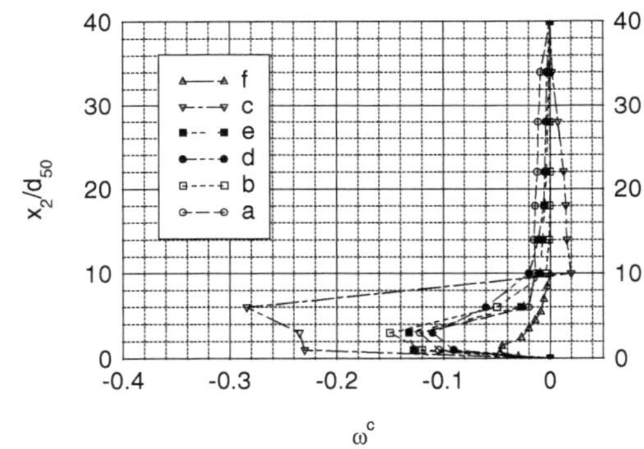

Fig. 5.29. Distribution of Cosserat rotation ω^c across the normalised height x_2/d_{50} in the middle of the specimen after 5 shear cycles: a) e_o=0.60, d_{50}=0.5 mm, k=250 kPa/mm, b) e_o=0.70, d_{50}=0.5 mm, k=250 kPa/mm, c) e_o=0.90, d_{50}=0.5 mm, k=250 kPa/mm, d) e_o=0.70, d_{50}=0.5 mm, k=25 kPa/mm, e) e_o=0.70, d_{50}=0.5 mm, k=2500 kPa/mm, f) e_o=0.70, d_{50}=2.0 mm, k=250 kPa/mm

The results of FE modeling of shear resistance degradation in granular materials during cyclic shearing under CNS condition indicate a satisfactory qualitative agreement between numerical and experimental results:

- The horizontal shear and vertical normal stresses show a significant loss during cyclic shearing due to a global material densification. Thus, the material contractancy is responsible for the shear resistance fatigue. Locally, the granulate is subject to dilatancy in the wall shear zone while experiences contractancy in the remaining region of the specimen.
- Locally, the granulate is subject to pronounced dilatancy in the wall interface shear zone (except of an initially loose specimen) while experiences contractancy in the

remaining region of the specimen. Thus, the volume change of the specimen is not homogeneous and the average value can strongly deviate from the local value. As the average value shows contractancy, the zone of dilatancy must be much smaller than the region of contractancy. The global contractancy during shearing increases with increasing initial void ratio and decreasing vertical normal stiffness.

- The shear zone thickness diminishes during cyclic shearing due to the global material contractancy. The final thickness during cyclic shearing is larger with increasing mean grain diameter and initial void ratio and decreasing vertical normal stiffness.

During each reversal shearing, contractancy always occurs in granular material. The normal stress ratios become equal to 1 in the course of cyclic shearing.

The degradation of shear resistance during cyclic shearing increases with increasing initial void ratio and vertical normal stiffness.

5.3 Wall Boundary Conditions

Soil-structure interfaces are frequently encountered in geotechnical engineering (foundations, tunnels, retaining walls, anchors, silos, piles, geotextiles etc.). Compared with interfaces e.g. between metals and rocks, the behaviour of the soil-structure interface is entirely different due to the formation of a wall shear zone in soil, i.e. a thin zone of intense shearing and with pronounced volume changes. In this zone, significant grain rotations occur. The determination of the thickness of the wall shear zone is of importance for the estimation of the shear resistance and forces transferred from the surrounding soil to the structure, i.e. eventually for the determination of the structure bearing capacity.

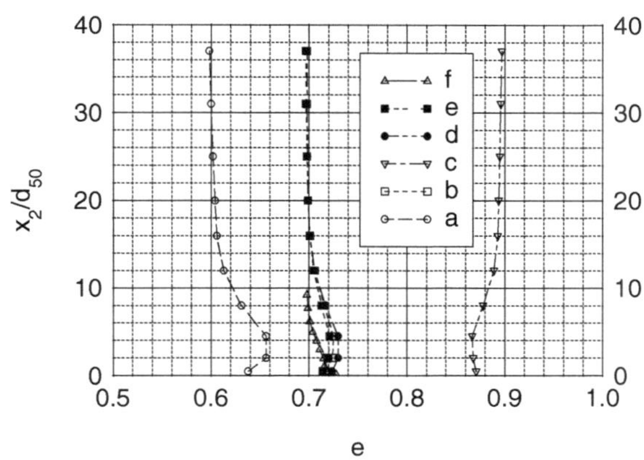

Fig. 5.30. Distribution of void ratio e across the normalised height x_2/d_{50} in the middle of the specimen after the initial shearing: a) e_o=0.60, d_{50}=0.5 mm, k=250 kPa/mm, b) e_o=0.70, d_{50}=0.5 mm, k=250 kPa/mm, c) e_o=0.90, d_{50}=0.5 mm, k=250 kPa/mm, d) e_o=0.70, d_{50}=0.5 mm, k=25 kPa/mm, e) e_o=0.70, d_{50}=0.5 mm, k=2500 kPa/mm, f) e_o=0.70, d_{50}=2.0 mm, k=250 kPa/mm

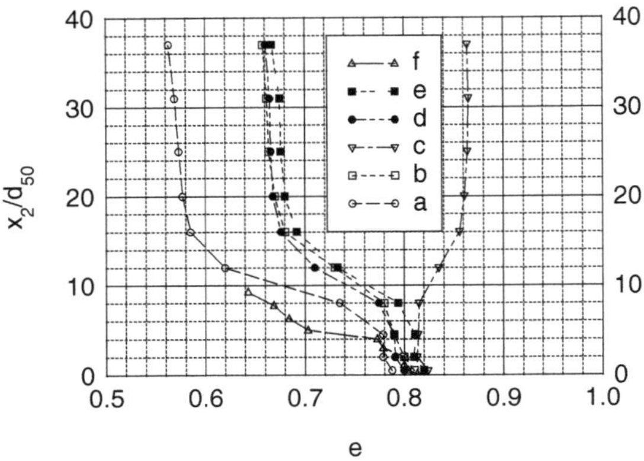

Fig. 5.31. Distribution of void ratio *e* across the normalised height x_2/d_{50} in the middle of the specimen after 5 shear cycles: a) e_o=0.60, d_{50}=0.5 mm, *k*=250 kPa/mm, b) e_o=0.70, d_{50}=0.5 mm, *k*=250 kPa/mm, c) e_o=0.90, d_{50}=0.5 mm, *k*=250 kPa/mm, d) e_o=0.70, d_{50}=0.5 mm, *k*=25 kPa/mm, e) e_o=0.70, d_{50}=0.5 mm, *k*=2500 kPa/mm, f) e_o=0.70, d_{50}=2.0 mm, *k*=250 kPa/mm

In classical FE analyses of soil-structures interfaces, Coulomb's friction law is usually used, which assumes a constant ratio between the shear wall stress and normal wall stress. However, such an assumption provides only an approximate description of the granular soil-structure interface and does not necessarily reflect the reality. First, the mobilised wall friction angle is not constant and can change significantly during shearing. An example is the pronounced softening and dilatancy behaviour for rough and very rough walls in contact with dense granular materials (Tejchman and Wu 1995). Second, the wall friction angle is not a state variable, since it depends on a number of factors, such as the boundary conditions, pressure level and initial stress state. Furthermore, the wall friction angle depends also on the specimen size and cannot be always directly transferred to other boundary value problems (Tejchman and Wu 1995, Tejchman and Bauer 2005). Moreover, classical finite element solutions cannot properly capture the formation of wall shear zones due to the lack of characteristic length of microstructure.

As we know, wall roughness plays an important role in characterizing interface behaviour. Traditionally, roughness is usually characterized qualitatively and left to experimental determination. The purpose of this FE study is to put the wall roughness of interfaces in numbers. This is furnished by specifying the wall boundary conditions based on a micro-polar constitutive model. The boundary conditions take into account the asperity of the wall surface and the rolling, sliding and shear strain of the granular material adjacent to the wall. In this way, the mobilized wall friction angle is obtained as the outcome of the calculation rather than as prescription. The soil-structure interface behaviour was investigated on a thin infinite layer of granular material between two rigid horizontal plates (walls) under plane strain conditions. The granular material was subjected to constant lateral pressure and free dilatancy between the walls is allowed.

The numerical calculations are performed on walls of different roughness. First, the material behavior in the neighborhood of very rough walls is investigated. Attention is paid to the ratio between the Cosserat rotation and slip and the ratio between between normal and shear stresses and horizontal couple stress close to a stationary and a moving boundary. Based on these initial studies, the boundary conditions along a stationary wall and a moving wall are investigated. While the stationary wall is representative for such problems like silos and retaining walls, the moving wall is relevant for problems involving piles and anchors. These boundary conditions take into account the different ratios between the Cosserat rotations and slip along the wall nodes to depict the different wall roughness. Due to the lack of experimental data with measurements of slip and rotation of granular materials along the wall, the numerical results cannot be verified quantitatively by laboratory experiments.

The interface between granular material and structure has been investigated using various testing devices and methods, e.g. direct shear apparatus ([8] Potyondy 1961, Sondermann 1983, Desai et al. 1985), improved direct shear device (Tejchman and Wu 1995, Boulon, 1988, Hassan, , 1995), torsional ring shear apparatus (Neuffer and Leibnitz 1964, Yoshimi and Kishida 1981, Huck and Saxena 1981, Löffelmann 1989), ring shear device (Brumund and Leonards 1973), simple shear apparatus(Kishida and Uesugi 1987, Uesugi 1988), plane strain apparatus (Tejchman and Wu 1995), Couette apparatus (Löffelmann 1989, Becker and Lippmann 1977), shear and wear tester (Haaker 1988), and in experiments with piles (Vesic 1973), anchors (Wernick 1978) and silos (Tejchman 1989). The experimental results show significant effect of wall roughness, grain size, grain distribution, pressure level, initial density, specimen size and velocity on the mobilized wall friction angle and wall shear zone thickness. The shear zone thickness is found to increase with increasing wall roughness (Uesugi 1988, Tejchman 1989), grain size (Tejchman 1989), pressure (Hassan 1995), velocity (Löffelmann 1989) and specimen size (Tejchman 1989), and to decrease with initial unit weight (Hassan 1995, Tejchman 1989). The wall friction angle increases with increasing wall roughness (Uesugi 1987, Tejchman 1989), grain size (Tejchman 1989), initial unit weight (Unterreiner et al. 1994, Tejchman 1989) and velocity (Löffelmann 1989) and decreases with pressure (Tejchman and Wu 1985, Hassan 1995) and specimen size (Löffelmann 1989). Moreover, large void fluctuations, grain mixing and grain segregation were observed in a wall shear zone (Löffelmann 1989). The magnitude of pure slip during shearing was measured by Uesugi (1987). In the experiments mentioned above, however, the ratio between grain rotations and slip was not measured. In addition to the experimental studies, numerical analyses have been carried out to investigate the interface behaviour in granular materials. The boundary conditions along interfaces with consideration of characteristic length of microstructure were investigated following different approaches, e.g. within micropolar elasto-plasticity (Tejchman 1989, Unterreiner 1994, Unterreiner et al. 1994), second-gradient elasto-plasticity (Vardoulakis et al. 1992) and micro-hypoplasticity (Tejchman 2001, Huang et al. 2003). However, the numerical results were obtained assuming fully developed shearing of the granular material along the wall. Obviously, this assumption is realistic only for very rough walls. For smooth and rough walls, however, further investigations are needed, where simultaneous slip and shear are expected.

An infinitely long granular strip with the height of h=10 mm between two rigid walls was considered (Section 4.2). The finite element calculations were performed on one element column with a width of b=10 cm, consisting of 20 quadrilateral horizontal elements composed of four diagonally crossed triangles (displacements and rotations along both sides of the column were constrained by the same amount, Section 4.2). The calculations are carried out with large deformation and curvature (Section 4.3.3).

As to the initial stress state, a K_o-stress state without polar quantities (σ_{22}=-1.0 kPa, $\sigma_{11}=\sigma_{33}$=-0.4 kPa, $\sigma_{12}=\sigma_{21}=m_1=m_2$=0) was assumed. Gravity was not considered. A vertical uniform pressure of p=200 kPa was applied to the specimen.

An initially dense sand specimen of e_0=0.60 was subject first to the vertical uniform pressure p and then to shearing with free dilatancy.

Shearing between two very rough walls

First, calculations of fully developed shearing of sand between two rigid and very rough walls under free dilatancy are performed (u_1=0, u_2=0) to study the ratios between Cosserat rotation and slip and the ratios between stresses and horizontal couple stress near the walls (Section 4.2). The results serve as a limit for interfaces with varying wall roughness.

Two different set of boundary conditions with respect to the Cosserat rotation are assumed along the bottom:

a) with fully constrained Cosserat rotation (ω^c=0, $m_2\neq0$),
b) with a unconstrained free Cosserat rotation ($\omega^c\neq0$, m_2=0).

In the first case, the boundary conditions along the top are: $u_1=n\Delta u$, ω^c=0, and $\sigma_{22}=p$ (the parameter n denotes the number of time steps and Δu is the constant displacement increment in one step). In the second case, the boundary conditions along the top are: $u_1=n\Delta u$, m_2=0, and $\sigma_{22}=p$. Note that the second case is not physically realistic for very rough walls, where no grain rotations are expected (Tejchman 1989, 1997). However, the FE results for this case will show the importance of the constrained Cosserat rotation along the wall and allow to determine the sign of the ratio between the Cosserat rotation and slip along the wall in the micro-polar boundary conditions.

Constrained Cosserat rotation along the bottom (ω^c=0)

Fig.5.32 shows the evolution of normalised stress components σ_{ij}/h_s at the mid-point of the strip and at the walls, the evolution of the normalised couple stress $m_2/(h_s d_{50})$ at the walls and the evolution of the mobilised wall friction angle $\varphi_w=arctan(\sigma_{12}/\sigma_{22})$ with the normalised horizontal displacement at the top u_1^t/h. A superimposed t is used to denote the variable at the top. The wall friction angle φ_w is calculated over the entire granular layer, since the stresses σ_{12} and σ_{22} are constant along the layer height and layer length.

Fig.5.33 presents the distribution of the normalised horizontal displacement u_1^t/h, Cosserat rotation ω^c, void ratio e, normalised stresses σ_{ij}/h_s and couple stress $m_2/(h_s d_{50})$ along the normalised height x_2/d_{50} in the residual state. The sign of the Cosserat rotation is positive, when it is counter clockwise as shown in Fig.3.28. The evolutions of the horizontal u_1 and vertical displacement u_2, Cosserat rotation ω^c and ratio $(\omega^c d_{50})/u_1$ in the neighborhood of walls (node at the mid-point of the quadrilateral element next to

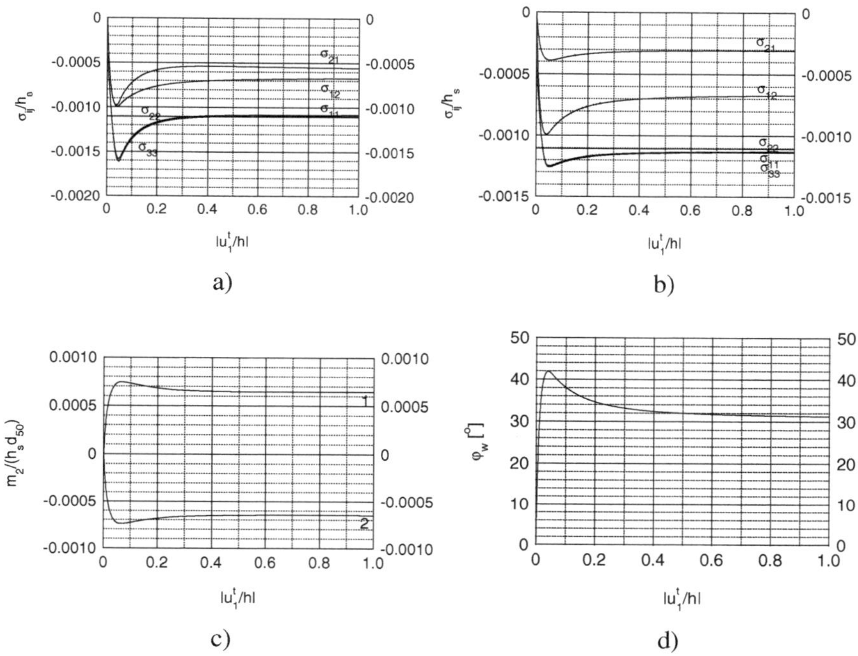

Fig. 5.32. Shearing of dense sand between two very rough boundaries(ω^f=0): a) evolution of normalised stresses σ_{ij}/h_s at the mid-point versus shear deformation u_1^t/h, b) evolution of normalised stresses σ_{ij}/h_s at the wall versus u_1^t/h, c) evolution of normalised wall couple stress $m_2/(h_s d_{50})$ versus u_1^t/h (1 – bottom wall, 2 – top wall), d) evolution of wall friction angle φ_w versus u_1^t/h

both horizontal boundaries) are given in Fig.5.34 The evolutions of the different stress quantities: $m_2/(\sigma_{12}d_{50})$, $m_2/(\sigma_{21}d_{50})$, $m_2/[0.5(\sigma_{12}+\sigma_{21})d_{50}]$ and $m_2/(\sigma_{22}d_{50})$ close to the boundaries (mean value of four triangular elements next to the wall) are shown in Fig.5.35.

All variables (stress, couple stress, void ratio and wall friction angle, displacements and rotations) tend to some asymptotic values (Fig.5.32). The Cosserat rotation, curvature, couple stress and non-symmetry of the stress tensor ($\sigma_{12}\neq\sigma_{21}$) are noticeable during shearing. In the middle of the layer, a shear zone is characterised by the appearance of the Cosserat rotation (Fig.5.33b) and strong increase of the void ratio (Fig.5.33c). As can be seen from the displacement profile in Fig.5.33a, there is a boundary layer adjacent to the walls, where the material assumes the wall displacement. Only at a distance of about $5\times d_{50}$ away from the wall, the displacement starts to increase steeply. The maximum normalized wall couple stress at the shear zone edges is about $m_2/(h_s d_{50})\pm0.0075$. The following sign convention applies: the couple stress is positive at the bottom and negative at the top (Fig.5.33c). The thickness of the shear zone, as ascertained from the Cosserat rotation, stress jump of σ_{21} and couple stress jump of m_2 at the shear zone edges, is about $10\times d_{50}$. The distribution of stresses σ_{11},

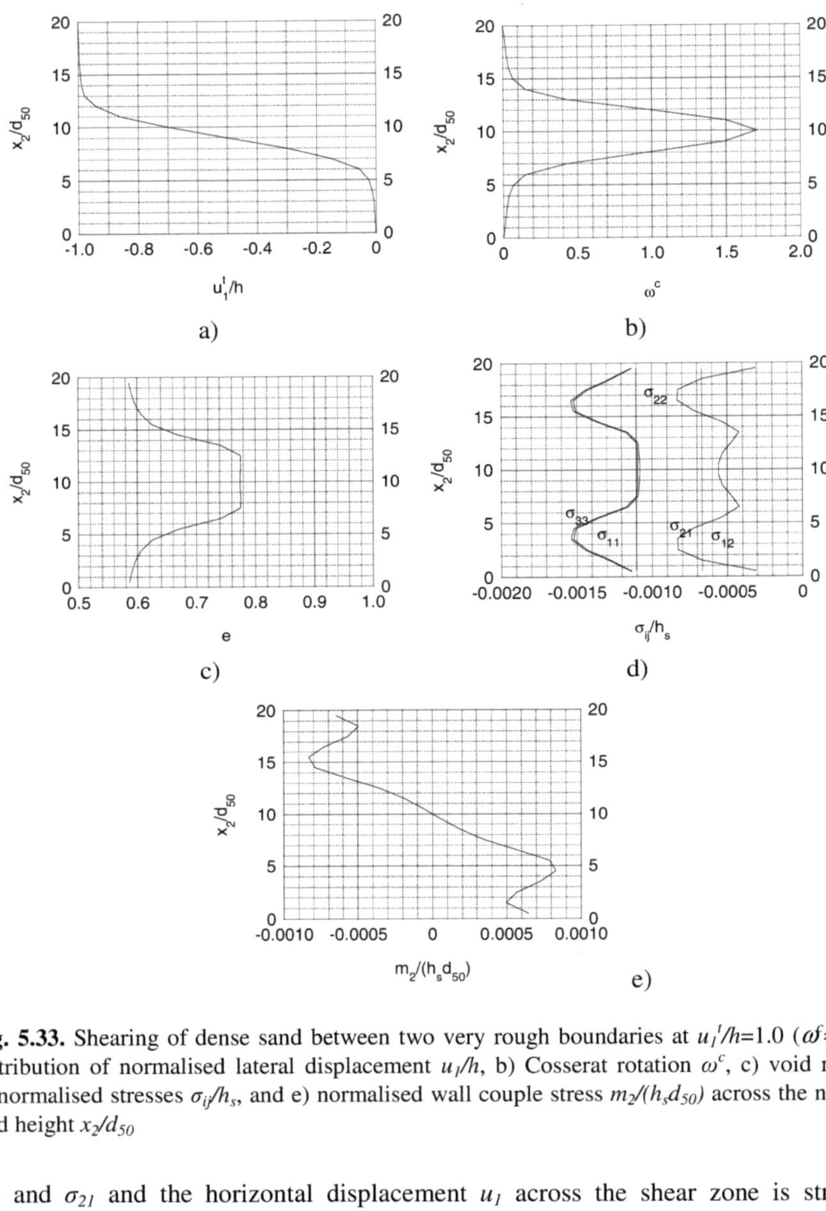

Fig. 5.33. Shearing of dense sand between two very rough boundaries at u_1^t/h=1.0 (ω^c=0): a) distribution of normalised lateral displacement u_1/h, b) Cosserat rotation ω^c, c) void ratio e, d) normalised stresses σ_{ij}/h_s, and e) normalised wall couple stress $m_2/(h_s d_{50})$ across the normalised height x_2/d_{50}

σ_{33} and σ_{21} and the horizontal displacement u_1 across the shear zone is strongly non-linear (Figs.5.33a and 5.33d). The stresses σ_{11}, σ_{33} and σ_{21} in the shear zone show parabolic distribution (Fig.5.33d). The stresses σ_{11} and σ_{33} have their minima and the stress σ_{21} their maxima in the middle of the shear zone. The stress ratios σ_{11}/σ_{22} and σ_{11}/σ_{33} in residual state approaches unity in the shear zone (Figs.5.32a and 5.33b). The mobilized wall friction angle φ_w is about 42° at peak and about 31.5° in residual state (Fig.5.33d). The void ratio in the middle of the shear zone e_c is about 0.78 in residual state (Fig.5.33c). Outside the shear zone, the void ratio lies in the range 0.59-0.64. In

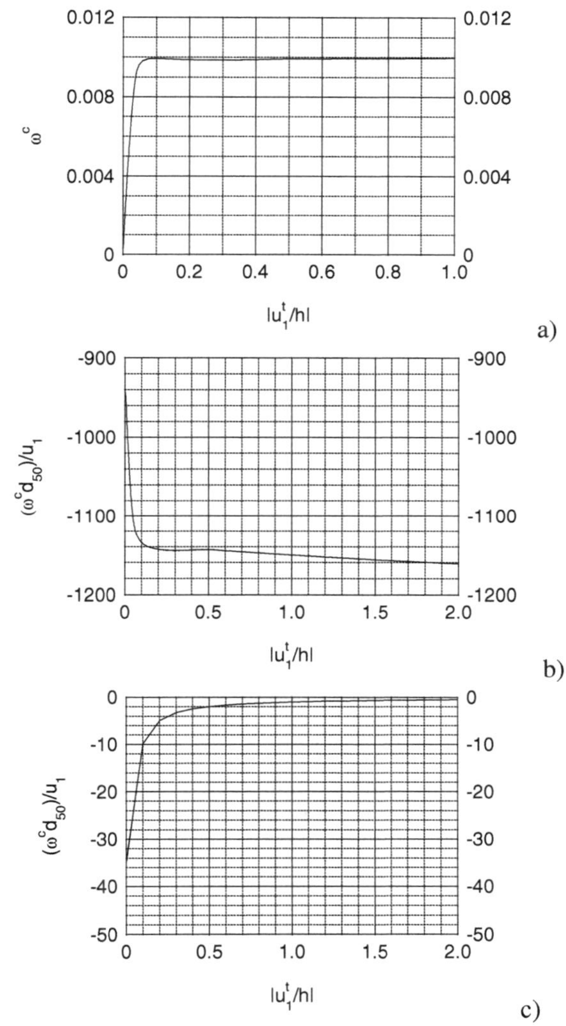

Fig. 5.34. Shearing of dense sand between two very rough boundaries (ω^c=0): a) evolution of Cosserat rotation ω^c near the walls, b) evolution of ratio $(\omega^c d_{50})/u_1$ near the bottom, c) evolution of ratio $(\omega^c d_{50})/u_1$ near the top versus u_1^t/h

the vicinity of the walls, the void ratio changes insignificantly and lies between 0.585 and 0.586.

The evolution of Cosserat rotations are shown in Fig.5.34. The normalised ratio between the Cosserat rotation and horizontal displacement $(\omega^c d_{50})/u_1$ close to the walls are nearly linear in the residual state. The normalized ratio is $(\omega^c d_{50})/u_1 \approx$-1160 at $u_1^t/h \approx$2.0 near the fixed bottom and $(\omega^c d_{50})/u_1 \approx$–0.50 at u_1^t/h=2.0 near the moving top (Fig.5.34c). This is due to small horizontal displacement near the fixed boundary and large horizontal displacement near the moving boundary. The normalized ratios between the couple stress m_2 and shear stresses σ_{ij} and the couple stress m_2 and normal stress σ_{22} near the walls remain virtually unchanged in residual state (Fig.5.35) with $m_2/(\sigma_{12}d_{50})\approx-1.0$, $m_2/(\sigma_{21}d_{50})\approx-2.1$, $m_2/[0.5(\sigma_{12}+\sigma_{21})d_{50}]\approx-1.3$ and $m_2/(\sigma_{22}d_{50})\approx0.6$.

Free Cosserat rotation along the bottom (m_2=0)

Fig.5.36 presents the evolution of the normalised stress components σ_{ij}/h_s at the bottom and mobilised wall friction angle $\varphi_w=arctan(\sigma_{12}/\sigma_{22})$ with the normalised horizontal displacement at the top u_1^t/h. In Fig.5.37, the distribution of the normalised horizontal displacement u_1/h, Cosserat rotation ω^c, normalised stresses σ_{ij}/h_s and couple stress $m_2/(h_s d_{50})$ along the normalised height x_2/d_{50} at the residual state is demonstrated.

A free Cosserat rotation along the bottom changes drastically the interface behaviour. In contrast to the boundary condition with ω^c=0 (Fig.5.33), the shear zone is not created in the mid-height but immediately above the bottom wall (Figs.5.37a-5.37e). The thickness of the shear zone is about $7\times d_{50}$ and slightly smaller than the thickness in Fig.5.33. Similar to the last section, the thickness of the shear zone is estimated based on the Cosserat rotation, shear stress and couple stress. The maximum Cosserat rotation is about twice as large as in Fig.5.33. This is due to the free rotation and the smaller shear zone thickness. The maximum void ratio in the wall shear zone is about 0.78 and is comparable with Fig.5.33. The evolution of the mobilized wall friction angle is also similar to Fig.5.32. The maximum wall friction angle is somewhat smaller by about 0.4-1.0° ($\varphi_w\approx41.6°$ at peak and $\varphi_w\approx30.6°$ at residual state).

The evolution of the normalized ratio $(\omega^c d_{50})/u_1$ close to both boundaries is also similar. However, the absolute value of this ratio is significantly smaller than that with the constrained Cosserat rotation, i.e. $(\omega^c d_{50})/u_1$=-0.40 (near the fixed bottom) and $(\omega^c d_{50})/u_1$=-0.002 (near the moving top).

The results of shearing of the granular material between two very rough boundaries show that the constraint of the Cosserat rotation strongly affects the interface behaviour and indicate the possibility to use novel micro-polar boundary conditions. These

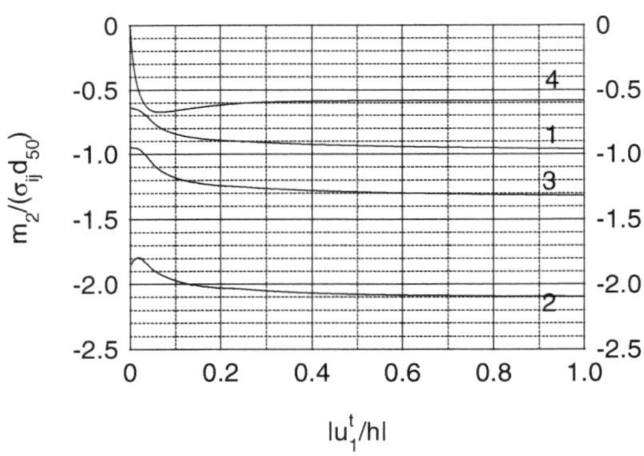

Fig. 5.35. Shearing of dense sand between two very rough boundaries (ω^c=0): evolution of ratios: $m_2/(\sigma_{12}d_{50})$ (1), $m_2/(\sigma_{21}d_{50})$ (2), $m_2/0.5[(\sigma_{12}+\sigma_{21})d_{50}]$ (3), $m_2/(\sigma_{22}d_{50})$ (4) near the wall versus u_1^t/h

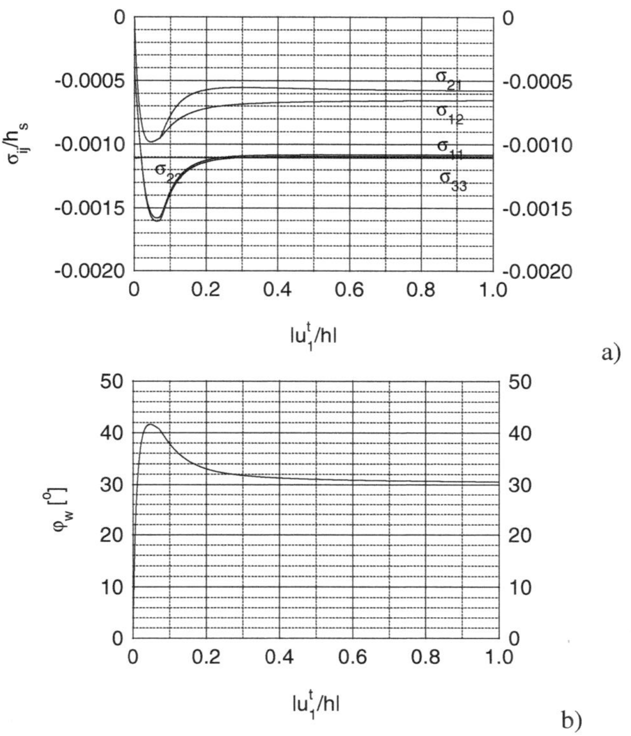

Fig. 5.36. Shearing of dense sand between two very rough boundaries(m_2=0 along the top): a) evolution of normalised stresses σ_{ij}/h_s at the wall versus u_1^t/h, b) evolution of wall friction angle φ_w versus u_1^t/h

new boundary conditions make use of sliding and rolling expressed either by the ratio between the Cosserat rotation and slip or by the ratio between couple stress and stresses.

Shearing along the wall with different roughness

Two sets of boundary conditions are considered by assuming different ratio of $(\omega^c d_{50})/u_1$ close to the fixed bottom and close to the horizontally moving top. The first set of boundary conditions is prescribed to the fixed bottom. This kind of boundary conditions can be found in silo flow and at retaining walls. The second set is prescribed to the moving wall. Such boundary conditions are representative for pile foundations and ground anchors. Since the ratio $(\omega^c d_{50})/u_1$ is found to be linear in the neighbourhood of very rough walls at residual state (Figs.5.34b and 5.35c), a kinematic boundary condition in the form of a constant ratio between the Cosserat rotation and displacement is proposed along the wall nodes to describe different interface roughness.

$$u_2 = 0\,, \qquad (\omega^c d_{50})/u_1 = \pm b_1\,, \qquad (\sigma_{12} d_{50})/m_2 = \mp b_1\,. \qquad (5.1)$$

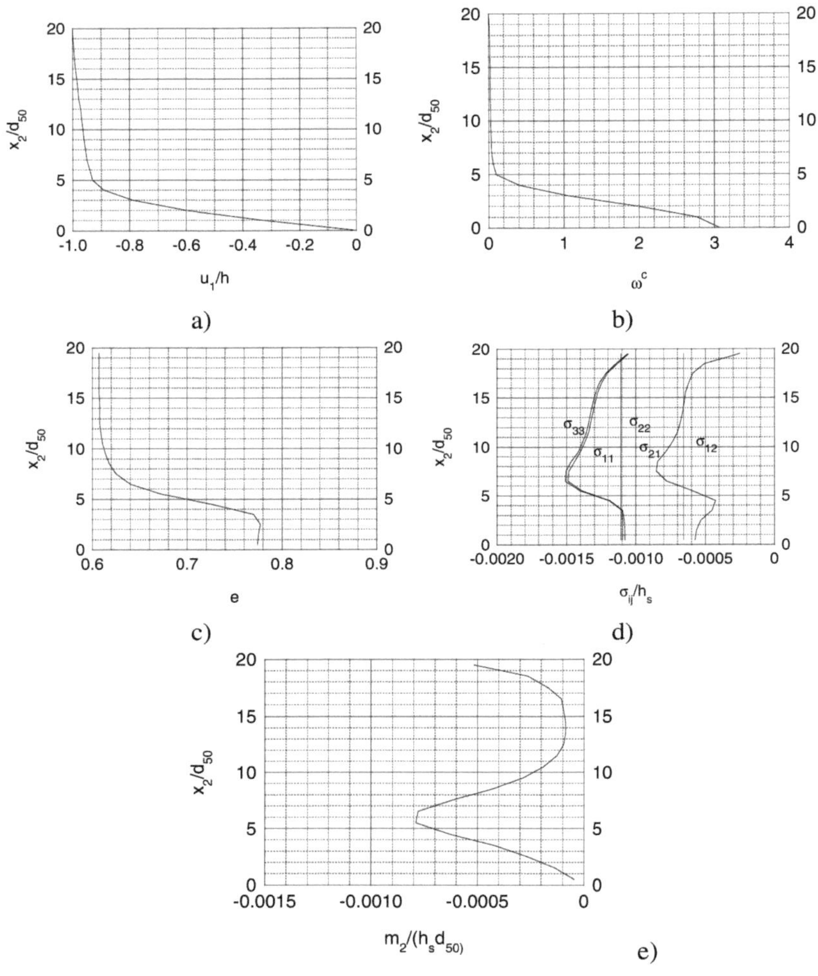

Fig. 5.37. Shearing of dense sand between two very rough boundaries at u_1^t/h=1.0 (m_2=0 along the top): a) distribution of normalised lateral displacement u_1/h, b) Cosserat rotation ω^c, c) void ratio e, d) normalised stresses σ_{ij}/h_s, and e) normalised wall couple stress $m_2/(h_s d_{50})$ across the normalised height x_2/d_{50}

Stationary fixed bottom wall

For problems where the wall remains stationary while the granular material moves relatively to the wall, the following boundary conditions were proposed along the rigid fixed bottom (Fig.5.38):

The sign convention for the Cosserat rotation and couple stress conforms to Figs.3.26 and 3.28. The first condition in Eq.5.1 denotes the lack of the vertical displacement of the rigid horizontal wall. The second condition in Eq.5.1 specifies a ratio between the Cosserat rotation and slip displacement. In turn, the third condition in

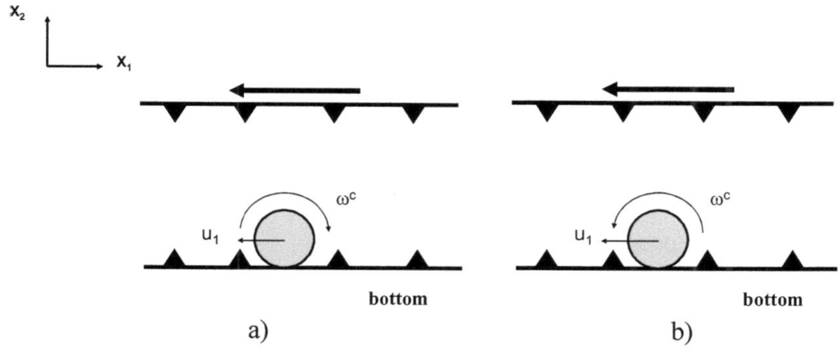

Fig. 5.38. Micro-polar boundary conditions along the bottom: a) clockwise Cosserat rotation, b) counter clockwise Cosserat rotation

Eq.5.1 determines a ratio between the horizontal shear stress and couple stress m_2. The parameter b_1 can be related to the ratio between the surface asperity of the wall r_w and the mean grain diameter of sand d_{50} (Eq.5.2). For a very rough wall, the asperity is assumed to be larger than the mean grain diameter, namely $r_w \geq d_{50}$. The sand grains are trapped in the asperities. In this case, both the Cosserat rotation ω^c and the displacement u_1 approach zero. For a smooth wall, the asperity is much smaller than the mean grain diameter, namely $r_w << d_{50}$. The sand grains slide along the smooth wall surface with decreasing shearing and rotation, i.e. $u_1 \neq 0$ and $\omega^c \to 0$.

Clockwise Cosserat rotation $(\omega^c \times d_{50})/u_1 = b_1, (\sigma_{12} d_{50})/m_2 = -b_1$

Figs.5.39 and 5.40 show the results with the boundary conditions in Eq.5.1 (with d_{50}=0.5 mm). The parameter b_1 is assumed to be positive and varies between 0.005 and 2.0. The following results are presented in Fig.5.39: the evolutions of the normalised stress components σ_{ij}/h_s, the normalised wall couple stress $m_2/(h_s d_{50})$ at the bottom and the evolution of the mobilised wall friction angle with the normalised horizontal displacement at the top u_1^t/h. The numerical results in the residual state in Fig.5.40 contain the distributions of the normalised horizontal displacement u_1/h, Cosserat rotation ω^c, void ratio, normalised stresses σ_{ij}/h_s and couple stress $m_2/(h_s d_{50})$ along the normalised height x_2/d_{50}. The results in Figs.5.39 and 5.40 can be compared to the results with very rough walls (ω^c=0).

The evolution of stresses, couple stress and wall friction angle and the distribution of stresses, horizontal displacement, Cosserat rotation and void ratio for the case b_1=1 are not much different from the results in Figs.5.32 and 5.33. Further numerical calculations show that the parameter b_1 should be equal to 0.8 in order to obtain identical results; the parameter b_1 depends on a_c (Eq.3.86).

The maximum and residual wall friction angle are similar in the range of b_1=0.5-2.0. For b_1<0.5, however, both friction angles decrease almost linearly with decreasing b_1.

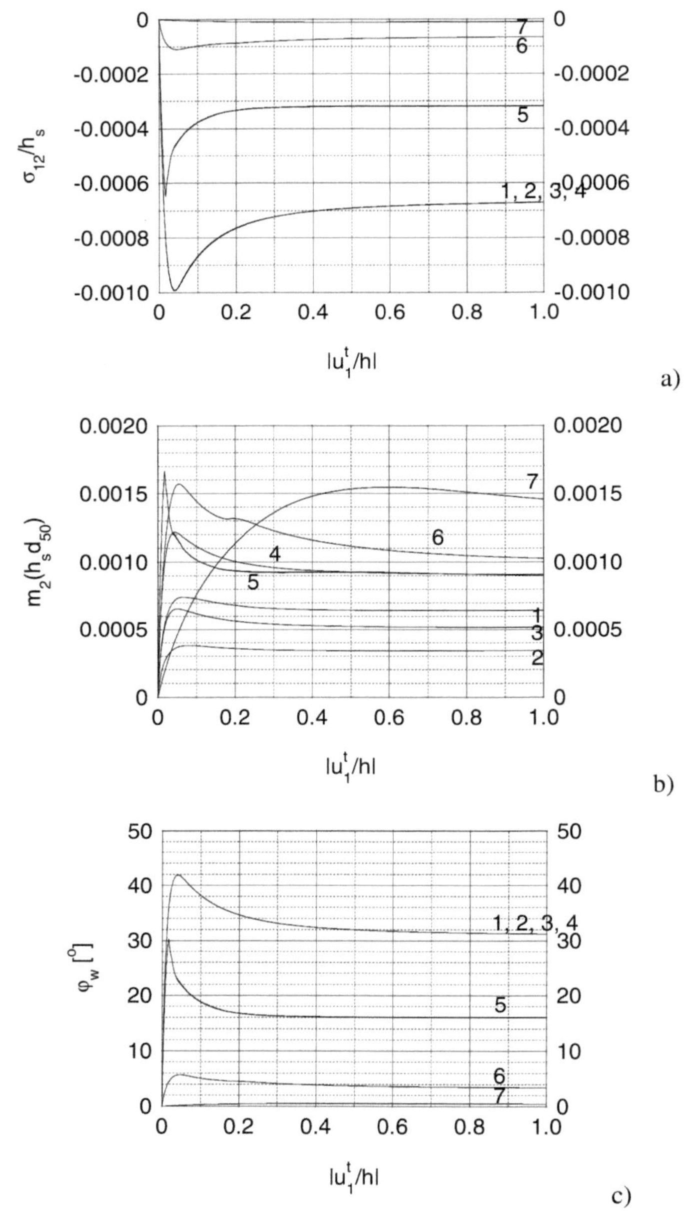

Fig. 5.39. Shearing of dense sand between two boundaries $((\omega^f d_{50})/u_1=b_1)$: a) evolution of normalised stress σ_{12}/h_s at the wall versus u_1^t/h, b) evolution of normalised couple stress $m_2/(h_s d_{50})$ at the bottom versus u_1^t/h, c) evolution of wall friction angle φ_w versus u_1^t/h: 1. very rough wall $(\omega^f=0)$, 2) $b_1=2$, 3) $b_1=1$, 4) $b_1=0.5$, 5) $b_1=0.25$, 6) $b_1=0.05$, 7) $b_1=0.005$

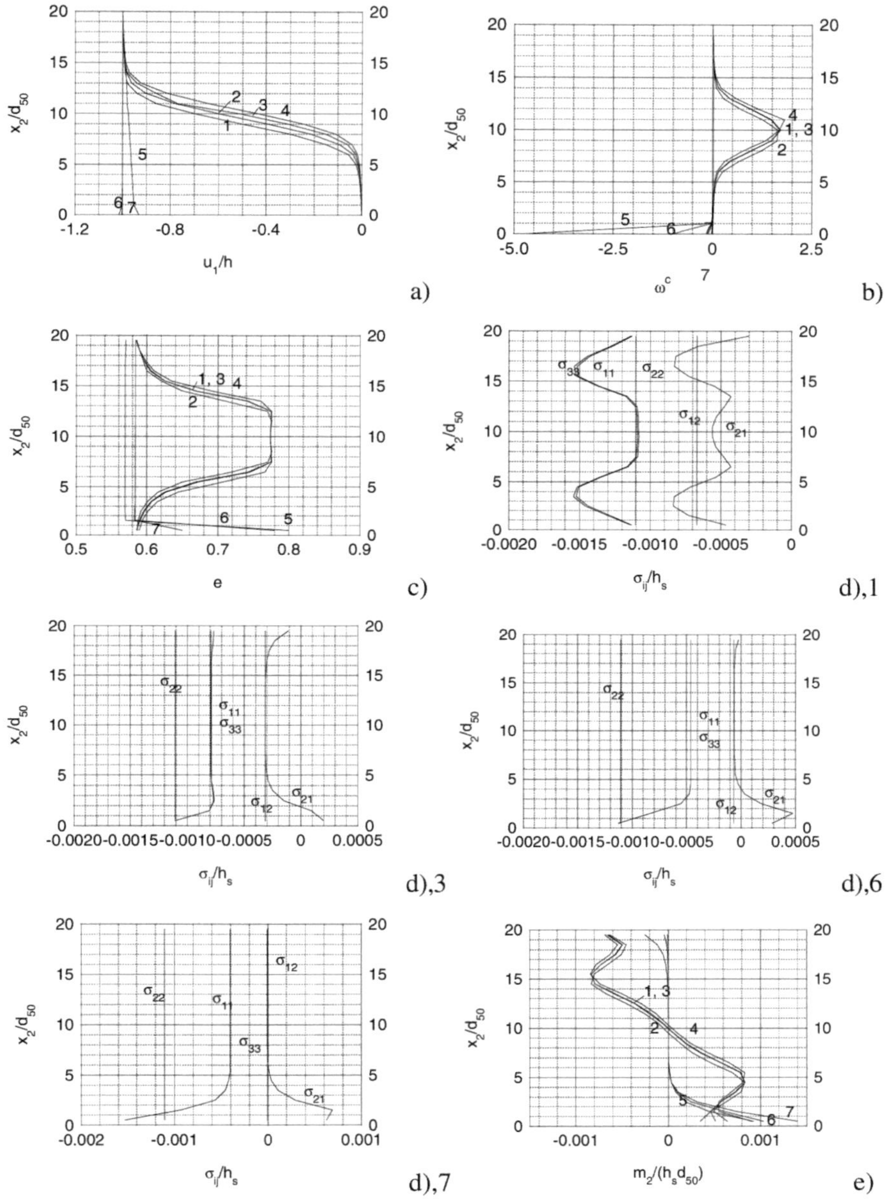

Fig. 5.40. Shearing of dense sand between two very rough boundaries at $u_1^t/h=1.0$ $((\omega^f d_{50})/u_1=b_1)$: a) distribution of normalised lateral displacement u_1/h, b) Cosserat rotation ω^c, c) void ratio e, d) normalised stresses σ_{ij}/h_s, and e) normalised wall couple stress $m_2/(h_s d_{50})$ across the normalised height x_2/d_{50}: 1. very rough wall ($\omega^f=0$), 2) $b_1=2$, 3) $b_1=1$, 4) $b_1=0.5$, 5) $b_1=0.25$, 6) $b_1=0.05$, 7) $b_1=0.005$

The residual wall friction angle is about 31.5° (b_1=0.5-2.0), 16° (b_1=0.25), 5° (b_1=0.05) and 0.5° (b_1=0.005), respectively (Fig.5.39c). It is always smaller than the peak friction angle. In the case of a very rough and rough bottom (b_1=0.25-2.0), this difference is about 10°-14°. In turn, for a smooth wall (b_1=0.005-0.05), the difference is 0.5°-2°.

The evolution of the wall friction angle φ_w with the normalized horizontal displacement u_1^t/h is qualitatively in good agreement with experimental observation (Tejchman and Wu 1995). For the parameter b_1=0.5-2.0, the shear zone occurs at mid-height as in very rough walls with ω^c=0. The thickness of the shear zone is about 10×d_{50} (Fig.5.40), which can be compared with the thickness in Fig.5.33. For b_1<0.5, the shear zone forms along the bottom with an extremely small thickness of about d_{50}. The horizontal displacement along the bottom increases and the Cosserat rotation decreases for $b_1 \leq 0.25$.

Counter clockwise Cosserat rotation $(\omega^c d_{50})/u_1 = -b_1$, $(\sigma_{12} d_{50})/m_2 = b_1$

The numerical results with negative parameter b_1 in the boundary condition in Eq.5.1 are given in Figs.5.41 and 5.42. The parameter b_1 varies from -0.005 down to -2.0. Again, the numerical results can be compared to the results with very rough walls and free Cosserat rotation $\omega^c \neq 0$ (Figs.5.36 and 5.37).

With increasing parameter b_1, the results approach those with a very rough bottom and free Cosserat rotation at m_2=0 (Figs.5.41 and 5.42). The thickness of the wall shear zone at the bottom decreases with increasing b_1, from about 5×d_{50} (b_1=-2.0), 3×d_{50} (b_1=-1), 2×d_{50} (b_1=-0.25 - -0.5) down to 1×d_{50} (b_1 = -0.05 to -0.005). The maximum and residual wall friction angle decrease with decreasing $|b_1|$. The residual wall friction angle is about 29.5° (b_1=-2.0), 26.5° (b_1=-1), 20.5° (b_1=-0.5), 13° (b_1=-0.25), 3.5° (b_1=-0.05) and 0.5° (b_1=-0.005).

The numerical results in Figs.5.41-5.42 show that the proposed micro-polar boundary condition in Eq.5.1 is qualitatively consistent. The wall friction angle and the

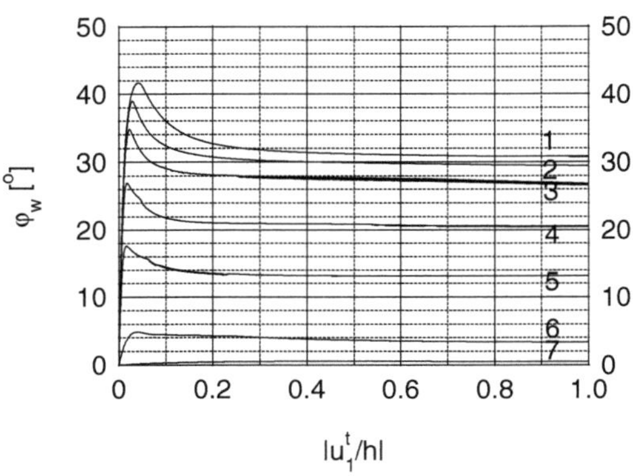

Fig. 5.41. Shearing of dense sand between two boundaries (($\omega^c d_{50}$)/u_1=-b_1): evolution of wall friction angle φ_w versus u_1^t/h: 1. very rough walls (m_2=0), 2) b_1=-2, 3) b_1=-1, 4) b_1=-0.5, 5) b_1=-0.25, 6) b_1=-0.05, 7) b_1=-0.005

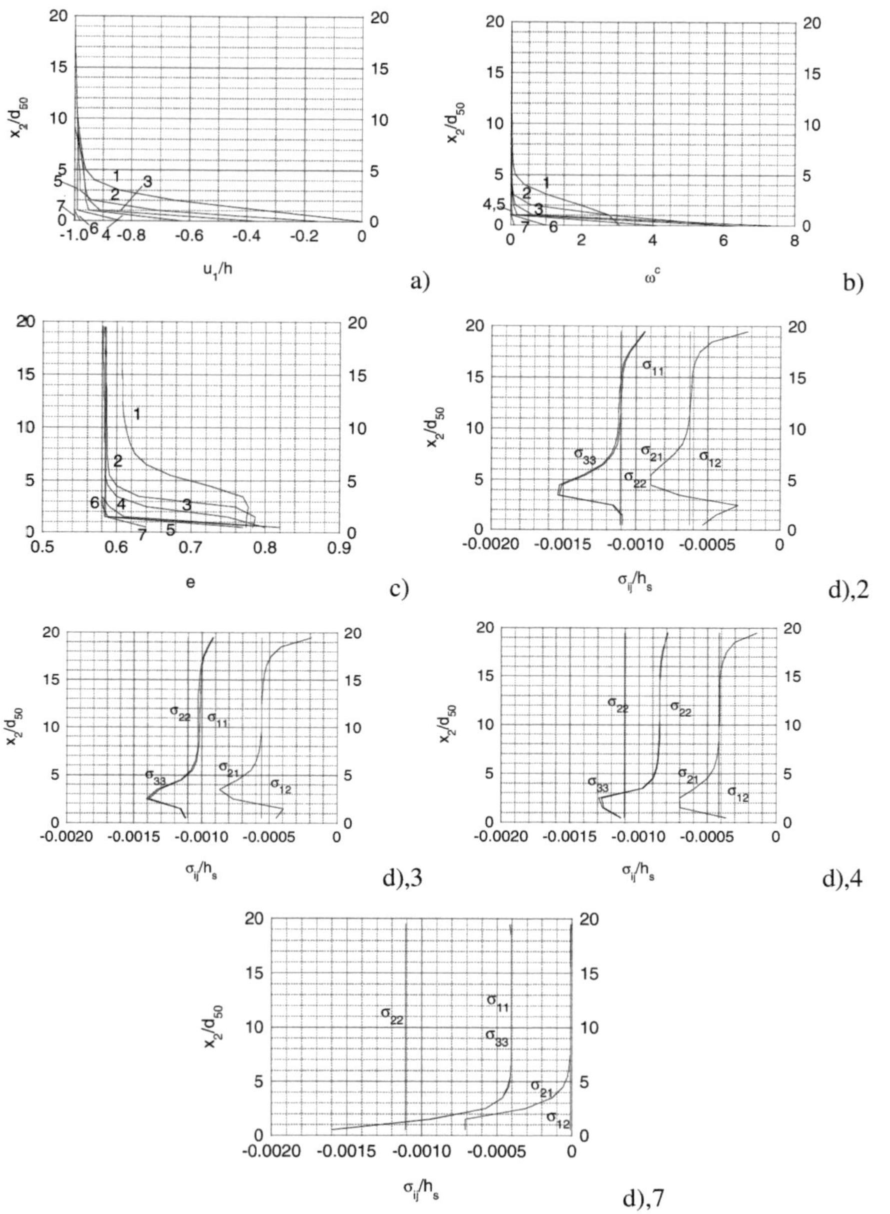

Fig. 5.42. Shearing of dense sand between two very rough boundaries at u_1^t/h=1.0 (($\omega^c d_{50}$)/u_1=-b_1): a) distribution of normalised lateral displacement u_1/h, b) Cosserat rotation ω^c, c) void ratio e and d) normalised stresses σ_{ij}/h_s across the normalised height x_2/d_{50}: 1. very rough walls (m_2=0), 2) b_1=-2, 3) b_1=-1, 4) b_1=-0.5, 5) b_1=-0.25, 6) b_1=-0.05, 7) b_1=-0.005

thickness of the wall shear zone decrease with decreasing $|b_1|$. Since the rotation of grains has not been experimentally observed during shearing along a very rough wall (Tejchman and Wu 1995, Tejchman 1997), the boundary condition $(\omega^c d_{50})/u_1=+b_1$ seems suitable for describing the roughness of stationary walls.

To gain perspective, the parameter b_1 can be further related to the wall roughness in the following way:

$$b_1 = c_1(r_w / d_{50}). \tag{5.2}$$

The ratio r_w/d_{50} lies in the range

$$0 < r_w / d_{50} \leq 1 . \tag{5.3}$$

In Eq.5.2, c_1 is a material parameter and r_w is a parameter describing the wall asperity, which denotes the mean distance between the highest point and the lowest point along the wall at the distance of a few grains ($3\times d_{50}$) (Uesugi 1987). According to this definition, the wall roughness is depicted, for the sake of simplicity, by only one constant parameter r_w. When the ratio r_w/d_{50} is equal to 1, the wall is said to be very rough (ω^c=0), and at $r_w/d_{50}\cong 0$, the wall is ideally smooth, i.e. slip occurs without shearing.

The parameter c_1 can be calibrated with wall friction experiments. Based on the numerical results, the parameter c_1=0.5 can be assumed.

Moving top wall

For problems where the wall moves relative to the granular material, the following boundary conditions are proposed along the horizontally moving top boundary:

$$u_1 = n\Delta u , \qquad \sigma_{22} = p , \qquad (\omega^c \times d_{50}) / u_1 = \pm b_2 , \tag{5.4}$$

where the parameter b_2 is again a constant related to the wall roughness (Eq.5.5). The calculations are carried out only with negative b_2 (counter clockwise Cosserat rotation) (Fig.5.43). The assumption of the positive sign for the parameter b_2 is not realistic since in this case the Cosserat rotation acts in the displacement direction.

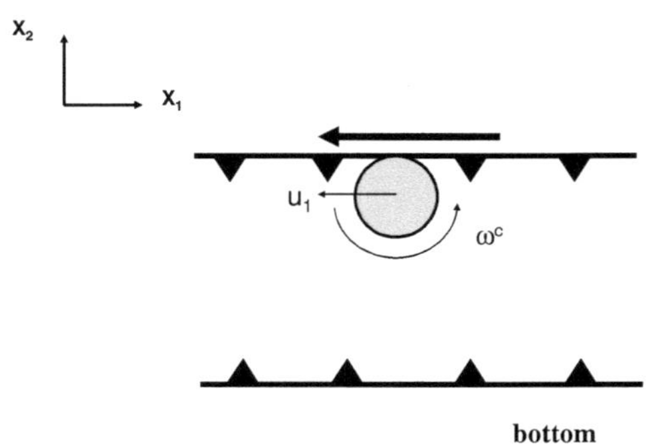

Fig. 5.43. Micro-polar boundary conditions along the bottom with a counter clockwise Cosserat rotation

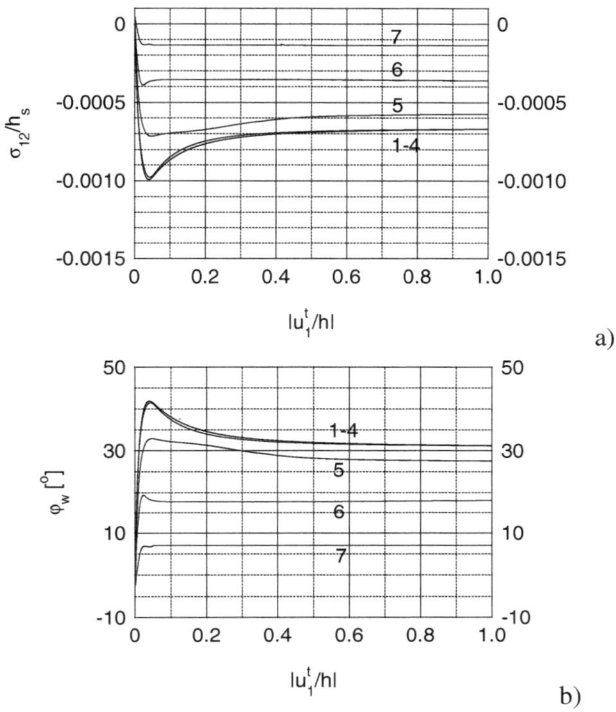

Fig. 5.44. Shearing of dense sand between two boundaries ($(\omega^f d_{50})/u_1$=-b_2): a) evolution of normalised stress σ_{12}/h_s at the wall versus u_1^t/h, b) evolution of wall friction angle φ_w versus u_1^t/h: 1. very rough walls (ω^f=0), 2) b_1=-0.000001, 3) b_1=-0.0005, 4) b_2=-0.05, 5) b_2=-0.5, 6) b_2=-1, 7) b_2=-2

Figs.5.44 and 5.45 show the results with the condition $(\omega^f d_{50})/u_1=-b_2$ along the top for the parameter b_2 changing from -0.000001 down to -2.0. The following results are presented in Fig.5.44: the evolution of the normalised stress components σ_{ij}/h_s, the evolution of the normalised wall couple stress $m_2/(h_s d_{50})$ at the bottom and the evolution of the mobilised wall friction angle with the normalised horizontal displacement at the top u_1^t/h.

In Fig.5.45, the following results in residual state are shown: the distribution of the normalised horizontal displacement u_1/h, Cosserat rotation ω^c, void ratio, normalised stresses σ_{ij}/h_s and couple stress $m_2/(h_s d_{50})$ along the normalised height x_2/d_{50}. These results can be compared to the results with very rough walls with ω^f=0 (Figs.5.32 and 5.33).

The evolution of stresses, couple stress and wall friction angle and distribution of stresses, horizontal displacement, Cosserat rotation and void ratio with b_2=-0.000001 are almost identical with the results for the conditions $u_1=n\Delta u$ and ω^f=0 (Figs.5.32 and 5.33). The shear zone is formed at mid-height for b_2=-0.000001 down to -0.0005. For b_2>-0.0005 the shear zone is formed along the top boundary (Fig.5.45). The

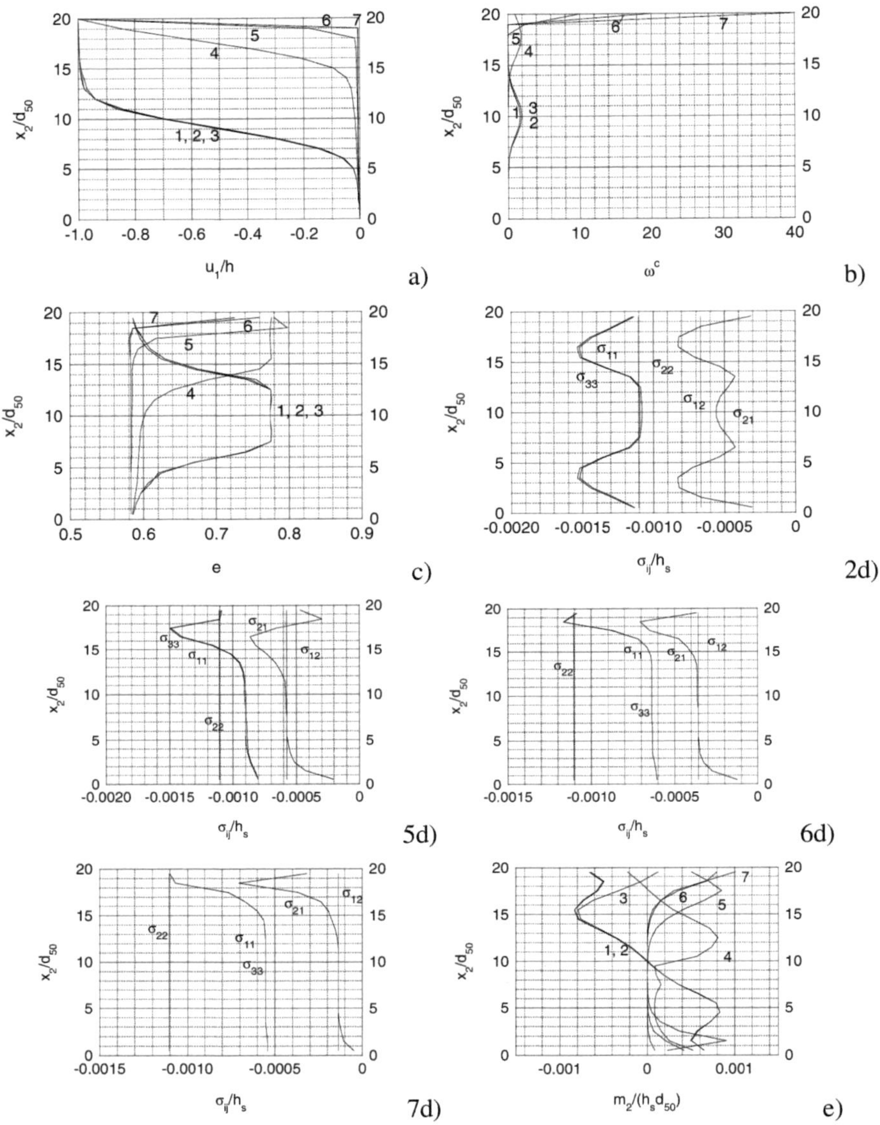

Fig. 5.45. Shearing of dense sand between two very rough boundaries at u_1^t/h=1.0 $((\omega^c d_{50})/u_1=-b_2)$: a) distribution of normalised lateral displacement u_1/h, b) Cosserat rotation ω^c, c) void ratio e, d) normalised stresses σ_{ij}/h_s, and e) normalised wall couple stress $m_2/(h_s d_{50})$ across the normalised height x_2/d_{50} : 1. very rough walls (ω^c=0), 2) b_1=-0.000001, 3) b_1=-0.0005, 4) b_2=-0.05, 5) b_2=-0.5, 6) b_2=-1, 7) b_2=-2

thickness of the wall shear zone at the bottom decreases with decreasing b_2 from about 10×d_{50} (b_2=-0.000001 to -0.0005), 7×d_{50} (b_2=-0.05), 2×d_{50} (b_2 =-0.5) to 1×d_{50} (b_1=-1 to-2).

The maximum and residual wall friction angle decrease almost linearly with decreasing b_2. The residual wall friction angle is about 31° (b_2=-0.000001 to -0.05), 27° (b_2=-0.5), 18° (b_2=-1) and 7° (b_2= -2), see Fig.5.44. For $b_2 \leq -2.5$, the residual wall friction angle is less than 1°.

Some ratios between stress and couple stress are given in Tab.5.1. When using the boundary condition $(\omega^c d_{50})/u_1=-b_2$, a relationship between the couple stress-shear stress ratio and parameter b_2 was not found.

In analogy to the boundary condition in Eq.5.2, the parameter b_2 can be related to the wall roughness as follows:

$$-b_2 = c_2 (r_w / d_{50} - 1) . \tag{5.5}$$

The ratio r_w/d_{50} lies again in the range

$$0 < r_w / d_{50} \leq 1 . \tag{5.6}$$

In Eq.5.5, c_2 is a material parameter, The parameter c_2 can be calibrated with wall friction experiments (e.g. c_2==2.5).

The following conclusions can be drawn:

- The constraint of the Cosserat rotation has a pronounced effect on the behaviour of the interface, in particular on the shear zone formation.
- When using micro-polar boundary conditions, the mobilised wall friction angle is obtained as natural outcome rather than prescribed. The additional boundary conditions enable us to account for the effect of the wall roughness on the mobilised wall friction angle and shear zone thickness during wall shearing. The residual wall friction angle changes nearly linearly with r_w/d_{50}.
- The interface behaviour along a stationary wall can be described by a kinematic condition expressed by the ratio between the Cosserat rotation and slip and a static condition expressed by the ratio between the shear stress and couple stress. The ratio can be related to the wall roughness.
- The interface behaviour along a moving wall can be described by a kinematic condition expressed by the ratio between the Cosserat rotation and slip. The ratio can be again related to the wall roughness.

Table 5.1. Different ratios $(\sigma_{ij} d_{50})/m_2$ near the top at residual state for the different wall roughness ($(\omega^c d_{50})/u_1=-b_2$)

B_2	$(\sigma_{12}d_{50})/m_2$	$(\sigma_{21}d_{50})/m_2$	$0.5[(\sigma_{12}+\sigma_{21})]d_{50}/m_2$	$(\sigma_{22}d_{50})/m_2$
-2	-0.14	-0.3	-0.2	-1.1
-1	-0.45	-0.5	-0.5	-1.4
-0.5	-1.1	-0.9	-1	-2
-0.05	3.3	2.5	2.5	5
-0.0005	-2.5	-0.4	-3.3	-5
-0.000001	1	-0.5	-0.7	1.7

5.4 Size Effects

One of the salient characteristics of the behaviour of granular and brittle materials is a size effect phenomenon, i.e. experimental findings vary with the size of the specimen. In general, the shear resistance in granular material (Wernick 1978, Tatsuoka et al. 1997, Tejchman 2004) and tensile strength in brittle ones (Bazant and Chen 1997, Bazant and Planas 1998, van Vliet 2000, Chen et al. 2001, Le Bellego et al. 2003), increase with decreasing specimen size during many experiments including strain localization. The specimen ductility (ratio between the energy consumed during a shearing or fracture process after and before the peak) also grows. Thus, the results from laboratory tests which are scaled versions of the actual structures cannot be directly transferred to them. Two main size effects can be defined: deterministic and statistical. The first one is caused by strain localization which cannot be appropriately scaled in laboratory tests. Thus, the specimen strength increases with increasing ratio l_c/L (l_c – characteristic length of microstructure influencing both the thickness and spacing of strain localization, L – specimen size). This feature is strongly influenced by the pressure level. The statistical effect is due to the presence of the randomness of the local material strength caused by number of weak spots whose amount usually grows with increasing specimen size. Thus, the specimen strength diminishes with increasing specimen size. In dynamic problems, this effect is also called a stochastic one. Up to now, the size effects are still not taken into account in the specifications of most of design codes for engineering structures.

For quasi-brittle and brittle materials, there exist only few reliable approaches to the size effect phenomenon. For example, two deterministic size effect laws by Bazant (Bazant and Chen 1997) allow us to take into account a size difference by determining the tensile strength for pre-notched structures and structures without an initial crack. The material strength is bound for small sizes by the plasticity limit whereas for large sizes the material follows linear elastic fracture mechanics. In the case of a statistical size effect, the most known is the Weibull or the weakest link theory (Weibull 1951) based on a distribution of flaws in materials. It postulates that a structure is as strong as its weakest component. When its strength is exceeded, the structure fails since the stress redistribution is not considered. This model is not able to account for a spatial correlation between local material properties. Another approach to size effect was proposed by Carpintieri et al. (1994) which was based on the multifractality of a fracture surface which increased with spreading disorder of the material in large structures. In this approach, the material strength is bound for small and large sizes by the plasticity limit. According to Bazant and Yavari (2005), the cause of size effect is energetic-statistical not fractal. The numerical calculations of a size effect in concrete specimens under tension with a microplane material model show (Bazant and Pang 2006), that a statistical strength (obtained from random sampling) can be larger than a deterministic one in small specimens in contrast to large specimens which rather obey the weakest link model. The difference between a deterministic material strength and a mean statistical strength grows with increasing size. The structural strength exhibits a gradual transition from Gaussian distribution to Weibull distribution at increasing size (Vorechovsky and Matesova 2006).

In the case of granular materials, no reliable size effect laws were proposed due to the fact that a determination of the distribution of material properties and a performance of laboratory tests are more complex than in brittle materials. In addition, the effect of the pressure level is more pronounced. Deterministic size effects have been studied in granular materials intensively by numerous researchers using a FE method based on enhanced continua including a characteristic length of micro-structure (e.g. Tejchman 2004) and a strong discontinuity approach (Regueiro and Borja 2001). In turn, in the case of statistical size effects, non-linear calculations are only few (Gutierrez and de Borst 1998, Fenton and Griffiths 2002, Niemunis et al 2005).

The intention of the numerical simulations was to investigate a deterministic and statistical effect in cohesionless granular materials with shear localization under quasi-static conditions. In the first step, these effects were investigated during a shearing of an infinite granular layer between two very rough boundaries under plane strain conditions, free dilatancy (without additional constraint) and constant uniform pressure (Tejchman and Górski 2007, 2008a). In the second case, the FE analyses were carried out during plane strain compression test (Tejchman and Górski 2008b, 2008c). Various properties of granular bodies may be considered as randomly distributed. In the present work, for the sake of simplicity, only fluctuations of void ratio are of primary interest as proposed by Gudehus and Nübel (2004). Due to the lack of experimental data, the FE results could not be compared with laboratory tests.

The insight into physical mechanisms of the size effects is of a major importance for civil engineers to extrapolate experimental findings at laboratory scale to results which can be used in real field situations. Since large structures are far beyond the range of failure testing, their design must rely on a realistic extrapolation of testing results with smaller specimens.

Stochastic calculations of discrete random fields

Continuous, second-order, real-valued, vector random field $\mathbf{X}(\mathbf{r},\omega)$ is specified by the expected (mean) value function

$$\bar{\mathbf{X}}(\mathbf{r}) = E(\mathbf{X}(\mathbf{r},\omega)) \tag{5.7}$$

and the correlation tensor function

$$\mathbf{K}_x(\mathbf{r}_1,\mathbf{r}_2) = E\left((\mathbf{X}(\mathbf{r}_1,\omega) - \bar{\mathbf{X}}(\mathbf{r}_1)) \otimes (\mathbf{X}(\mathbf{r}_2,\omega) - \bar{\mathbf{X}}(\mathbf{r}_2))\right), \tag{5.8}$$

where $E(\cdot)$ is the expectation operator, $\omega \in \Omega$ denotes an elementary event, the sign $\otimes$ denotes the tensor product and $\mathbf{r},\mathbf{r}_1,\mathbf{r}_2 \in \Re^3$. For the homogeneous random fields, $\bar{\mathbf{X}}(\mathbf{r}) = const$ and $\mathbf{K}_x(\mathbf{r}_1,\mathbf{r}_2) = \mathbf{K}_x(\mathbf{r}_1 - \mathbf{r}_2)$. A discrete, second-order, real-valued random field $\mathbf{X}(\omega)$ is defined by the expected value vector

$$\bar{\mathbf{X}} = E(\mathbf{X}(\omega)) \tag{5.9}$$

and the covariance matrix

$$\mathbf{K} = E\left((\mathbf{X}(\omega) - \bar{\mathbf{X}})(\mathbf{X}(\omega) - \bar{\mathbf{X}})^T\right). \tag{5.10}$$

Probabilistic foundations of a generation of arbitrary vector random variables have been formulated in the rejection theorem by Devroye (1986). Let $\mathbf{X} = (X_1, X_2, ..., X_m)$ be a random vector with density f on $\Re^m$ and let U be an independent uniform [0, 1] random variable. Then $(\mathbf{X}, cUf(\mathbf{X}))$ is uniformly distributed on $A = \{(\mathbf{x}, u) : \mathbf{x} \in \Re^m,\ 0 \le u \le cf(\mathbf{x})\}$, where $c > 0$ is an arbitrary constant. Vice versa, if $(\mathbf{X}, U)$ is a random vector in $\Re^{m+1}$ uniformly distributed on A, then $\mathbf{X}$ has a density f on $\Re^m$.

To apply the theorem to the random fields generation, the probability density function $f(\mathbf{X})$ was defined on a compact domain in $\Re^m$ and obeyed the condition $f(\mathbf{X}) < +\infty$. Therefore, one generates a random point Π uniformly distributed in $\Re^{m+1}$:

$$\Pi(\omega) = (\mathbf{x},\ u) = \begin{cases} x_i = a_i + u_i(\omega)(b_i - a_i), \\ u \ = u_{i+1}(\omega) c f_{\max}, \end{cases} \tag{5.11}$$

where (a_i, b_i), $i = 1, 2, ..., m$ are given intervals of the reals and $u_i(\omega)$, $u_{i+1}(\omega) \in U[0,1]$ are the values of independent, uniform random variables. If $\Pi \in A$ (i.e. $\mathbf{\Pi}$ is not rejected), then the generated random variable $X_i (i = 1, 2, ..., m)$ is

$$X_i = a_i + u_i (b_i - a_i). \tag{5.12}$$

The intervals (a_i , b_i) are defined for all points of the mesh and establish an envelope of the random field. The envelope specifies the characteristic features of the field under consideration, for example, the maximal and minimal values and the field boundary conditions. It can be given in the form of a function or by an experimental discrete data. The intervals (a_i , b_i) are connected with assumed standard deviation σ_i at the node i by the following equation

$$\left(\int_{a_i}^{b_i} (x_i - \bar{x}_i)^2 f(x_i) dx_i \right)^{1/2} = \sigma_i . \tag{5.13}$$

Experience shows that for a large value of m, the method is not feasible because of time inefficiencies. For that reason, a conditional distribution was proposed (Walukiewicz et al. 1997, Górski 2006). The method makes it possible to simulate any homogeneous or non-homogeneous truncated Gaussian random field described on regular or irregular spatial meshes. Here, only a short description of this method is provided. A standard algorithm concerns the following steps:

1. Determination of the local covariance matrix $\mathbf{K}$, the known part of the random vector $\mathbf{X}_k$ and the expected values vector $\bar{\mathbf{X}}$:

$$\mathbf{X}=\begin{Bmatrix}\mathbf{X}_u\\ \mathbf{X}_k\end{Bmatrix}\begin{matrix}n\\ p\end{matrix}\qquad \mathbf{K}=\begin{bmatrix}\mathbf{K}_{11} & \mathbf{K}_{12}\\ \mathbf{K}_{21} & \mathbf{K}_{22}\end{bmatrix}\begin{matrix}n\\ p\end{matrix},\qquad \bar{\mathbf{X}}=\begin{Bmatrix}\bar{\mathbf{X}}_u\\ \bar{\mathbf{X}}_k\end{Bmatrix}\begin{matrix}n\\ p\end{matrix}. \tag{5.14}$$

$$\qquad\qquad n \quad p$$

2. Determination of the conditional variance $\mathbf{K}_c$ and the conditional mean $\bar{\mathbf{X}}_c$:

$$\mathbf{K}_c=\mathbf{K}_{11}-\mathbf{K}_{12}\mathbf{K}_{22}^{-1}\mathbf{K}_{21}, \tag{5.15}$$

$$\bar{\mathbf{X}}_c=\bar{\mathbf{X}}_u+\mathbf{K}_{12}\mathbf{K}_{22}^{-1}\left(\mathbf{X}_k-\bar{\mathbf{X}}_k\right). \tag{5.16}$$

3. Determination of the maximal value of the density function $f_{\max}$

$$f_{\max}=(1-t)^{-n/2}\left(\det\mathbf{K}_c\right)^{-n/2}(2\pi)^{-n/2}\times\left(2\operatorname{erf}(s)\right)^{-n}, \tag{5.17}$$

where t is the truncation parameter and s defines the truncation level:

$$t=\frac{s\times\exp\left(-s^2/2\right)}{\sqrt{2\pi}\operatorname{erf}(s)},\qquad s\geq 0\quad\text{and}\quad \operatorname{erf}(s)=\frac{1}{2\pi}\int_0^s\exp\left(-\frac{x^2}{2}\right)dx\cdot \tag{5.18}$$

4. Generation of the vector of unknown values $\mathbf{X}_u$:

$$X_i=a_i+\left(b_i-a_i\right)u_i\qquad\qquad i=1,\ldots,n, \tag{5.19}$$

where u_i is a random variable uniformly distributed in the interval [0,1].

5. Calculation of the conditional density function $f\left(\mathbf{X}_u\right)$

$$f\left(\mathbf{X}_u\right)=f_{\max}\exp\left(-0.5J\left(\mathbf{X}_u\right)\right), \tag{5.20}$$

where

$$J\left(\mathbf{X}_u\right)=-\frac{1}{2(1-t)}\left(\mathbf{X}_u-\bar{\mathbf{X}}_c\right)^T\mathbf{K}_c^{-1}\left(\mathbf{X}_u-\bar{\mathbf{X}}_c\right). \tag{5.21}$$

6. Generation of the independent random variable u_{n+1} from the interval [0,1] and checking the condition

$$f_{\max}u_{n+1}\leq f\left(\mathbf{X}_u\right). \tag{5.22}$$

6. If this condition holds, the random value $\mathbf{X}_u$ is accepted; if not, the calculation returns to the point 4.

An additional improvement of the calculation was achieved by introducing a sequential type technique. A "base scheme" was defined (Walukiewicz et al. 1997, Górski 2006) which covered a limited mesh area (hundred points), and only these points

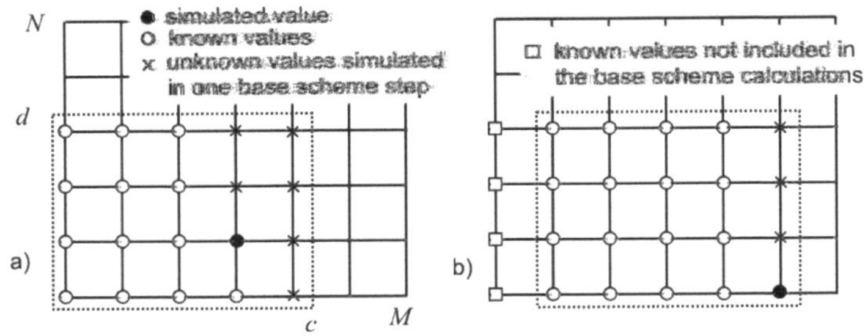

Fig. 5.46. Successive coverage of the field points with the moving propagation scheme (Walukiewicz et al. 1997)

were used in the calculations of the next random values (Fig.5.46). The simulation process was divided into three stages. First, the four-corner random values were generated using an unconditional method. Next, all random variables in the defined base scheme (dotted rectangle in Fig.5.46a) were generated, one by one, using a conditional method. In the third stage, the base scheme was appropriately shifted, and the next group of unknown random values was simulated (Fig.5.46b). The base scheme was translated so as to cover all the field nodes. The proposed approach allows for generation of practically unlimited random fields (thousands of discrete points).

To describe the discrepancies between the theoretical and generated fields, the following global G_{er} and local V_{er} (variance) errors were calculated (Walukiewicz et al. 1997):

$$G_{er}\left(\mathbf{K},\,\hat{\mathbf{K}}\right)=\frac{\left\|\|\mathbf{K}\|-\|\hat{\mathbf{K}}\|\right\|}{\|\mathbf{K}\|}\times 100\%\,, \tag{5.23}$$

$$V_{er}\left(k_{ii},\hat{k}_{ii}\right)=\sum_{i=1}^{m}\frac{\left(k_{ii}-\hat{k}_{ii}\right)}{\left(k_{ii}\right)}\times 100\%\,, \tag{5.24}$$

where $\hat{\mathbf{K}}$ is the estimator of the covariance matrix $\mathbf{K}$ calculated with the help of the generated fields

$$\hat{\mathbf{K}}=\frac{1}{NR-1}\sum_{i=1}^{NR}(\hat{\mathbf{X}}_i-\bar{\hat{\mathbf{X}}})(\hat{\mathbf{X}}_i-\bar{\hat{\mathbf{X}}})^{\mathrm{T}}\,,\qquad \bar{\hat{\mathbf{X}}}=\frac{1}{NR}\sum_{i=1}^{NR}\hat{\mathbf{X}}_i\,. \tag{5.25}$$

The parameter $\hat{\mathbf{X}}$ is the estimator of the random vector $\mathbf{X}$, and $\bar{\hat{\mathbf{X}}}$ is the estimator of its mean value, *NR* denotes the number of realizations, $\|\mathbf{K}\|=\sqrt{tr\left(\mathbf{K}\right)^2}$ is the matrix norm, k_{ii} and $\hat{k}_{ii}$ denote the diagonal element of the covariance matrix $\mathbf{K}$ and its estimators $\hat{\mathbf{K}}$, respectively.

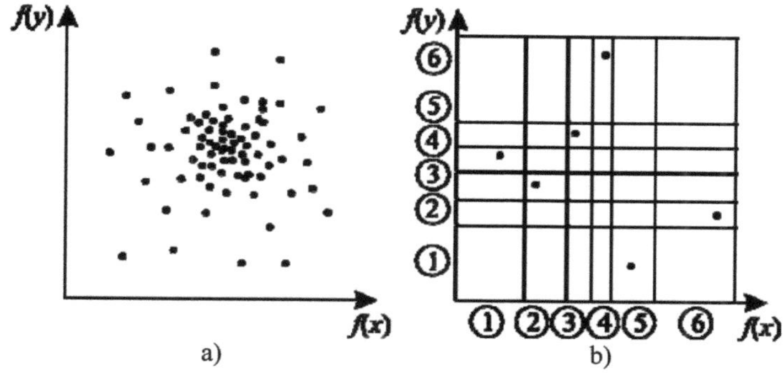

Fig. 5.47. Latin hypercube method – the scheme of random sampling

Contrary to stochastic finite element codes, the Monte Carlo method (Hurtado and Barbat 1998, Przewłócki and Górski 2001, Górski 2006) does not impose any restriction to the solved random problems. The only limitation of the Monte Carlo method is the time of calculations. To reproduce exactly the input random data at least 2000 random samples should be used (Górski 2006., Bielewicz and Górski 2002). Any nonlinear calculations for such number of initial data are, however, impossible due to excessive computation times. To determine a minimal, but sufficient number of samples (which allows to estimate the results with a specified accuracy), a convergence analysis of the outcomes has been proposed (Bielewicz and Górski 2002). It has been estimated that in case of various engineering problems only ca. 50 realizations have to be considered. A further decrease of sample numbers can be obtained using Monte Carlo reduction methods (e.g. stratified (Rubinstain 1981, Górski 2006) and Latin sampling methods (Bazant and Lin 1985, Florian 1992)). It should be pointed out that these methods were not used for the generation of two-dimensional random fields, but for their classification. For that reason the single realization was generated according to the initial data, i.e. the mean value and the covariance matrices were exactly reproduced. The calculations have shown that using these reduction methods the results can be properly estimated by several realizations (e.g.12-15). In the case of the stratified sampling, the whole space of the samples was divided into equal subsets or subsets of equal probability. From each subset only one sample was chosen for the analysis (Górski 2006). According to the Latin hypercube sampling method, the random field realizations were chosen in a strictly defined manner (Fig.5.47). First, an initial set of random samples was generated in the same way as in the case of the direct Monte Carlo method. Next, the generated samples were classified according to chosen parameters, for example norms of the random vectors, their mean values, changes of the vector signs and others. The samples were arranged according to this classification, in increasing order. On this basis, their distributions in the form of frequency histograms were specified. The whole space of the samples was divided into subsets of equal probability and numbered. The Latin hypercube sampling method combined at random each subset number with other subset numbers of the remaining variables only once. From each subset defined in this way, only one sample was chosen for the analysis. The calculations were performed only for these samples.

The input data of the considered problems were seta of random fields describing the distribution of the initial void ratio e_o. A truncated Gaussian random field was applied

$$e_o = \overline{e}_o \left(1 + \nu\beta(x_1, x_2)\right), \tag{5.26}$$

where $\overline{e}_o$ is the mean value of the initial void ratio, $\nu = s_{e_o} / \overline{e}_o$ is the coefficient of variation, s_{e_o} describes the standard deviation of the mean value and $\beta(x_1, x_2)$ stands for the normalized homogeneous random field. The initial void ratio scattering in the specimen was also limited by the pressure dependent void ratios e_{i0} (upper bound) and e_{d0} (lower bound) (Eqs.3.80 and 3.81) what is essential for large standard deviations e.g. when s_{e_0} >0.10 (Tejchman and Górski 2007). The truncation parameter t of the Gaussian field allows us to fulfill these conditions. The midpoint method was applied. The method approximated the random field in each finite element by a single random variable defined as the value of the field at its centre. Randomness of the initial void ratio e_o should be described by a correlation function. For lack of the appropriate data the correlation function is usually chosen arbitrarily. Here, the following second order, homogeneous correlation function was adopted (Bielewicz and Górski 2002)

$$K(x_1, x_2) = s_{e_o}^2 \times e^{-\lambda_{x1}\Delta x1} (1 + \lambda_{x1}\Delta x_1) e^{-\lambda_{x2}\Delta x2} (1 + \lambda_{x2}\Delta x_2), \tag{5.27}$$

where Δx_1 and Δx_2 is are the distances between two field points along the horizontal axis x_1 and vertical axis x_2, λ_{x1} and λ_{x2} are the decay coefficients (damping parameters) characterizing a spatial variability of the specimen properties (i.e. describe the correlation between the random field points), while the standard deviation s_{e_o} represents the field scattering. The mean value of the random field was assumed to be constant.

In finite element methods, continuous correlation function has to be represented by the appropriate covariance matrix. For this purpose, the procedure of local averages of the random fields proposed by Vanmarcke (1983) was adopted. After an appropriate integration of the function (Eq.6.27), the following expression describing the variances D_w and covariances K_w were obtained (Knabe et al. 1998):

$$\begin{aligned} D_w(\Delta x_1, \Delta x_2) = {} & \frac{2}{\lambda_{x_1}\Delta x_1} s_{e_o}^2 \left[2 + e^{-\lambda_{x_1}\Delta x_1} - \frac{3}{\lambda_{x_1}\Delta x_1}\left(1 - e^{-\lambda_{x_1}\Delta x_1}\right) \right] \times \\ & \times \frac{2}{\lambda_{x_2}\Delta x_2} \left[2 + e^{-\lambda_{x_2}\Delta x_2} - \frac{3}{\lambda_{x_2}\Delta x_2}\left(1 - e^{-\lambda_{x_2}\Delta x_2}\right) \right] \end{aligned}, \tag{5.28}$$

$$\begin{aligned} K_w(\Delta x_1, \Delta x_2) = {} & \frac{e^{\lambda_{x_1}\Delta x_1}}{\left(\lambda_{x_1}\Delta x_1\right)^2} s_{e_o}^2 \left\{ \left[\cos\left(\lambda_{x_1}\Delta x_1\right) - \sin\left(\lambda_{x_1}\Delta x_1\right)\right] + 2\lambda_{x_1}\Delta x_1 - 1 \right\} \times \\ & \times \frac{e^{\lambda_{x_2}\Delta x_2}}{\left(\lambda_{x_2}\Delta x_2\right)^2} s \left\{ \left[\cos\left(\lambda_{x_2}\Delta x_2\right) - \sin\left(\lambda_{x_2}\Delta x_2\right)\right] + 2\lambda_{x_2}\Delta x_2 - 1 \right\} \end{aligned}. \tag{5.29}$$

In the stochastic calculations, the mean initial void ratio $\overline{e}_0$ =0.60 (initially dense sand) was mainly assumed. One took into consideration a weak or strong correlation of the initial void ratio (λ_{x1}=1 and λ_{x2}=1) or (λ_{x1}=3 and λ_{x2}=3), and a low or high standard deviation (s_{e_o} =0.05 or s_{e_o} =0.10). The range of significant correlation was approximately 20-40 mm. The dimension of the random field was the same as the finite element mesh.

Shearing between two very rough boundaries

FE calculations of plane strain shearing under vertical pressure and free dilatancy were performed for an infinite granular layer with a height of h_o=5 mm, h_o=50 mm, h_o=500 mm and h_o=2000 mm. The calculations were performed with a section of an infinite shear layer with a width of $b=h_o$, discretised by 20 (h_o=5 mm and h_o=50 mm), 200 (h_o=500 mm) and 800 (h_o=2000 mm) quadrilateral elements. The height of the elements was always $5\times d_{50}$ (with d_{50}=0.5 mm). The behaviour of an infinite shear layer was modelled by lateral boundary conditions, i.e. displacements and rotations along both sides of the column were constrained by the same amount (Section 4.2). Consequently, the evolution of state quantities was independent of the direction of shearing (and layer length). As the initial stress state in the granular strip, a K_o-state without polar quantities (σ_{22}=-1.0 kPa, $\sigma_{11}=\sigma_{33}$=-0.3 kPa, $\sigma_{12}=\sigma_{21}=m_1=m_2=0$, K_o=0.45) was assumed. The influence of the gravity was neglected. A quasi-static shear deformation was initiated through constant horizontal displacement increments prescribed at the nodes along the top of the layer. Both bottom and top surfaces were assumed to be very rough, i.e. sliding and rotation of particles against the bounding surface were excluded. The boundary conditions were along the bottom: u_1=0, u_2=0 and ω^c=0, and along the top: $u_1=n\Delta u$, ω^c=0, and $\sigma_{22}=p$ (Section 4.2). Due to the fact that the numerical effect of the vertical surface pressure on shear localization during shearing of an infinite granular layer is not so significant (Section 4.2), the calculations were carried out only with uniform constant pressure p=200 kPa. In the FE investigations of a deterministic size effect, initially dense sand was used (with the initial void ratio of e_0=0.60).

Deterministic size effect

Figures 5.48-5.50 show results for an initially dense sand layer with a different initial height h_o=5-2000 mm (e_o=0.60, p=200 kPa). Fig.5.48 presents the evolution of the mobilized internal friction angle $\phi=\arctan(\sigma_{12}/\sigma_{22})$ against the normalized horizontal displacement of the top boundary u_1^t/h_o. The mobilized friction angle ϕ is related to the entire granular layer since the stresses σ_{12} and σ_{22} are independent of both the height and length of the layer. The deformed FE meshes for different h_o at residual state are shown in Fig.5.49. In turn, Fig.5.50 presents the distribution of the Cosserat rotation ω^c and void ratio e across the normalized height x_2/d_{50} (for higher layers the distribution is shown only in the region of shear localization).

All state variables (stress, couple stress, mobilized friction angle and void ratio) tend to asymptotic values. The mobilized friction angle at peak ϕ_p slightly decreases with increasing h_o/d_{50} reaching its asymptote for h_o=50 mm (h_o/d_{50}=100): ϕ_p=46.2° (h_o=5 mm), ϕ_p=45.2° (h_o=50 mm), ϕ_p=45.1° (h_o=500 mm) and ϕ_p=45.0° (h_o=2000 mm). The residual friction angle similarly behaves: ϕ_{cr}=32° (h_o=5 mm), ϕ_{cr}=31.7° (h_o=50 mm), ϕ_{cr}=31.2°

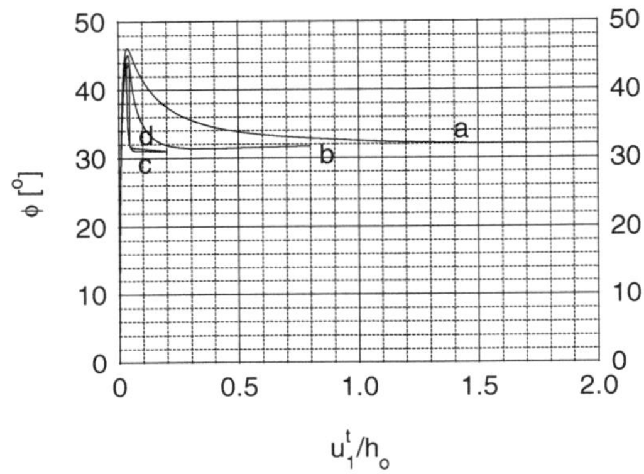

Fig. 5.48. Evolution of mobilized internal friction angle ϕ versus u_1^t/h_o during shearing with dense sand (e_0=0.60): a) h_o=5 mm, b) h_o=50 mm, c) h_o=500 mm, d) h_o=2000 mm

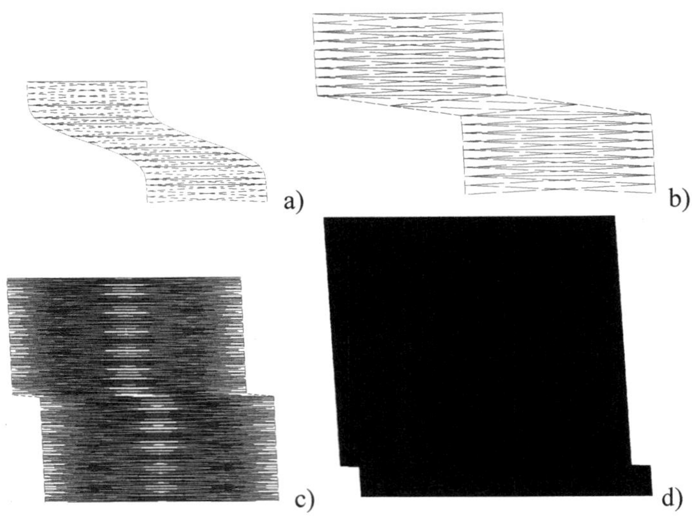

Fig. 5.49. Deformed FE meshes at residual state during shearing with dense sand (e_o=0.60): a) h_o=5 mm, a) h_o=50 mm, a) h_o=500 mm, a) h_o=2000 mm

(h_o=500 mm) and ϕ_p=31.0° (h_o=2000 mm). The rate of softening increases with increasing h_o (material becomes more brittle). For h_o/d_{50}=4000, the material softening is almost vertical. One could expect that for h_o/d_{50}>4000, a stress-strain curve of a snap-back type will occur after the peak.

The width of the shear zone appearing inside of the layer (characterized among others by the presence of Cosserat rotation and an increase of the initial void ratio, Fig.5.50) is $5\times d_{50}$ (h_o=5 mm) and $16\times d_{50}$ (h_o=50-2000 mm) on the basis of ω^c,

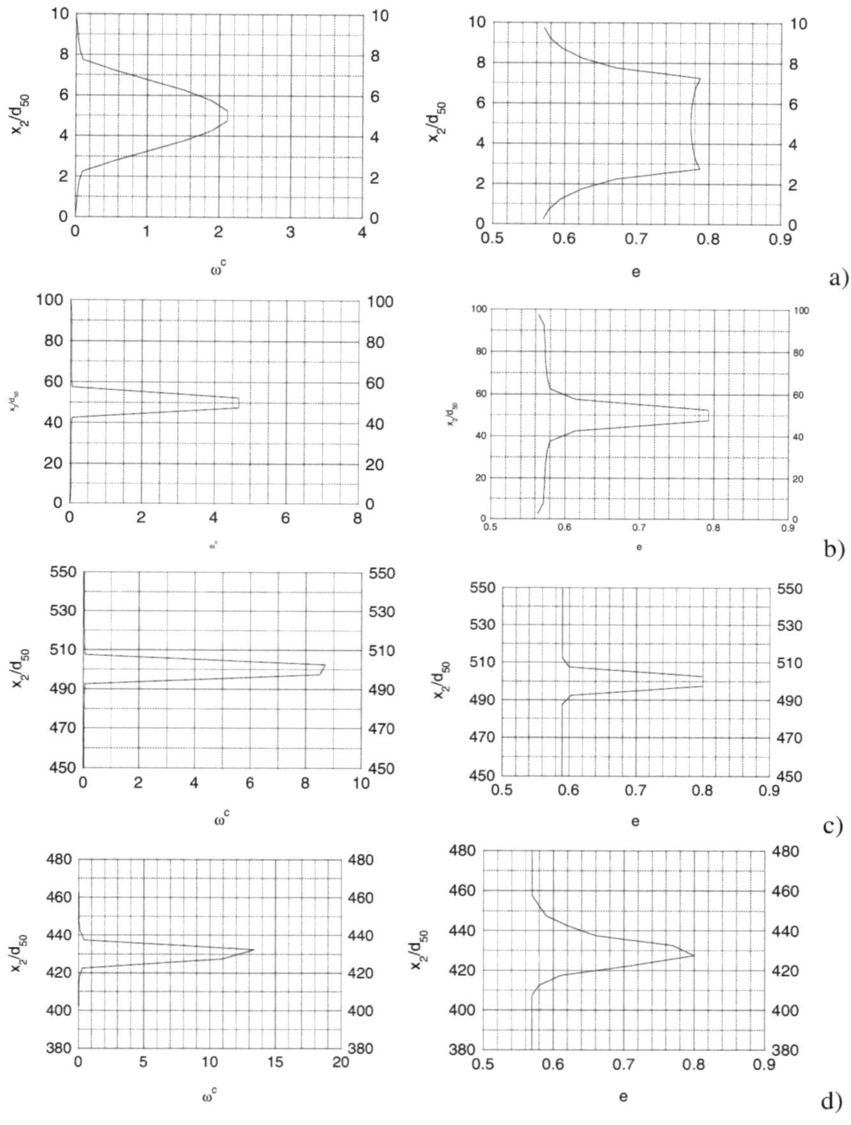

Fig. 5.50. Distribution Cosserat rotation ω^c and void ratio e across the normalized height x_2/d_{50} in dense sand (e_o=0.60) at residual state: a) h_o=5 mm, b) h_o=50 mm, c) h_o=500 mm, d) h_o=2000 mm

respectively. The thickness of the shear zone on the basis of void ratio is slightly larger since each dense granulate undergoes dilatancy before shear localization occurs.

The ratios between the mobilized friction angle at peak and normalized shear zone thickness against the parameter h_o/d_{50} are summarized in Fig.5.51.

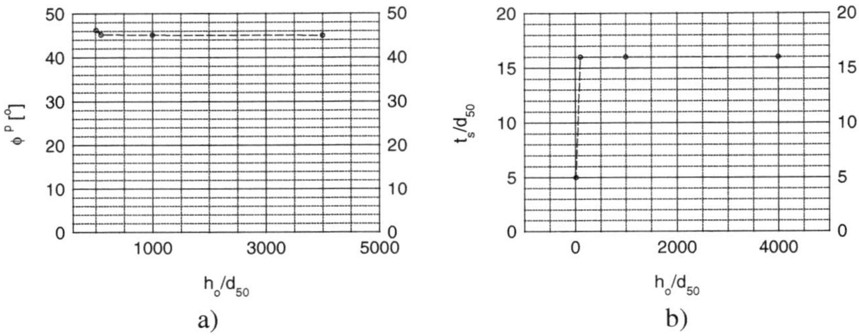

Fig. 5.51. Ratio between mobilized internal friction angle at peak ϕ_p (a) and normalized shear zone thickness t_s at residual state (b) and parameter h_o/d_{50} (e_o=0.60)

Statistical size effect (layer height h_o=50 mm (h_o/d_{50}=100)

The calculations were carried out with a mean initial void $\overline{e}_0$ =0.60 (initially dense sand), small standard deviation s_{e_o} =0.05 and strong correlation of the initial void ratio e_o in a vertical direction (λ_{x1}=0 and λ_{x2}=1).

First, the statistical size effect was analysed using the direct Monte Carlo method. According to this method the FE calculations were performed for succeeding realizations in order as they had been generated. The set of 2000 realizations of the initial voids was used. The outcomes were analysed in every step.

Figs.5.52-5.56 present the results obtained for the first N=50 random fields of the set. In Figs.5.52-5.55 the expected values and standard deviations of the friction angle at peak ϕ^p and at residual state ϕ_{res}, the normalized horizontal displacement of the top boundary u_1^t/h_o corresponding to the peak and the normalized shear zone thickness t_s/d_{50} are presented versus the number of realizations. In addition, the distribution of the Cosserat rotation ω^c across the normalized height x_2/d_{50} is shown in Fig.5.56 for one wide and one thin shear zone. Enclosed is also the evolution of the vertical displacement of the top boundary during initial compression and further shearing and the evolution of the mobilized friction angle.

Analyzing the results one notices that 50 samples are the sufficient number of realizations (the outcomes display stabilizations starting from 20 realizations). It is expected that the next calculations performed for the successive samples would not change the results visibly.

The friction angle at peak ϕ_p varies between 36.51° and 46.89°, the friction angle at residual state ϕ_{res} between 30.09° and 31.72°, the normalized horizontal displacement of the top boundary corresponding to the peak u_1^t / h_0 between 0.022 and 0.035 and the normalized shear zone thickness t_s/d_{50} between 15 and 25.

The estimated expected value and standard deviations were respectively: for the friction angle at peak $\hat{\phi}_p$ =42.89° and $\hat{s}_{\phi_p}$ =2.12° (Fig.5.52), for the residual friction

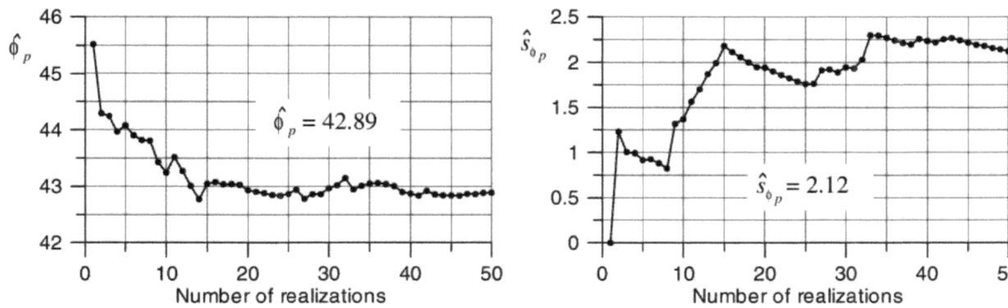

Fig. 5.52. The mobilized friction angle at peak ϕ_p: expected values and standard deviation

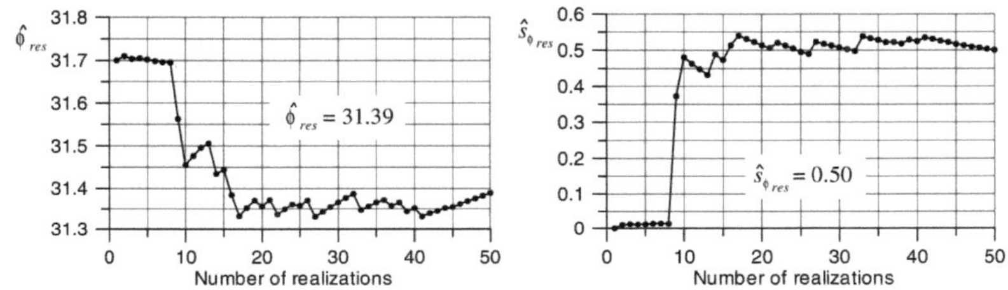

Fig. 5.53. The mobilized friction angle at residual state ϕ_{res}: expected values and standard deviation

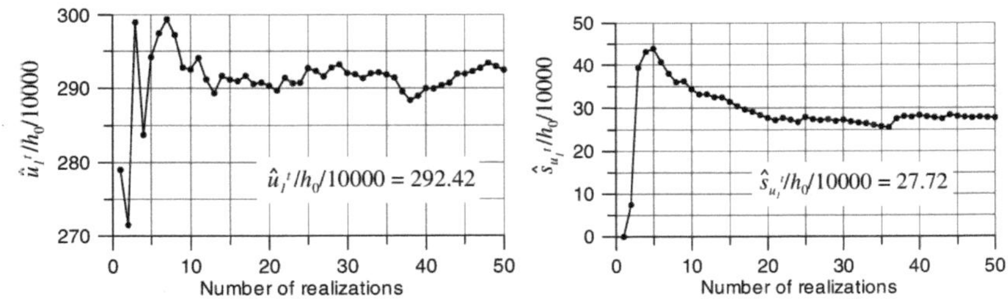

Fig. 5.54. The normalized horizontal displacement of the top boundary $u_1^t/h_o/10000$ corresponding to the peak: expected values and standard deviation

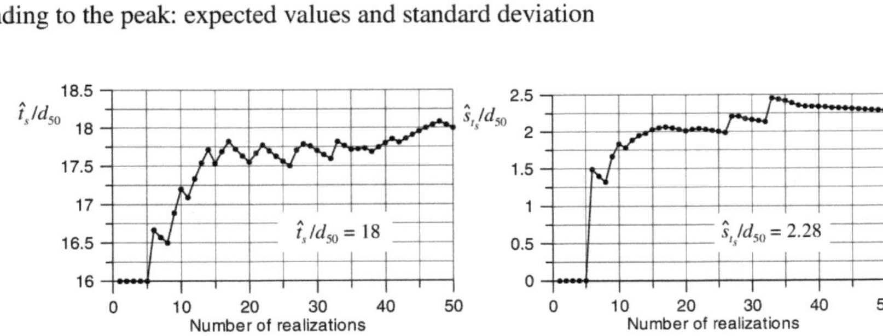

Fig. 5.55. The normalized shear zone thickness t_s/d_{50}: expected values and standard deviation

angle $\hat{\phi}_{res}$ =31.89° and $\hat{s}_{\phi_{res}}$ =1.12° (Fig.5.53), the normalized horizontal displacement of the top boundary u_1^t/h_o corresponding to the peak $\hat{u}_1^t / h_o$ =0.0029 and $\hat{s}_{u_1^t} / h_o$ = 0.0028 (Fig.5.54) and the normalized shear zone thickness $\hat{t}_s / d_{50}$ =18.00 and $\hat{s}_{t_s} / d_{50}$ =2.28 (Fig.5.55).

It is important to estimate the confidence interval of the calculated values. The friction angle at peak ϕ_p was chosen for the analysis. It follows from the central limit theorem that the variable $(\hat{\phi}_p - \bar{\phi}_p)$ has a normal distribution with the zero mean value and unit variance and $\hat{s}_{\phi_p}^2$ is the chi-square distributed with $(N-1)$ degrees-of-freedom. Then the random variable

$$t = \frac{(\hat{\phi}_p - \bar{\phi}_p)\sqrt{N-1}}{\hat{s}_{\phi_p}^2} \tag{5.30}$$

has the Student's t-distribution with $(N-1)$ degrees-of-freedom. Therefore for $(1-\alpha)100\%$ $(0 \le \alpha \le 1)$, the confidence interval on the mean value $\bar{\phi}_p$ is (Benjamin and Cornell 1970)

$$\hat{\phi}_p - t\left(\frac{\alpha}{2}\right)\frac{\hat{s}_{\phi_p}}{\sqrt{N-1}} < \bar{\phi}_p < \hat{\phi}_p + t\left(\frac{\alpha}{2}\right)\frac{\hat{s}_{\phi_p}}{\sqrt{N-1}}, \tag{5.31}$$

where $t(\alpha/2)$ is the value of Student's t-distribution for the assumed parameter α. The confidence interval of variance $\bar{s}_{\phi_p}$ is given by the following inequalities

$$\frac{(N-1)\hat{s}_{\phi_p}^2}{\chi_{\alpha/2}^2} < \bar{s}_{\phi_p}^2 < \frac{(N-1)\hat{s}_{\phi_p}^2}{\chi_{1-\alpha/2}^2}, \tag{5.32}$$

where $\chi_{\alpha/2}^2$ and $\chi_{1-\alpha/2}^2$ are the values of a random variable having a chi-square distribution that cuts off $\alpha/2\times100\%$ of the right tail of the distributions and $\alpha/2\times100\%$ of the left tail, respectively. Assuming 99% confidence interval $(\alpha = 0.01)$ and $N = 50$, one can obtain according to Eqs.5.30 and 5.31:

$$42.89 - 2.69\frac{2.12}{\sqrt{50-1}} < \bar{\phi}_p < 42.89 + 2.69\frac{2.12}{\sqrt{50-1}}, \tag{5.33}$$

$$\frac{(50-1)\cdot 2{,}12^2}{79.06} < \bar{s}_{\phi_p}^2 < \frac{(50-1)\cdot 2{,}12^2}{26.83}. \tag{5.34}$$

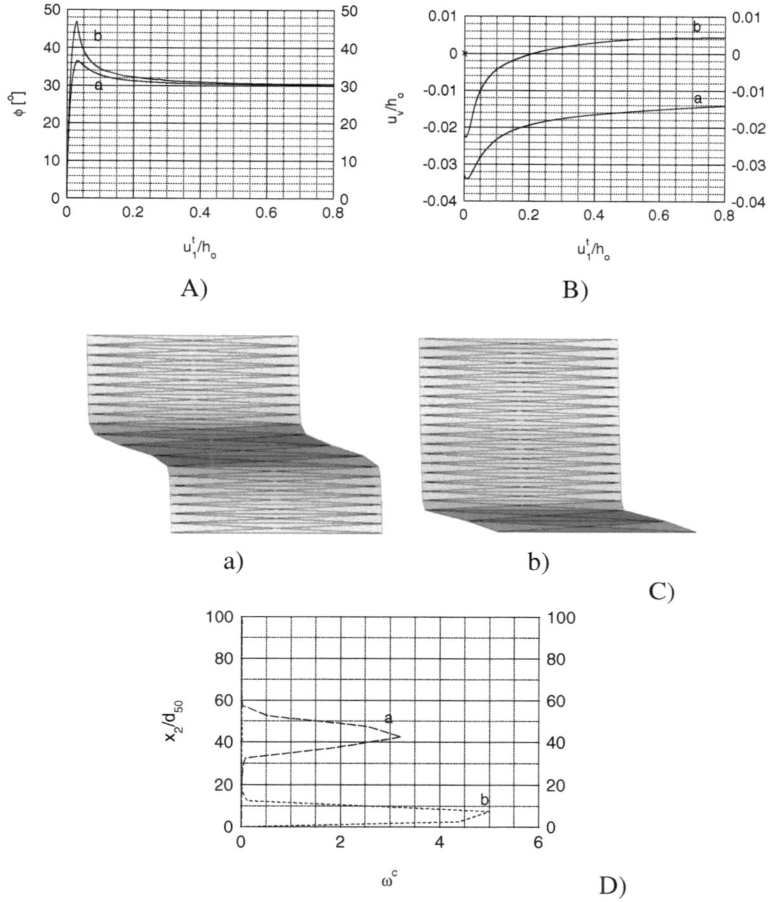

Fig. 5.56. Evolution of mobilized friction angle ϕ (A) and vertical displacement of the top boundary (B) versus u_1^t/h_o (A), deformed meshes with the distribution of void ratio *e* (C) and distribution Cosserat rotation ω^c across the normalized height x_2/d_{50} (D) at residual state in dense random sand (h_o=50 mm): a) ϕ_p=36.51°, b) ϕ_p=46.89°

Therefore, it can be expected that the admissible (with 99% confidence) values of the mean value of $\overline{\phi}_p$ and the standard deviation $\overline{s}_{\phi_p}$ belong to the following intervals:

$$42.07 < \overline{\phi}_p < 43.70 \qquad \text{and} \qquad 1.67 < \overline{s}_{\phi_p} < 2.87\,. \tag{5.35}$$

It is easy to notice that the mean value converges to the exact solution very fast. An adequate estimation of the standard deviation is more difficult.

Additionally it can be found out that the shear zone develops horizontally at the weakest element layer (Fig.5.56). The larger is the material softening rate (larger stiffness after the peak), the narrower the shear zone. The shear zone width changes between 15×d_{50} ($\hat{e}_o = 0.57$, ϕ^p=46.89°) and 25×d_{50} ($\hat{e}_o = 0.68$, ϕ^p=36.51°). The dilatancy increases with increasing ϕ^p. The mean random internal friction angle at peak ($\hat{\phi}_p$=42.89°) is by 2.5° smaller than this with the uniform initial void ratio ϕ_p=45.2° (Fig.5.51). The mean random shear normalized zone thickness (t_s/d_{50}=18.0) is by 15% larger than this with the uniform initial void ratio t_s/d_{50}=16.0 (Fig.5.51) due to a smaller softening rate.

To decrease the number of realizations, the stratified sampling and Latin hypercube sampling method were applied. First, the generated 2000 realizations (random vectors) of the initial void ratio had to be appropriately classified. The following classification parameters were taken into consideration: the mean values of the initial void ratio e_o, the norms of the random vectors, the gap between the lowest and the highest value of the void ratio, and the number of waves formed by the random vector values. In the case of the stratified sampling reduction method only one parameter, namely the mean value of the random vectors of the initial void ratio was chosen. It was expected that this parameter influenced the results more than the others. The generated set of 2000 realizations was divided into 10 equal intervals, and on this basis, the probability distribution (histogram) of the void ratio mean values was obtained (Fig.5.57). In each interval, one representative vector of the void ratio distribution (close to the middle of each interval) was chosen.

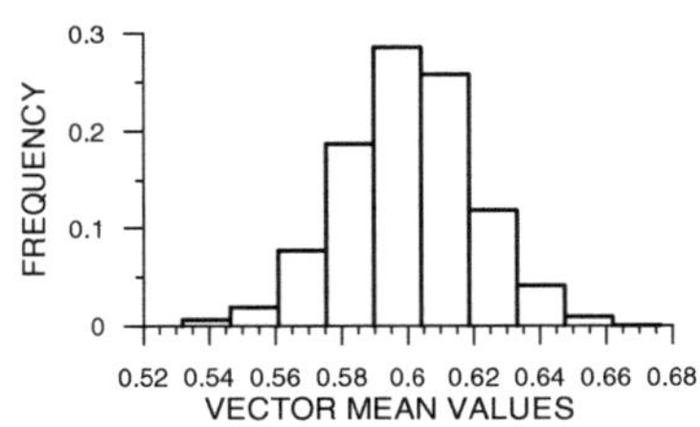

Fig. 5.57. Stratified sampling with 10 equal intervals (frequency against initial void ratio)

The probabilities of these representatives were described by the following formula

$$p_i = \frac{N_i}{NI}, \tag{5.36}$$

where N_i is the number of random vectors belonging to the interval "i" and NI is the number of all variables (in this case NI=2000). Using the chosen ten realizations, the nonlinear FE calculations were performed. To examine the efficiency of the reduction methods, the results for the friction angle at peak ϕ^p were analysed. The expected value and the standard deviations were calculated according to the following formulas

$$\hat{\phi}^p = \sum_{i=1}^{NRI} p_i \phi_i^p, \tag{5.37}$$

$$\hat{\sigma}_{\phi_p} = \sqrt{\sum_{i=1}^{NRI} p_i (\phi_i^p - \hat{\phi}^p)^2}, \tag{5.38}$$

where *NRI* is the number of intervals (*NRI*=10). The results for ϕ^p are presented in Tab.5.2. The estimated expected value and standard deviations for the friction angle at peak were: $\hat{\phi}^p$ =43.71° and $\hat{s}_{\phi^p}$ =2.35°, respectively (Tab.5.2). The same stratified sampling calculations were also performed for 5 equal intervals of the random set. The outcomes were compared with the results obtained using the direct Monte Carlo method for 50, 10 and 5 realizations (Tab.5.2).

An alternative version of the stratified Monte Carlo method was also examined. Use was made of intervals of the void ratio mean values with the same probability. Assuming, as previously 10 intervals, the probability distribution was obtained (Fig.5.58). The results for ϕ^p are presented in Tab.5.2 (the estimated expected value and standard deviations were, respectively, for the friction angle at peak $\hat{\phi}_p$ =43.21° and $\hat{s}_{\phi_p}$ =1.94°). In this case, the outcome has the best agreement with the direct Monte Carlo method (discrepancies are slightly smaller compared to the results of the stratified sampling method with equal intervals). But using 5 realizations, much worse results were, however, obtained (Tab.5.2).

Table 5.2. Expected values and standard deviation for the friction angle at peak ϕ_p versus the reduction method and number of realizations (h_o/d_{50}=50 mm)

Method	Number of realizations n	Expected value of ϕ_p [°]	Standard deviation of ϕ_p [°]
direct Monte Carlo	50	42.89	2.12
direct Monte Carlo	10	43.24	1.37
stratified sampling – equal intervals	10	43.71	2.35
stratified sampling – equal probabilities	10	43.21	1.94
Latin sampling	10	42.97	1.39
direct Monte Carlo	5	44.08	0.92
stratified sampling – equal intervals	5	43.77	1.87
stratified sampling – equal probabilities	5	43.13	2.90
Latin sampling	5	43.87	2.42

Another possibility for the reduction of the number of realizations is the application of the Latin hypercube sampling. Here, in contrary to the procedure which applied the Latin hypercube sampling to the generation of random fields (Niemunis et al. 2005), the method was used to select the realizations which were taken in the calculations. In this way, the considered field was generated strictly according to the initial assumptions. Two classification parameters of the realizations were chosen. The first was the same as in the case of the stratified sampling, i.e. the mean value of the initial void ratio. The second parameter was assumed to be the gap between the lowest and the highest value of the initial void ratio. The joint probability distribution (so called "ant hill") is presented in Fig.5.59. One dot represents one random vector described by its mean value and the difference between its extreme values. The distribution of the vector max-min gaps were also divided in ten intervals with equal probabilities. Next, according to the Latin hypercube sampling assumptions, ten random numbers in the range 1-10 were generated (one number appeared only once) using the uniform distribution. Here, the following pairs of numbers were obtained: 1 – 8, 2 – 10, 3 – 7, 4 – 6, 5 – 1, 6 – 4, 7 – 9, 8 – 2, 9 – 5, 10 – 3. According to these pairs, the generated set of 2000 realizations was appropriate divided. The areas (subfields) corresponding to the selected pairs of numbers are presented as rectangles in Fig.5.59. From each subfield only one realization was chosen and used as the input data to the FEM calculations. The results using 10 and 5 samples for ϕ^p are also presented in Tab.5.2. The agreement with the direct Monte Carlo method (50 samples) is worse as in the case of the stratified sampling method (the estimated expected value and standard deviations were, respectively, for the friction angle at peak $\hat{\phi}_p$ =42.97° and $\hat{s}_{\phi_p}$ =1.39). The reasons can be two: first, the vectors used in the calculation were not selected from the middle of the intervals and second, the Latin hypercube sampling is a powerful method when different, uncorrelated initial parameters are subject to an analysis (the application of this method to random field selection is not so obvious).

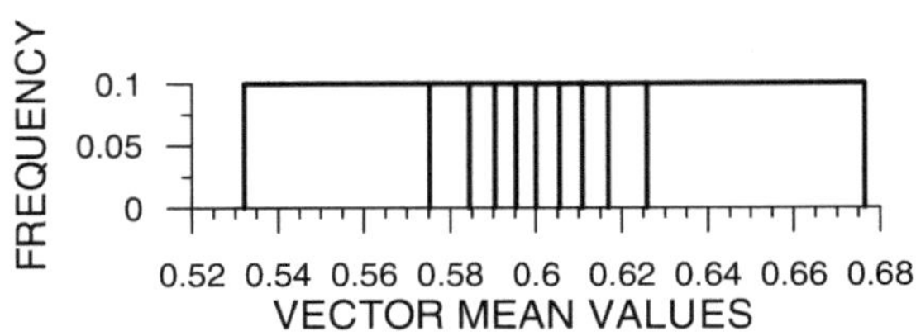

Fig. 5.58. Stratified sampling with 10 intervals with equal probabilities (frequency against initial void ratio)

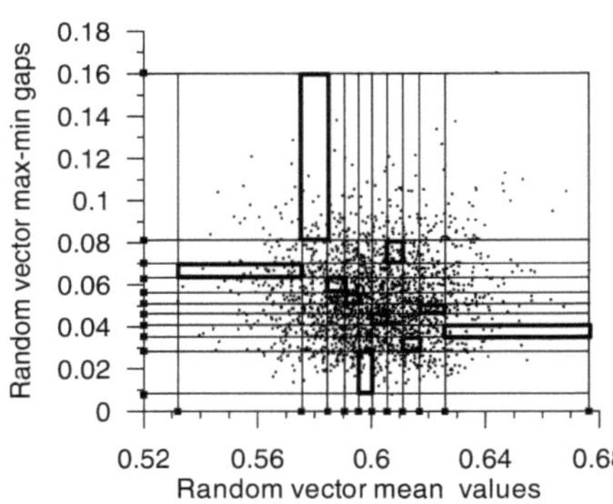

Fig. 5.59. Latin hypercube method: selection of 10 pairs of random samples: 1 – 8, 2 – 10, 3 –7, 4 – 6, 5 – 1, 6 – 4, 7 – 9, 8 – 2, 9 – 5, 10 – 3

Statistical size effect (layer height h_o=500 mm (h_o/d_{50}=1000))

The calculations were performed only with the stratified sampling method using 10 intervals of the void ratio mean values with the equal probability. As it was established earlier, this reduction method was the most accurate.

The estimated expected value and standard deviations were respectively: for the friction angle at peak: 39.81° and 0.72, for the residual friction angle: 30.80° and 0.35, for the normalized horizontal displacement of the top boundary u_1^t/h_o corresponding to the peak: 0.0022 and 0.0016 and for the normalized shear zone thickness t_s/d_{50}: 16.0 and 2.29. The mean random internal friction angle at peak (ϕ^p=39.81°) is by 3.1° smaller than this with h_o=50 mm (ϕ^p=42.89°). Thus, the random shear resistance diminishes with increasing layer height.

During initial shearing, several parallel shear zones occurred (Fig.5.60). Next, only one developed horizontally at the weakest strip.

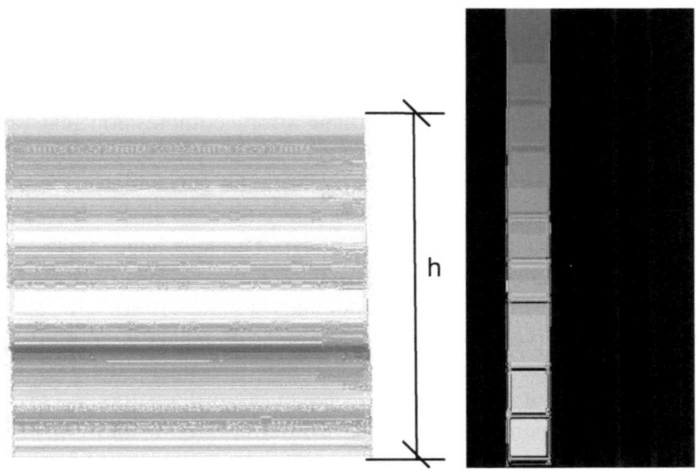

Fig. 5.60. Distribution of void ratio *e* during initial shearing with dense sand along layer height of 500 mm (darker color denotes an increase of *e*)

The following conclusions can be drawn from simplified non-linear FE investigations of a deterministic and statistical size effect during quasi-static shearing of an infinite granular layer:

a) with respect to the deterministic size effect:

- The deterministic size effect is rather small. The differences in the mobilized internal friction angle are maximum 1.0-1.5°. The shear resistance at peak and at residual state decrease with increasing h_o/d_{50} up to $h_o/d_{50}\leq 100$.
- The material brittleness strongly increases with increasing h_o/d_{50}. For the layer height h_o/d_{50}=4000, the material softening is almost vertical.
- The thickness of the shear zone increases with increasing initial void ratio e_o (due to a decreasing rate of softening) and ratio between the layer height and mean grain diameter h_o/d_{50}. (due to a decreasing effect of boundaries). It reaches a constant

value of $16\times d_{50}$ for initially dense granulates and $25\times d_{50}$ for initially medium dense granulates at $h_o/d_{50}>100$.

b) with respect to the statistical size effect:

- The solution of random nonlinear problems on the basis of several samples is possible. The most efficient reduction method is the stratified sampling with intervals described by equal probabilities.
- The mean random shear resistance at peak is always smaller as compared to this with the deterministic uniform distribution. It diminishes with increasing h_o/d_{50}. Thus, the statistical size effect is smaller than the deterministic one. The weakest link principle always applies here due to that shear localization forms in a horizontal weak layer.
- The thickness of the shear zone in the random layer can be larger by 15% (h_o/d_{50}=100) than this at the deterministic uniform distribution of the initial void ratio.
- In the case of $h/d_{50}\geq 1000$, several horizontal shear zones occur during initial shearing. Later, only one dominates.

Plane strain compression

The deterministic FE calculations of a plane strain compression test (assuming an uniform distribution of the initial void ratio e_o) were performed with six different sand specimen sizes $b_o\times h_o$ (b_o – initial width, h_o – initial height) which were geometrically similar: 10×35 mm^2, 20×70 mm^2, 40×140 mm^2, 80×280 mm^2, 160×560 mm^2 and 320×1120 mm^2 (Tejchman and Górski 2008b). The specimen depth was equal to l=1.0 m with plane strain condition. The specimen dimensions of 40×140 mm^2 were similar as in the experiments by Vardoulakis (1977, 1980). In all cases, 896 quadrilateral elements divided into 3584 triangular elements were used. To properly capture shear localization inside of the granular specimen, the size of the finite elements s_e should be not larger than five mean grain diameters d_{50} (Tejchman and Bauer 1996). For specimen sizes changing from 10×35 mm^2 up to 40×140 mm^2 this condition was fulfilled ($s_e\leq 5\times d_{50}$). However, for specimen sizes larger than 40×140 mm^2 this condition was violated (e.g $s_e=40\times d_{50}$ for 320×1120 mm^2). Thus, these latter FE results were mesh-dependent; the mesh-dependence increased with increasing specimen size. A further increase of the amount of finite elements would increase the computation time drastically. A remedy in the form of remeshing was not the aim of these FE analyses.

A quasi-static deformation in sand was imposed through a constant vertical displacement increment Δu prescribed at nodes along the upper edge of the specimen. The boundary conditions implied no shear stress imposed at the smooth top and bottom of the specimen. To preserve the stability of the specimen against horizontal sliding, the node in the middle of the top edge was kept fixed. To simulate a movable roller bearing in the experiment (Vardoulakis 1977), the horizontal displacements along the specimen bottom were constrained to move by the same amount. Thus, no imperfections were used to induce shear localization with an uniform distribution of e_o. Comparative calculations were also performed with a very rough top and bottom boundary. In this case, the horizontal displacement and Cosserat rotation along both horizontal edges were assumed to be equal to zero. The vertical displacement increments were chosen as $\Delta u/h_o$=0.0000025. About 3000 steps were performed.

As an initial stress state, a K_0-state with $\sigma_{22}=\gamma_d x_2$ and $\sigma_{11}=K_0\gamma_d x_2$ was assumed in the specimen; x_2 is the vertical coordinate measured from the top of the specimen, γ_d=16.5 kN/m^3 denotes the initial volumetric weight and K_0=0.50 is the earth pressure coefficient at rest. Then, a uniform confining pressure of σ_c=200 kPa was prescribed.

When a stochastic distribution of the initial void ratio was assumed, two different specimen sizes were used: 40×140 mm^2 (medium size) and 320×1120 mm^2 (large size). The mean value of the initial void ratio was $\overline{e}_0$ =0.60 (initially dense sand). First, we assumed a strong correlation of the initial void ratio e_o in both directions (λ_{x1}=1 and λ_{x2}=1 in Eqs.5.33) and a low standard deviation s_{e_o} =0.05 to simulate a careful specimen preparation method (Vardoulakis 1977). In this case, the range of significant correlation was approximately 40 mm. Next, the calculations were performed for a weak correlation of e_o in both directions (λ_{x1}=3 and λ_{x2}=3 in Eqs 5.33) and a large standard deviation s_d =0.10 (the correlation range was about 20 mm). Next, The generated fields were classified according to two parameters: the mean value of the initial void ratio and the gap between the lowest and the highest value of the initial void ratio. The joint probability distribution (so-called "ant hill") is presented in Fig.5.61 (λ_{x1}=1, λ_{x2}=1, s_d =0.05). One dot represents one random vector described by its mean value and the difference between its extreme values. Two variable domains were divided in 12 intervals of equal probabilities. Next, according to the Latin hypercube sampling assumptions, 12 random numbers in the range 1-12 were generated (one number appeared only once) using the uniform distribution. The generated numbers formed the following 12 pairs: 1 – 2, 2 – 3, 3 – 7, 4 – 9, 5 – 8, 6 – 4, 7 – 6, 8 – 1, 9 – 11, 10 – 12, 11 – 5 and 12 – 10. According to these pairs, the appropriate areas (subfields) were selected (they are presented as rectangles in Fig.5.61). From each subfield only one realization was chosen and used as the input data to the FEM calculations. In this way the results of 12 realizations were analyzed.

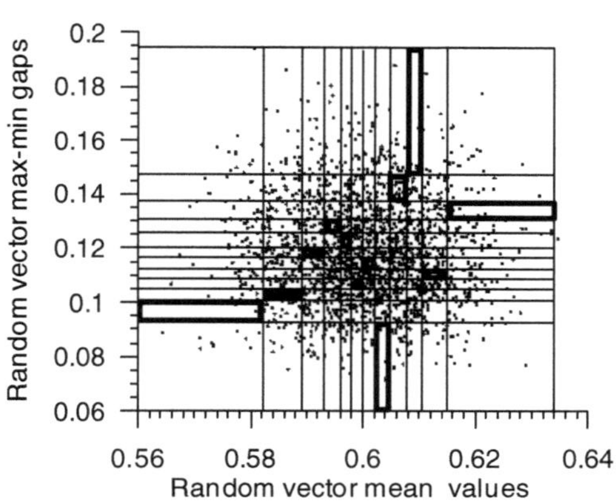

Fig. 5.61. Latin hypercube method: selection of 12 pairs of random samples: 1 – 2, 2 – 3, 3 – 7, 4 – 9, 5 – 8, 6 – 4, 7 – 6, 8 – 1, 9 – 11, 10 – 12, 11 – 5 and 12 – 10 (strong correlation, low standard deviation, initially dense specimen)

Deterministic size effect (smooth boundaries)

Fig.5.62 shows results for an initially dense sand during plane strain compression for six different specimen sizes $b_o \times h_o$ (same b_o/h_o): 10×35 mm^2, 20×70 mm^2, 40×140 mm^2, 80×280 mm^2, 160×560 mm^2 and 320×1112 mm^2, with a uniform distribution of the initial void ratio e_o=0.60 and the lateral confining pressure p=200 kPa. Presented is the evolution of the normalized vertical force $P/(\sigma_c b_o l)$ versus the normalized vertical displacement of the top boundary u_2^t/h_o (Tejchman and Górski 2008b). In addition, the deformed FE meshes for different sizes at u_2^t/h_o=0.075 with the distribution of the distribution of equivalent total strain $\bar{\varepsilon} = \sqrt{\varepsilon_{ij}\varepsilon_{ij}}$ are additionally shown in Fig.5.62. Tab.6.3 presents overall internal friction angles at peak ϕ^p and residual friction angles ϕ_{cr} (at u_2^t/h_o=0.10), vertical strain corresponding to the maximum force u_2^t/h_o and normalized shear zone thickness t_s/d_{50}. The overall friction angle was calculated from the formula including the principal stresses on the basis of the Mohr's circle (Eq.4.1).

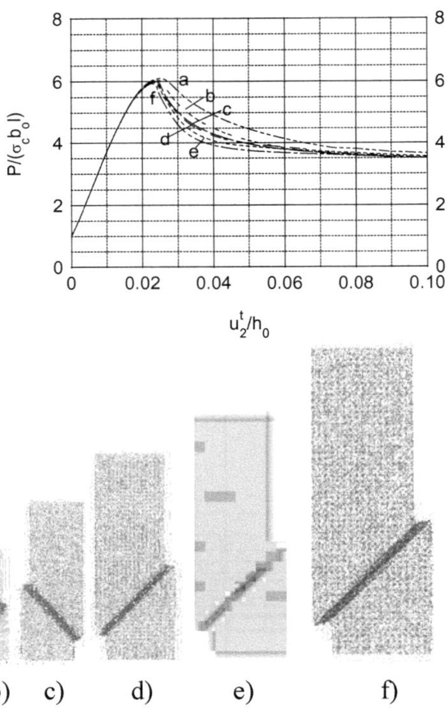

Fig. 5.62. Evolution of mobilized internal friction angle φ versus normalized vertical displacement of the top edge u_2^t/h_o and deformed FE mesh with the distribution equivalent total strain measure during plane strain compression with an initially dense specimen and uniform distribution of initial void ratio (e_o=0.60): a) 10×35 mm^2, b) 20×70 mm^2, c) 40×140 mm^2, d) 80×280 mm^2, e) 160×560 mm^2, f) 320×1120 mm^2

Table 5.3. The values of peak friction angles ϕ^p, residual friction angles ϕ_{cr}, vertical strain corresponding to the maximum vertical force u_2^t/h_o, normalized shear zone thickness t_s/d_{50} (uniform distribution of initial void ratio with e_o=0.60, smooth boundaries)

Specimen size $b_o \times h_o$	ϕ^p [°]	ϕ_{cr} [°]	u_2^t/h_o	t_s/d_{50}
10×35 mm^2	45.92	35.9	0.0250	7
20×70 mm^2	45.71	34.7	0.0240	11
40×140 mm^2	45.62	34.8	0.0236	14
80×280 mm^2	45.55	35.4	0.0233	28
160×560 mm^2	45.48	35.2	0.0231	46
320×1120 mm^2	45.38	34.5	0.0229	46

The resultant vertical force on the specimen top increases first, shows a pronounced peak, drops later and reaches then a residual value. The strength increases and the softening rate decreases with decreasing specimen size. The peak friction angle ϕ_p insignificantly increases, only from ϕ_p=45.38° (32×1120 mm^2) to ϕ_p=45.92° (10×35 mm^2), due to a similar failure mechanism and pressure level. The vertical strain corresponding to the peak grows with decreasing specimen size (from 2.29% up to 2.50%). The residual friction angle is similar for all specimen sizes, it changes between 34.5° and 36° at u_2^t/h_o=7.5%.

The thickness of the shear zone at mid-point of the specimen increases with increasing specimen size; it varies between 7×d_{50} up to 46×d_{50}. The thickness of the shear zone for the specimen sizes larger than 40×140 mm^2 is certainly influenced by the mesh discretisation (as it was noted) – it is too large. The thickness of the shear zone was determined on the basis of shear deformation and Cosserat rotation. To define the edges of the shear zone, we assumed that the Cosserat rotation larger than 0.1 occurred in the shear zone. The calculated thickness for the specimen of 40×140 mm^2, t_s=14×d_{50}, is similar as the observed thickness, t_s=15×d_{50}, during plane strain compression tests with dense sand (e_o=0.60) at σ_c=200 kPa (Vardoulakis 1977). The calculated inclination of the shear zone (about θ=53°-54°) is also close to the experiment with dense sand (55°-60°).

Deterministic size effect (very rough boundaries)

Results for an initially dense sand during plane strain compression for 3 different specimen sizes $b_o \times h_o$ (geometrically similar): 10×35 mm^2, 40×140 mm^2 and 320×1112 mm^2 with a uniform distribution of the initial void ratio e_o=0.60, lateral confining pressure of 200 kPa and very rough horizontal boundaries are shown in Fig.5.63 and Tab.5.4.

The evolution of the normalized vertical force is similar as for smooth plates. The friction angle at peak ϕ_p increases from: ϕ_p=45.4° (32×1120 mm^2) to ϕ_p=45.7° (10×35 mm^2) (the values are almost identical as for smooth plates). The residual friction angle changes between 39.5° and 40.4° at u_2^t/h_o=10%. It is significantly higher (by about 5°) than for smooth plates.

Table 5.4. The values of peak friction angles ϕ^p, residual friction angles ϕ_{cr}, vertical strain corresponding to the maximum vertical force u_2^t/h_o, normalized shear zone thickness t_s/d_{50} (uniform distribution of initial void ratio with e_o=0.60, very rough boundaries)

Specimen size $b_o \times h_o$	ϕ^p [°]	ϕ_{cr} [°]	u_2^t/h_o	t_s/d_{50}
10×35 mm^2	45.70	39.40	0.0235	9
40×140 mm^2	45.60	40.40	0.0239	13
320×1120 mm^2	45.40	39.70	0.0234	57

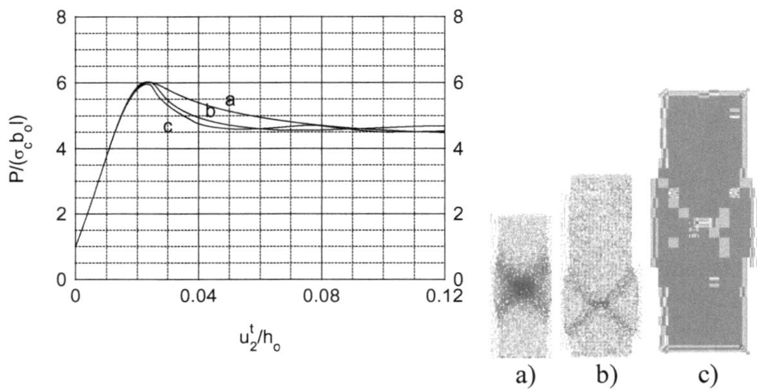

Fig. 5.63. Evolution of mobilized internal friction angle φ versus normalized vertical displacement of the top edge u_2^t/h_o and deformed FE mesh with the distribution equivalent total strain measure during plane strain compression with an initially dense specimen and uniform distribution of initial void ratio: a) 10×35 mm^2, b) 40×140 mm^2, c) 320×1120 mm^2

In contrast to smooth plates, two symmetric intersecting shear zones always appear inside of the specimen.

Statistical size effect (strong correlated field, low standard deviation)

a) medium sand specimen (smooth boundaries, specimen 40×140 mm^2)

12 selected random samples using Latin hypercube sampling are shown in Fig.5.61. The evolution of the normalized vertical force with respect to the initial specimen width b_o, $P/(\sigma_c b_o l)$ (Fig.5.64a), and with respect to the actual specimen width b, $P/(\sigma_c bl)$ (Fig.5.64b), versus the normalized vertical displacement of the top boundary u_2^t/h_o and the deformed FE meshes at u_2^t/h_o=0.075 with the distribution of equivalent total strain $\bar{\varepsilon} = \sqrt{\varepsilon_{ij}\varepsilon_{ij}}$ are shown in Fig.5.65, respectively.

The evolution of the normalized vertical force depends of the location of the shear zone. If this hits the top boundary, the residual normalized vertical force $P/(\sigma_c b_o l)$ has a increasing tendency (caused by an increase of the specimen's width). If the shear zone intersects the vertical sides the evolution of the residual normalized vertical

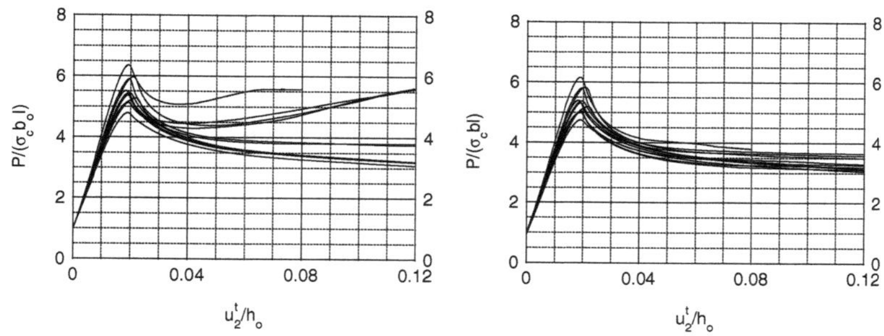

a) b)

Fig. 5.64. Evolution of normalized vertical force $P/(\sigma_c b_o l)$ (a) and $P/(\sigma_c bl)$ (b) versus normalized vertical displacement of the upper edge u_2^t/h_o during plane strain compression with a dense specimen for 12 different random fields of e_o ($\bar{e}_o$ =0.60, specimen size 40×140 mm^2, smooth boundaries, λ_{x1}=1, λ_{x2}=1, s_{eo}=0.05)

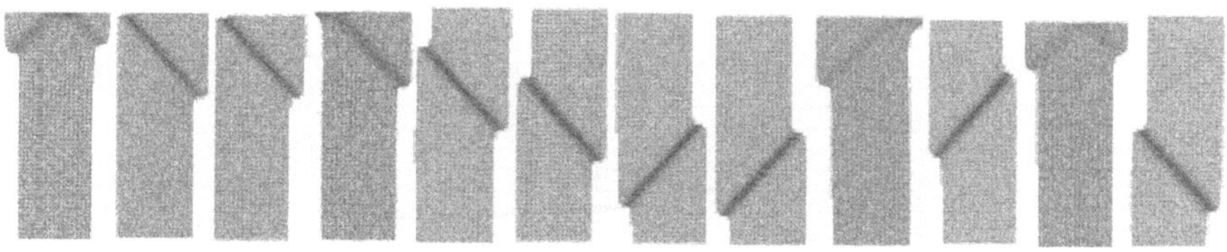

a) b) c) d) e) f) g) h) i) j) k) l)

Fig. 5.65. Deformed FE mesh with the distribution equivalent total strain measure during compression with a dense specimen for 12 different random fields of e_o ($\bar{e}_o$ =0.60, specimen size 40×140 mm^2, smooth boundaries, λ_{x1}=1, λ_{x2}=1, s_{eo}=0.05)

force $P/(\sigma_c b_o l)$ is similar as for a uniform distribution of e_o. The shear zone develops inside of the specimen somewhere at the weakest spot, depending on the initial distribution of e_o (Fig.5.65). The shear zone width changes between 11×d_{50} and 17×d_{50}.

The estimated mean values and standard deviations were: for the peak friction angle $\hat{\phi}_p$ =43.67° and $\hat{s}_{\phi}$ =1.52, the normalized horizontal displacement of the top boundary u_1^t/h_o corresponding to the peak $\hat{u}_1^t / h_o$ =0.0195 and $\hat{s}_{u_1^t} / h_o$ =0.00092 and the normalized shear zone thickness $\hat{t}_s / d_{50}$ =14 and $\hat{s}_{t_s} / d_{50}$ =1.7. The peak friction angle ($\hat{\phi}_p$ =43.74°) was by 2° smaller than with a uniform initial void ratio (ϕ_p =45.62°), whereas the shear zone thickness was similar. The residual friction angle (ϕ_{cr}=33.4° at u_2^t/h_o=7.5%) was by 1°-2° smaller. The residual friction angle was slightly larger if the shear zone hit the top boundary (Fig.5.64b).

b) large sand specimen (smooth boundaries, specimen 320×1120 mm^2)

The FE results (Figs.5.66 and 5.67) for a large sand specimen of (320×1120 mm^2) are qualitatively the same as for a medium sand specimen (however, their scattering is smaller). For 3 cases, a shear zone hits a top boundary, and the residual normalized vertical force $P/(\sigma_c b_o l)$ is higher.

The estimated expected values and standard deviations were respectively: for the peak friction angle $\hat{\phi}_p$ =44.22° and $\hat{s}_{\phi}$ =0.37°, the normalized horizontal displacement of the top boundary u_1^t/h_o corresponding to the peak $\hat{u}_1^t / h_o$ =0.0020 and $\hat{s}_{u} / h_o$ =0.00062 and the normalized shear zone thickness $\hat{t}_s / d_{50}$ =70.3 and $\hat{s}_{t} / d_{50}$ =14.59. The peak friction angle ($\hat{\phi}_p$ =44.23°) is by 1.0° smaller than this with the uniform initial void ratio (ϕ_p =45.38°). The shear zone thickness (t_s/d_{50}=70) is larger than with a uniform initial void ratio t_s/d_{50}=34. The residual friction (ϕ_{cr}=34.5° at u_2^t/h_o=7.5%) is by 1.5° smaller.

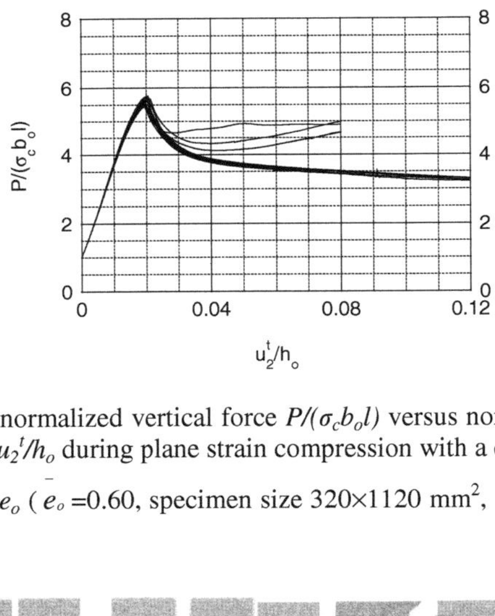

Fig. 5.66. Evolution of normalized vertical force $P/(\sigma_c b_o l)$ versus normalized vertical displacement of the upper edge u_2^t/h_o during plane strain compression with a dense specimen for 12 different random fields of e_o ($\bar{e}_o$ =0.60, specimen size 320×1120 mm^2, smooth boundaries, λ_{x1}=1, λ_{x2}=1, s_{eo}=0.05)

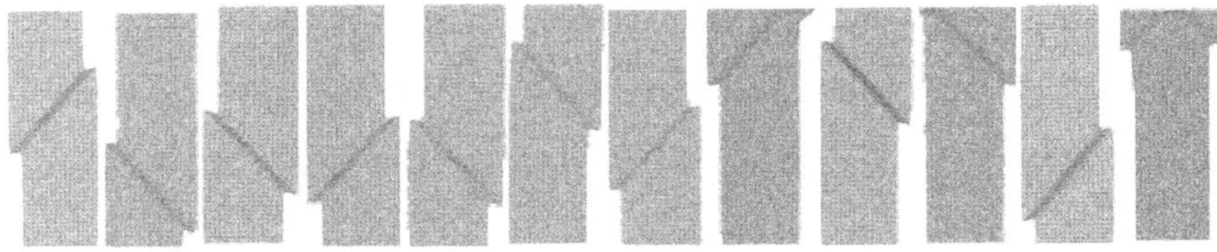

a) b) c) d) e) f) g) h) i) j) k) l)

Fig. 5.67. Deformed FE mesh with the distribution equivalent total strain measure during compression with a dense specimen for 12 different random fields of e_o ($\bar{e}_o$ =0.60, specimen size 320×1120 mm^2, smooth boundaries, λ_{x1}=1, λ_{x2}=1, s_{eo}=0.05)

c) medium sand specimen (very rough boundaries, specimen 40×140 mm^2)
The results are demonstrated in Figs.5.68 and 5.69 (for the same random fields of e_o as in the case of smooth boundaries).

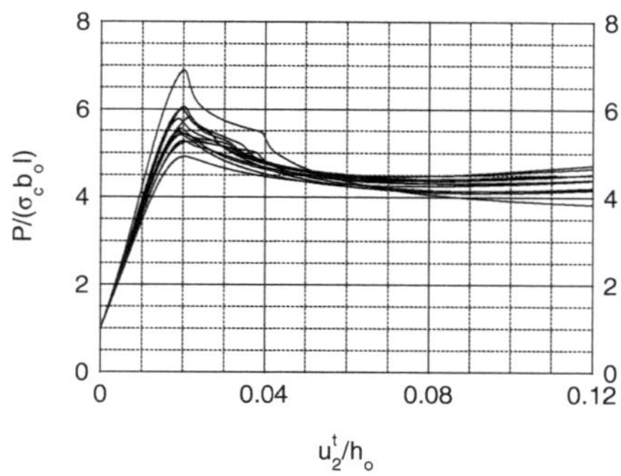

Fig. 5.68. Evolution of normalized vertical force $P/(\sigma_c b_o l)$ versus normalized vertical displacement of the upper edge u_2^t/h_o during plane strain compression with a dense specimen for 12 different random fields of e_o ($\bar{e}_o$ =0.60, specimen size 40×140 mm^2, very rough boundaries, λ_{x1}=1, λ_{x2}=1, s_{eo}=0.05)

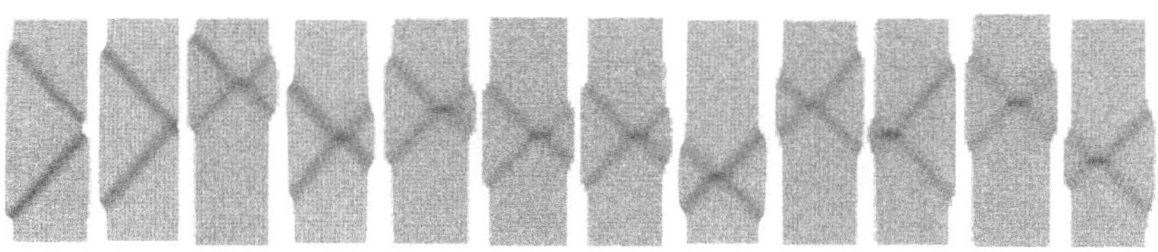

a) b) c) d) e) f) g) h) i) j) k) l)

Fig. 5.69. Deformed FE mesh with the distribution equivalent total strain measure during compression with a dense specimen for 12 different random fields of e_o ($\bar{e}_o$ =0.60, specimen size 40×140 mm^2, very rough boundaries, λ_{x1}=1, λ_{x2}=1, s_{eo}=0.05)

In all cases, two non-symmetric shear zones appear inside of the specimen. The peak friction angle ($\hat{\phi}_p$ =45.24°) is by 0.5° smaller than with a uniform initial void ratio (ϕ_p =45.60°). The shear zone thickness (t_s/d_{50}=13.3) is slightly smaller than this with the uniform initial void ratio t_s/d_{50}=14. The residual friction angle (ϕ_{cr}=38.77° at u_2^t/h_o=10%) is smaller by 1.7°.

Statistical size effect (weak correlated field, high standard deviation)
The calculations were carried out with initially dense sand specimens, and smooth and very rough boundaries (specimen size was 40×140 mm2). 12 selected random samples using Latin hypercube sampling are shown in Fig.5.70. The evolution of the

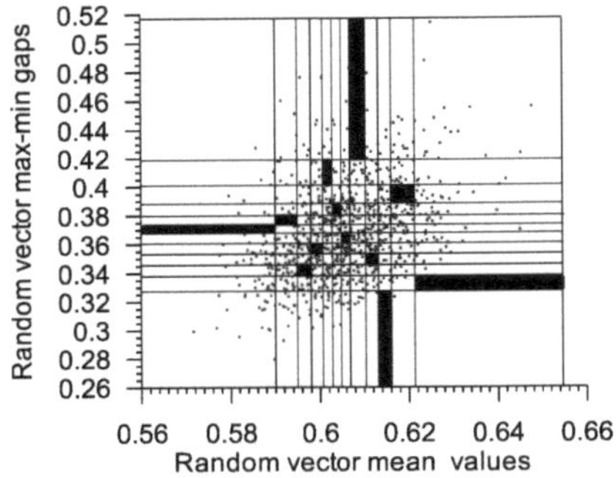

Fig. 5.70. Selection of 12 pairs of random samples using Latin hypercube sampling (initially dense sand, weak correlation, high standard deviation)

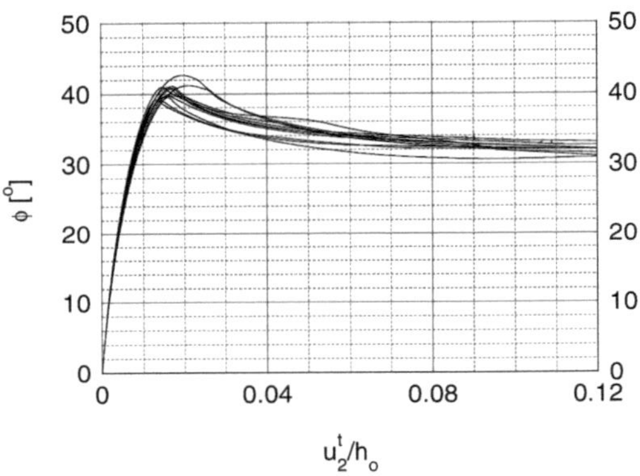

Fig. 5.71. Evolution of mobilized internal friction angle φ versus normalized vertical displacement of the upper edge u_2^t/h_o during plane strain compression with an initially dense specimen for 12 different random fields of e_o at smooth horizontal boundaries

mobilized friction angle with respect versus the normalized vertical displacement of the top boundary and the deformed FE meshes at u_2^t/h_o=7.5% with the distribution of the equivalent total strain $\bar{\varepsilon}=\sqrt{\varepsilon_{ij}\varepsilon_{ij}}$ are shown in Figs.5.71 and 5.72 (smooth horizontal boundaries), and in Figs.5.73 and 5.74 (very rough horizontal boundaries), respectively. In addition, the deformed FE meshes after the peak at u_2^t/h_o=2.15% with the distribution of the equivalent total strain $\bar{\varepsilon}=\sqrt{\varepsilon_{ij}\varepsilon_{ij}}$ are demonstrated in Fig.5.75 for very rough horizontal boundaries.

During deformation, first, a pattern of shear zones can be observed in the specimen independently of the boundary roughness. Next, deformation continues to localize within an inclined single shear zone for smooth boundaries (just before the peak on the load-displacement curve), Fig.5.75. This single shear zone always develops inside of the specimen somewhere at the weakest spot, depending on the initial distribution

a) b) c) d) e) f) g) h) i) j) k l)

Fig. 5.72. Deformed FE mesh with the distribution equivalent total strain measure during compression with an initially dense specimen for 12 different random fields of e_o at smooth horizontal boundaries (u_2^t/h_o=7.5%)

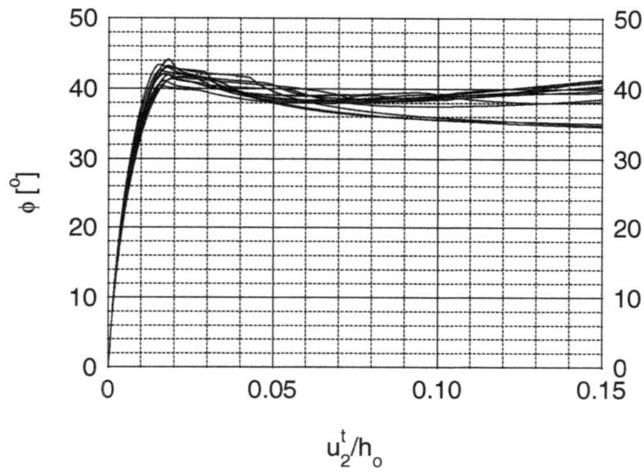

Fig. 5.73. Evolution of mobilized internal friction angle φ versus normalized vertical displacement of the upper edge u_2^t/h_o during plane strain compression with an initially dense specimen for 12 different random fields of e_o at very rough horizontal boundaries

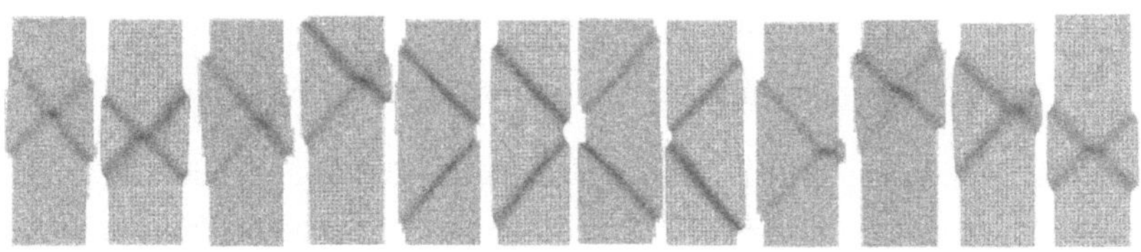

a) b) c) d) e) f) g) h) i) j) k) l)

Fig. 5.74. Deformed FE mesh with the distribution equivalent total strain measure during compression with an initially dense specimen for 12 different random fields of e_o at very rough horizontal boundaries (u_2^t/h_o=7.5%)

of e_o (Fig.5.75). In some cases (in contrast to a uniform distribution of e_o), it may hit the top boundary (cases '*a*', '*b*', '*d*', '*e*', '*g*', '*j*' and '*l*' in Fig.5.75). The shear zone thickness changes between t_s/d_{50}=11 and t_s/d_{50}=17. The mean shear zone thickness (t_s/d_{50}=11.6) is smaller by 15% than with a uniform initial void ratio (t_s/d_{50}=14). The mean shear inclination to the bottom is 53°.

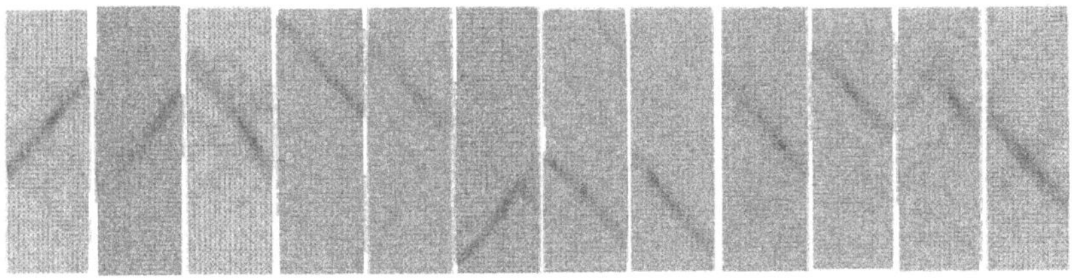

a) b) c) d) e) f) g) h) i) j) k) l)

Fig. 5.75. Deformed FE mesh with the distribution equivalent total strain measure during compression with an initially dense specimen for 12 different random fields of e_o at very rough horizontal boundaries (u_2^t/h_o=2.15%)

For very rough horizontal plates, two non-symmetric shear zones are obtained inside of the specimen at residual state. The shear zones are created one after another (it can be concluded from Figs.5.72 and 5.73). The mean shear zone thickness (t_s/d_{50}=18) is smaller by 10% than with a uniform initial void ratio (t_s/d_{50}=20). Both shear zones can have a different thickness. Their mean inclination to the bottom is 50°.

The evolution of the mobilized internal friction angle is similar for smooth boundaries. The friction angle increases first, shows a pronounced peak, drops later and reaches then a residual value. However, in the case of very rough boundaries, the mobilized friction angle drops after the peak and next can decrease reaching its asymptote (cases '*e*', '*g*' and '*h*' in Fig.5.72) or can increase (in remaining cases in Fig.5.72). The peak and residual friction angles are: ϕ^p=45.2° and ϕ_{res}=34.5° (smooth plates), and ϕ^p=45.6° and ϕ_{res}=40.5° (very rough plates), respectively. The mean peak friction angle for smooth plates ($\bar{\phi}_p$ =40.7°) is by 4.5° smaller than with a uniform initial void ratio (ϕ_p =45.2°). The mean residual friction angle ($\bar{\phi}_{cr}$ =32.0°) is smaller by 2.5° (smooth walls). In turn, for very rough plates, the mean peak friction angle ($\bar{\phi}_p$ =42.2°) is by 3.5° smaller than with a uniform initial void ratio (ϕ_p =45.6°). The mean residual friction angle ($\bar{\phi}_{cr}$ =38.5°) is smaller by 2.0° (very rough plates).

The calculated results of the peak friction angles ϕp, residual friction angles ϕcr, normalized shear zone thickness t_s/d_{50} and shear zone inclination to the bottom θ are summarized in Tab.5.5.

Table 5.5. The values of peak friction angles ϕ^p, residual friction angles ϕ_{cr}, normalized shear zone thickness t_s/d_{50} and shear zone inclination to the bottom θ

Initial density	Boundary roughness	ϕ^p [°]	ϕ_{cr} [°]	t_s/d_{50}	θ [°]
uniform dense	smooth	45.2	34.5	14	53
uniform dense	very rough	45.6	40.5	20	50
dense (random)	smooth	40.7	32.0	12	53
dense (random)	very rough	42.2	38.5	18	50

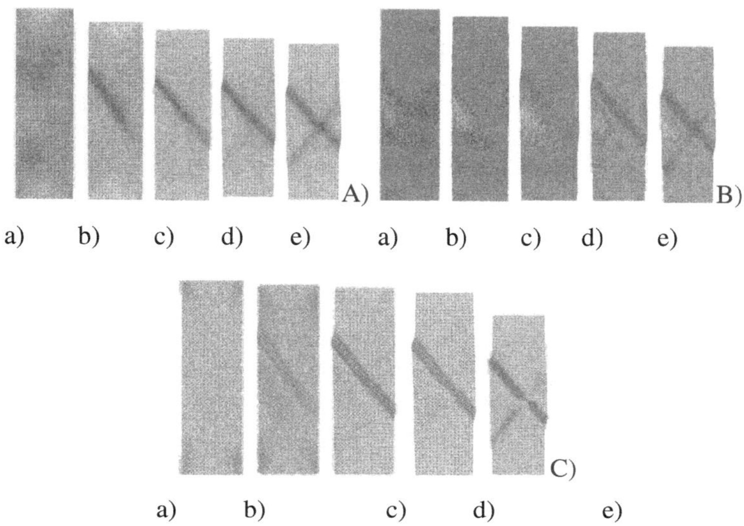

Fig. 5.76. Evolution of equivalent total strain measure (A), void ratio (B) and Cosserat rotation (C) field during plane strain compression (random field of Fig.5.220d) at: a) u_2^t/h_o=0.011, b) u_2^t/h_o=0.021, c) u_2^t/h_o=0.028, d) u_2^t/h_o=0.036, e) u_2^t/h_o=0.042 (e_o =0.60, very rough boundaries, specimen size 40×140 mm^2)

Finally, the evolution of deformation, void ratio and Cosserat rotation is shown for one selected random field with 2 intersecting shear zones (medium specimen size of Fig.5.69d, very rough boundaries), Fig.5.76. During deformation, first, a pattern of shear zones can be observed in the specimen (strain localization starts in corners). Next, deformation continues to localize within an inclined single shear zone (just before the peak at u_2^t/h_o=2%). This shear zone becomes well visible after the peak at u_2^t/h_o=2.1%. Next, a second intersecting shear zone occurs at u_2^t/h_o=3.5% (it causes a small jump on the load-displacement curve, Fig.5.76). In the specimen region beyond the shear zone, small changes of void ratio are visible, what is also in agreement with experiments by Yoshida et al. (1994).

The following conclusions can be drawn from our non-linear FE investigations of the effect of the stochastic distribution of the initial void ratio and roughness of horizontal boundaries on the shear localization in dry sand specimens during plane strain compression under constant lateral pressure:

- Shear localization always occurs during deformation. A stochastic distribution of the initial void ratio influences the location of shear zones and type of shear localization (in particular for very rough boundaries).
- Shear localization can be created in the form of a single shear zone if horizontal boundaries are smooth or in the form of two shear zones if horizontal boundaries are very rough. A single shear zone can reflect from a smooth top boundary if the specimen is initially dense. In turn, double shear zones can branch or intersect each other if the specimen has very rough horizontal boundaries. They are created successively or simultaneously slightly before the peak.

- The distribution of the initial void ratio has a large effect on the mobilized friction angle in initially dense specimen and an insignificant effect in initially loose specimens. The mean peak shear resistance with a stochastic distribution of the initial void ratio is always smaller than with the uniform distribution of the initial void ratio. For a large standard deviation and weakly correlated fields of the initial void ratio in both directions, the differences in the peak internal friction angle at peak and residual state are significant in initially dense specimens, i.e. 4.5° (at peak) and 2.5° (at residual state) for smooth plates and 3.5° (at peak) and 2.0° (at residual state) for very rough plates. The thickness of the shear zone is smaller by 10%-15% when the distribution of the initial void ratio is stochastic.

The obtained types of shear localization are qualitatively in agreement with experimental results of the Grenoble University (Desrues and Viggiani 2004) wherein single and multiple shear zones were observed in sand specimens during plane strain compression tests (Tab.2.2).

5.5 Non-coaxiality and Stress-Dilatancy Rule

In conventional plasticity theory, the stress and the plastic strain rate are assumed to be coaxial, i.e. with the same principal axes. The assumption of coaxiality is controversial for tests involving the rotation of the principal stress, e.g. simple shear tests (Cole 1967). As pointed out by Hill (1950), coaxiality can be seen as a consequence of material isotropy. As shown by Wu (1998), anisotropy (inherent or induced) leads inevitably to non-coaxiality for problems involving rotation of the principal stress axes. The simple shear tests reported by Roscoe et al. (1967) showed that coaxiality can be expected for fairly wide range of stress and strain. Their experimental finding was then confirmed by Arthur and Menzies (1972). However, non-coaxiality was observed when shear bands were involved in the tests (Arthur and Assadi (1977). Afterwards, there are extensive experimental data showing that plastic flow in granular materials is non-coaxial in loadings involving rotation of the principal stress and shear localization (Gutierrez and Vardoulakis 2007, Miura et al. 1986, Gutierrez et al. 1993, Gutierrez and Ishihara 2000, Vardoulakis and Georgopoulos (2005). The degree of non-coaxiality seems to depend on the magnitude of rotation. The angles of non-coaxiality can be as high as 30° in laboratory tests, e.g. in a hollow cylindrical apparatus (Gutierrez and Ishihara 2000). Such rotation tends to have influence on the post-bifurcation response.

The non-coaxiality can be expressed as the deviation angle between the principal stress and the principal plastic strain-rate directions

$$\xi = |\alpha - \beta|, \tag{5.39}$$

where the angle α is the orientation of the major principal stress σ_1

$$\tan 2\alpha = \frac{\sigma_{12} + \sigma_{21}}{\sigma_{11} - \sigma_{22}} \tag{5.40}$$

and β is the orientation of the major principal plastic strain rate $\dot{\varepsilon}_1^p$ ('p' denotes plastic)

$$\tan 2\beta = \frac{\dot{\varepsilon}_{12}^p + \dot{\varepsilon}_{21}^p}{\dot{\varepsilon}_{11}^p - \dot{\varepsilon}_{22}^p}. \tag{5.41}$$

The rate of energy dissipation can be expressed as follows (Gutierrez and Vardoulakis 2007):

$$\dot{W}^p = s \times \dot{v}^p + c \times t \times \dot{\gamma}^p \tag{5.42}$$

with the mean stress s and the shear stress t as

$$s = 0.5(\sigma_1 + \sigma_3), \tag{5.43}$$

and the volumetric strain rate $\dot{v}^p$ and shear strain rate $\dot{\gamma}^p$

$$\dot{v}^p = \dot{\varepsilon}_1^p + \dot{\varepsilon}_3^p, \qquad \dot{\gamma}^p = \dot{\varepsilon}_1^p - \dot{\varepsilon}_3^p, \tag{5.44}$$

and the non-coaxiality parameter c related to the deviation angle ξ (Eq.6.39)

$$c = \cos 2\xi. \tag{5.45}$$

The stresses σ_1 and σ_3 are the principal stresses and the strain rates $\dot{\varepsilon}_1^p$ and $\dot{\varepsilon}_3^p$ are the principal plastic strain rates. The dilatancy angle is defined as

$$\sin\psi = \frac{\dot{v}^p}{\dot{\gamma}^p}. \tag{5.46}$$

The following non-coaxial stress-dilatancy relationship can be obtained from Eq.5.42 (Gutierrez and Ishihara 2000)

$$\sin\psi = \sin\phi_{cx} - c \times \sin\phi \tag{5.47}$$

with

$$\sin\phi = \frac{t}{s}, \tag{5.48}$$

and ϕ_{cx} as the friction angle at vanishing dilatancy, ψ as the mobilized dilatancy angle and ϕ as the mobilized friction angle. According to Gutierrez and Vardoulakis (2007), ϕ_{cx} is the friction angle at the transition from initial contractancy to dilatancy. The parameter c serves as a measure of the effect of the deviation angle ξ on the energy dissipation and the stress-dilatancy rule. The parameter c can vary from about 1.0 at the point of bifurcation to about 0.5 (ξ=30°) at the residual state (Gutierrez and Vardoulakis 2007). The parameter c is not constant but changes with stress level and the direction of the stress increment. According to Eq.5.47, the dilatancy angle is first negative (material undergoes contractancy), then the material shows a significant dilatancy which is the highest at the peak friction angle (the dilatancy angle is negative). Next, the dilatancy angle diminishes down to zero in the course of shearing reaching a critical (residual) state. Gutierrez and Vardoulakis (2007) separated ϕ_{cx} in Eq.5.47 into two different parts since the stress-dilatancy relationship consists of two distinct parts (before and after bifurcation) and intersects the *t/s* axis at two different points.

The stress-dilatancy rule can be also expressed according to Vardoulakis and Georgopoulos (2005) as a modified Taylor's stress-dilatancy hypothesis (Taylor 1948) in the following manner

$$\tan\psi = \tan\phi_{cx} - c\times\tan\phi \ . \tag{5.49}$$

In turn, Wan and Guo (2004) proposed a stress-dilatancy rule without taking into account non-coaciality

$$\tan\psi = \frac{4}{3}\frac{(\sin\phi_{cx} - \sin\sin\phi)}{(1-\sin\phi_{cx}\sin\sin\phi)} \ . \tag{5.50}$$

Their rule was extended later by the effect of fabric.

The FE calculations of a plane strain compression test were performed with the specimen size $b_o \times h_o$=40×140 mm^2 (b_o – initial width, h_o – initial height) (Tejchman and Wu 2008b). Two sets of boundary conditions were assumed. First, the boundary conditions implied no shear stress imposed at the smooth top and bottom of the specimen. To preserve the stability of the specimen against horizontal sliding, the node in the middle of the top edge was kept fixed. To simulate a movable roller bearing in the experiment (Vardoulakis 1980), the horizontal displacements along the specimen bottom were constrained to move by the same amount from the beginning of compression. Unfortunately, this boundary condition cannot describe the initial rectangular deformations (since the displacements along the bottom moved in one direction). Comparative calculations were also performed with a very rough top and bottom boundary. In this case, the horizontal displacement and Cosserat rotation along both horizontal boundaries were assumed to be equal to zero.

As an initial stress state, a K_0-state with $\sigma_{22}=\gamma_d x_2$ and $\sigma_{11}=K_0\gamma_d x_2$ was assumed in the specimen; x_2 is the vertical coordinate measured from the top of the specimen, γ_d=16.5 kN/m^3 denotes the initial volumetric weight and K_0=0.50 is the earth pressure coefficient at rest. Then, a uniform confining pressure of σ_c=200 kPa was prescribed.

The initial void ratio in a granular specimen was assumed to be stochastically distributed to model volume fluctuations observed during experiments (Vardoulakis 1977). It was assumed to be spatially correlated using a truncated Gaussian random field (Section 5.3). One generated random field of e_o for a specimen of medium dense sand with a mean initial void ratio of 0.60 with a strong correlation of e_o in both directions and a low standard deviation (λ_{x1}=1, λ_{x2}=1, s_{e_o} =0.05) (Section 5.3) was chosen for the FE analysis.

Smooth boundaries

The numerical results for an initially dense specimen are shown in Figs.5.77-5.81 (smooth boundaries). Presented in Fig.5.77 is the evolution of the mobilized overall and local friction angle ϕ versus the normalized vertical displacement of the top boundary u_2^t/h_o and the deformed FE mesh at u_2^t/h_o=0.075 with the distribution of the distribution of equivalent total strain $\bar{\varepsilon} = \sqrt{\varepsilon_{ij}\varepsilon_{ij}}$. The overall friction angle was obtained from Eq.4.1. The local friction angle was determined at the mid-point of the shear zone (at $x_1=b_o/2$) based on the local principal stresses σ_1 and σ_3. The evolution of the Cosserat rotation at 3 points along the shear zone and the evolution of displacements in the mid-point of the shear zone (at $x_1=b_o/2$) are presented in Fig.5.78.

Fig.5.79 demonstrates the evolution of the mobilized global and local dilatancy angle ψ (mid-point of the shear zone) versus u_2^t/h_o. The global dilatancy angle was calculated either by Eq.5.46 ($\psi = \arcsin(\dot{\nu}/\dot{\gamma})$); plastic strain rates were replaced by total strain rates) or using the following formula (Gutierrez and Vardoulakis 2007) assuming that all deformation occurs inside the shear zone:

$$\psi = \theta - \tan^{-1}(\dot{u}_2/\dot{u}_1)\,, \tag{5.51}$$

where θ is the shear zone inclination against the bottom, and u_2 and u_1 are the vertical and horizontal displacements of the specimen. The relationship between the global dilatancy angle ψ (Eq.5.46) and the global friction angle ϕ (Eq.4.1) is shown in Fig.5.238. Finally, the evolution of the deviation angles α (Eq.6.40), β (Eq.5.41) and $\xi=\alpha-\beta$ (Eq.5.42) in the mid-point of the shear zone (at $x_1=b_o/2$) is depicted in Fig.5.81.

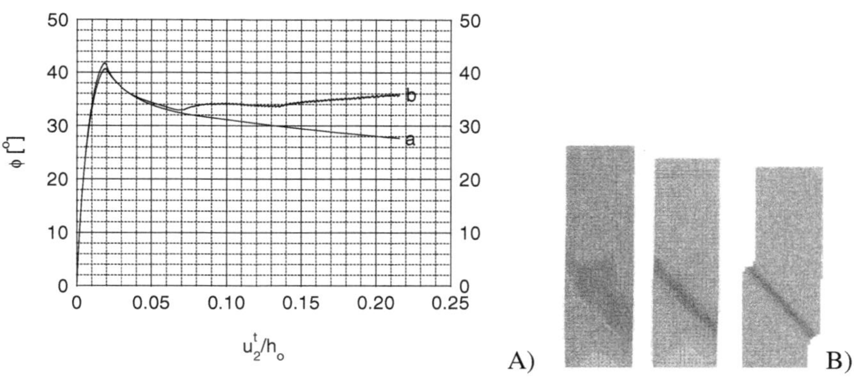

Fig. 5.77. FE results with smooth boundaries: A) evolution of mobilized overall friction angle (a) and mobilized local friction angle in the mid-point of the shear zone versus normalized vertical displacement of the upper edge u_2^t/h_o and B) deformed FE mesh with the distribution equivalent total strain at: a) u_2^t/h_o=0.018, b) u_2^t/h_o=0.021, c) u_2^t/h_o=0.071

Along with the deformation, the mobilized friction angle increases to reach a pronounced peak, drops gradually and reaches a residual state at large deformation. The overall peak friction angle is about ϕ_p=40.7° at u_2^t/h_o=0.019 (curve '*a*' in Fig.5.77A). For comparison, the overall residual friction angle is about ϕ_{res}=27.7° at u_2^t/h_o=0.21 The local peak friction angle in the mid-point of the shear zone is slightly higher than the global one (ϕ_p=41.8° at u_2^t/h_o=0.019). However, the local residual friction angle of the shear zone is found to be significantly higher, viz. ϕ_{res}=35° (curve '*b*' in Fig.5.77A).

Along with the deformation, a pattern of shear zones can be first observed in the specimen (Section 5.3). Next, the deformation continues to localize within an inclined single shear zone (Fig.5.77B). The shear zone is created certainly before the peak of u_2^t/h_o=0.019 on the basis of an increase of the Cosserat rotation (Figs.5.78A and 5.80 B). The thickness of the internal shear zone appearing inside of the specimen (at mid-point of the specimen) is about 14×d_{50} (Fig.5.77B). Its inclination against the bottom is about θ=54°. A Cosserat rotation of 0.1 was used to discriminate between points inside or outside the shear zone. The side boundaries of the specimen have a slight influence on the shear zone formation since the distribution of the Cosserat rotation is not uniform along the shear zone (Fig.5.78).

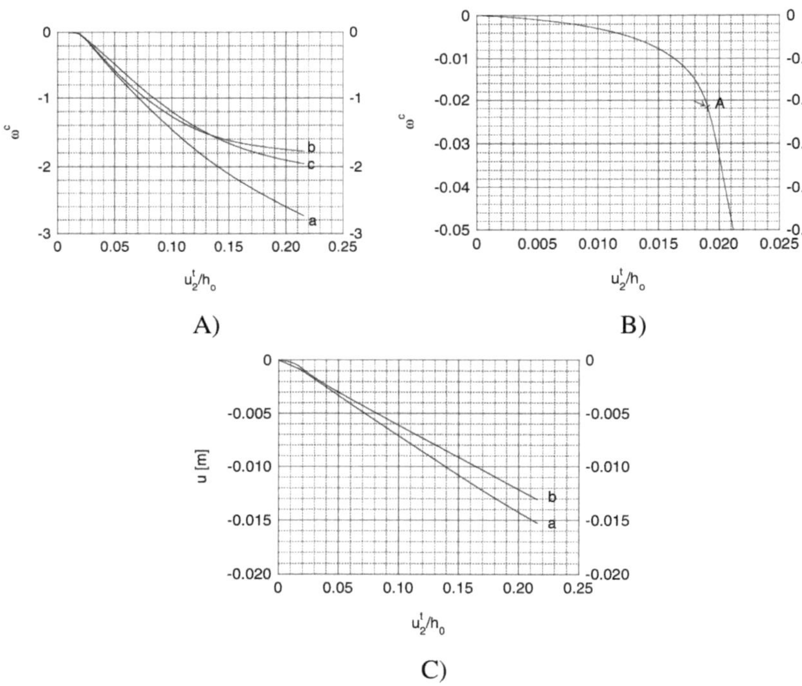

Fig. 5.78. FE results with smooth boundaries: A) evolution of Cosserat rotation in the shear zone: a) mid-point, b) left side, c) right side, B) evolution of Cosserat rotation in the mid-point of the shear zone at the beginning of loading (point 'A' – location of the peak friction angle), and C) evolution of displacements in the mid-point of the shear zone: a) u_2, b) u_1 versus normalized vertical displacement of the upper edge u_2^t/h

Globally (Eq.5.46), the specimen undergoes first contractancy (up to u_2^t/h_o=0.014) then dilatancy and finally contractancy (curve 'a' in Fig.5.79). The maximum global dilatancy angle is about ψ=8° (at u_2^t/h_o=0.02) and corresponds to the peak friction angle. Next, the dilatancy angle decreases down to ψ=-15° at u_2^t/h_o=0.20. A local peak dilatancy angle of a similar value of about 9° can be observed in the middle of the shear zone (curve 'c' in Fig.5.79). At residual state, the local dilatancy angle slightly increases. In the case of calculations of the global dilatancy angle by Eq.5.51 (curve 'b' in Fig.5.79), the behaviour is similar to the local behaviour in the shear zone. However, the peak global dilatancy angle is somewhat higher, viz. 20° (curve 'c' in Fig.5.79). It indicates that the assumption that all deformation occurs inside the shear zone cannot be fully justified at peak.

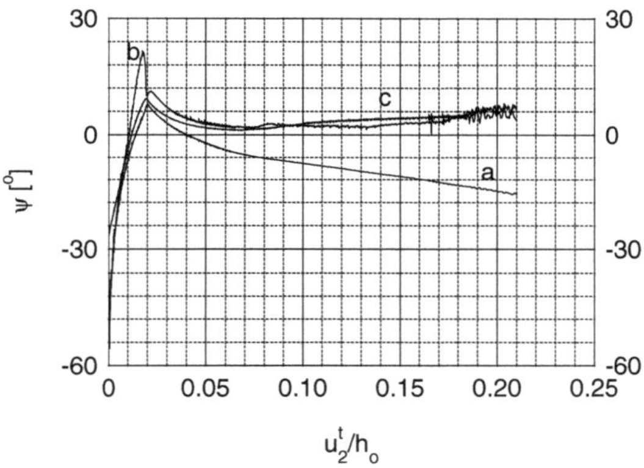

Fig. 5.79. FE results with smooth boundaries: evolution of mobilized global dilatancy angle (Eq.5.46) (a), mobilized global dilatancy angle (Eq.5.51) (b) and mobilized local dilatancy angle (Eq.5.46) in the mid-point of the shear zone (c) versus normalized vertical displacement of the upper edge u_2^t/h_o

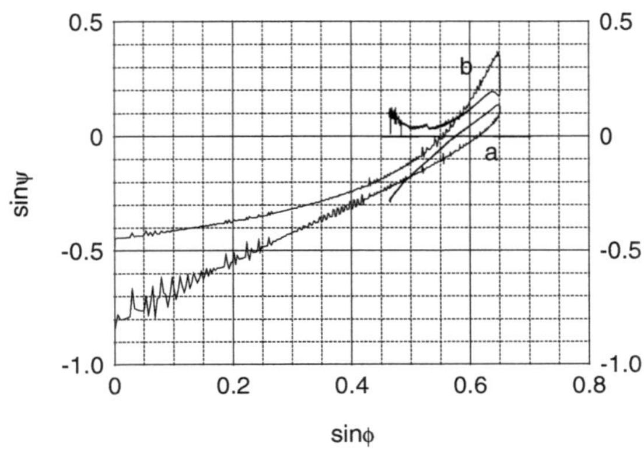

Fig. 5.80. FE results with smooth boundaries: relationship between the global dilatancy angle ψ and global friction angle ϕ (Eq.5.1): a) Eq.6.46, b) Eq.6.51

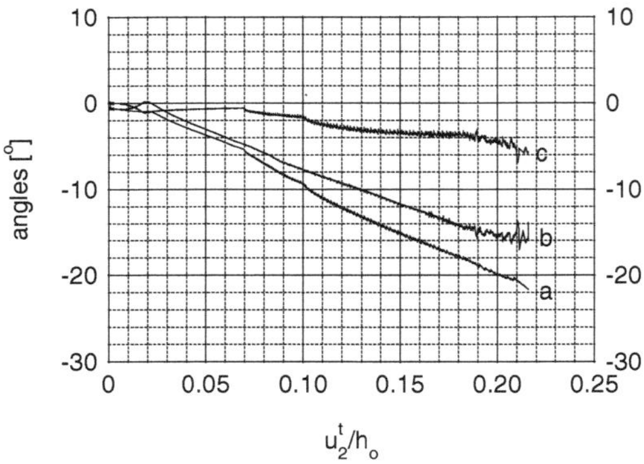

Fig. 5.81. FE results with smooth boundaries (mid-point of the shear zone): a) evolution of orientation of the major principal stress α, b) orientation of the major principal strain rate β (b) and c) non-coaxiality angle $\xi=\alpha-\beta$ versus normalized vertical displacement of the upper edge u_2^t/h_o

The friction angle ϕ_{cx} is ϕ_{cx}=38° (curve 'a' in Fig.5.80) and ϕ_{cx}=35° (curve 'b' in Fig.5.80). The stress-dilatancy behaviour (global and local in the shear zone) at residual state does not exactly follow the rules by Eqs.5.47, 6.49 and 6.50 at $u_2^t/h_o>0.07$ since the dilatancy angle can have an increasing tendency or a decreasing one reaching the contractancy region. The plot describing the evolution of the global dilatancy angle (Eqs.5.46) versus the global friction angle (Eq.4.1) in Fig.5.80 shows that two nearly linear curves exist (as in the experiments by Gutierrez and Vardoulakis 2007): one in the pre-peak region (the plot intersects the horizontal line of ψ=0° at about ϕ_{cx}=38°) and the second in the post-peak regime (the plot intersects the vertical line of ψ=0° at about ϕ_{cx}=34° (curve 'a' in Fig.5.80). In the case of the plot in Fig.5.80 describing the evolution of the global dilatancy angle versus the global friction angle (Eq.56.51), two non-linear curves exist. However, the plot intersects once the horizontal line of ψ=0° at about ϕ_{cx}=34° only in the pre-peak region.

The values of the dilatancy angle ψ calculated from Eqs.5.47, 5.49 and 5.50 (with ϕ_{cx}=38°, ϕ=40.7° and c=1) (curve 'a' in Fig.6.80) are significantly smaller (ψ=2°-5°). In turn, the values of ψ calculated from Eqs.5.47, 5.49 and 5.50 with ϕ_{cx}=34°, ϕ=40.7° and c=1 (curve 'b' in Fig.5.80) are more realistic (ψ=5°-10°).

The rotation of principal stresses and the principal strain-rate directions appears from the beginning of deformation and it increases almost linearly up to 16°-22° at u_2^t/h_o=0.21 (Fig.5.81). At peak, the deviation angle between the principal stress and the principal strain-rate directions is insignificant with ξ=1°. At the residual state, the deviation angle is not very much pronounced (ξ=6° at u_2^t/h_o=0.21) due to the moderate dilatancy. Thus, the effect of the non-coaxiality parameter, c=0.978-1.0, is found to be rather small during shearing. This numerical result agrees well with the experiments described by Vardoulakis and Georgopoulos (2005) on the same sand. However, this effect is much smaller than in the experiments on Nevada sand (c=0.5-1.0) (Gutierrez and Vardoulakis 2007).

A perusal of the experimental and numerical results (Vardoulakis 1977, 1980) shows, however, that the calculated stiffness is too high before the peak (in the hardening regime). The computed thickness of the shear zone is slightly larger (by 10-15%) than the thickness in experiments. Moreover, the calculated inclination of the shear zone to the horizontal is smaller than the experimental by 2°-4°. The evolution of the calculated global dilatancy angle by Eq.5.51 is qualitatively the same as in the experiments. However, significant volume fluctuations during shear zone evolution in laboratory were not numerically obtained in spite of the assumed initial spatially correlated fluctuations of the initial void ratio. It is necessary to investigate the effect of other spatially correlated random fields in initially very dense granular specimens. The initial evolution of displacements in the specimen (Fig.5.78C) differs slightly from the experiment due to the fact that the assumed boundary conditions along the bottom boundary do not allow for initial rectangular deformations at the beginning of loading (since horizontal displacements along the bottom move in one direction).

Very rough boundaries

The boundary roughness is found to have strong influence on the pattern of shear localization (Section 5.3). Instead of one shear zone, two intersecting shear zones occur with very rough boundaries (Fig.5.82B). They are created almost simultaneously before the peak (u_2^t/h_o=0.019) on the basis of the development of the Cosserat rotation (Fig.5.83). The thickness of the internal shear zones appearing inside of the specimen is again about $14\times d_{50}$ (Fig.5.82B). Their inclination against the bottom is about θ=50°.

As compared to smooth boundaries, the evolution of the mobilized friction angle is quite different. Along with deformation, the global friction angle increases to a pronounced peak, drops gradually and then increases slightly again (the residual state has not been reached). The global, peak friction angle is about ϕ_p=41.4° at u_2^t/h_o=0.020 (curve '*a*' in Fig.5.82A). This can be compared to a global friction angle of about ϕ_{res}=41° at large deformation of u_2^t/h_o=0.20; this is significantly higher than the global residual friction angle with smooth boundaries. The local peak friction angles in the shear zones are: ϕ_p=39.8° at u_2^t/h_o=0.018 (intersection point of two shear zones at point 'A', curve '*b*' in Fig.5.82A), ϕ_p=41.4° at u_2^t/h_o=0.020 (mid-point of one shear zone at point 'B', curve '*c*' in Fig.5.82A) and ϕ_p=42.3° at u_2^t/h_o=0.019 (mid-point of second shear zone at point 'C', curve '*d*' in Fig.5.82A), respectively. The local friction angles in the shear zones at large deformation of u_2^t/h_o=0.020 are significantly smaller, viz. 34°-36°. However, the local friction angle in the intersecting point of two shear zones (at large deformation) of 41° is comparable to the global one.

Globally (Eq.5.51), the specimen undergoes first contractancy (up to u_2^t/h_o=0.014) and then dilatancy. The dilatancy angle after reaching its maximum value peak continuously decreases down to 2° (Fig.5.84A). The maximum global dilatancy angle (ψ=8.5° at u_2^t/h_o=0.022) is similar as in the case of smooth boundaries. In the middle of the shear zones, the peak local dilatancy angle is also similar (Fig.5.235B). However, the local values at large deformation are different. Neither the stress-dilatancy rule of Eqs.5.52, 5.54 nor 5.55 is suitable for very rough boundaries in the case of the point 'A' (Fig.5.84B). The plot (Fig.5.85) describing the evolution of the global dilatancy angle (Eq.5.51) versus global friction angle (Eq.5.51) is different from the

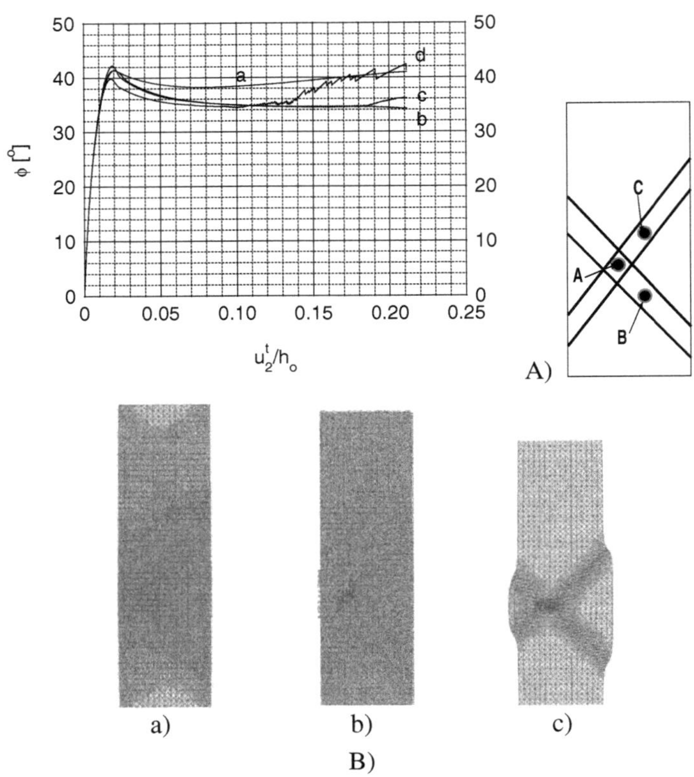

Fig. 5.82. FE results with very rough boundaries: A) evolution of mobilized overall friction angle: a) global friction angle, b) local friction angle in the mid-point of two intersecting shear zones (point *A*), c) local friction angle in the middle of one shear zone (point *B)*, d) local friction angle in the middle of second shear zone (point *C*) versus normalized vertical displacement of the upper edge $u_2^t/h_{o,}$ and B) deformed FE mesh with the distribution equivalent total strain at: a) u_2^t/h_o=0.018, b) u_2^t/h_o=0.021, c) u_2^t/h_o=0.071

results for smooth boundaries (due to increasing friction angle at residual state); the plot consists namely of three linear parts and intersects the horizontal line of ψ=0° only once in the pre-peak regime at about ϕ=38.3°.The local dilatancy angle can fluctuate in the shear zone at large deformation.

The maximum rotation of principal stresses and the principal strain-rate directions is about 12° at u_2^t/h_o=0.20 (Figs.5.86 and 5.87). The deviation angle between the principal stress and the principal strain-rate directions is again insignificant at peak (ξ=1°) (Fig.5.88). At large deformation, it becomes larger (ξ=12° for point 'A' in Fig.5.88) depending upon the position in the shear zone.

The numerical results collaborate well with the experimental finding by Arthur and Assadi (1977), where the non-coaxiality of 6-12° was observed in the shear band.

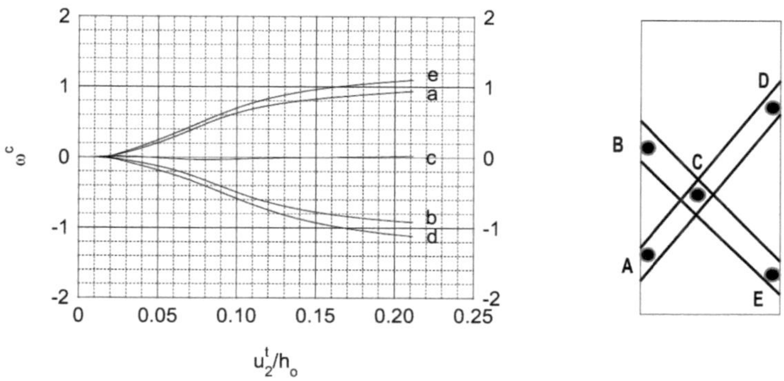

Fig. 5.83. Evolution of Cosserat rotation in the shear zones: a) point A, b) point B, c) point C, d) point D, e) point E (very rough boundaries)

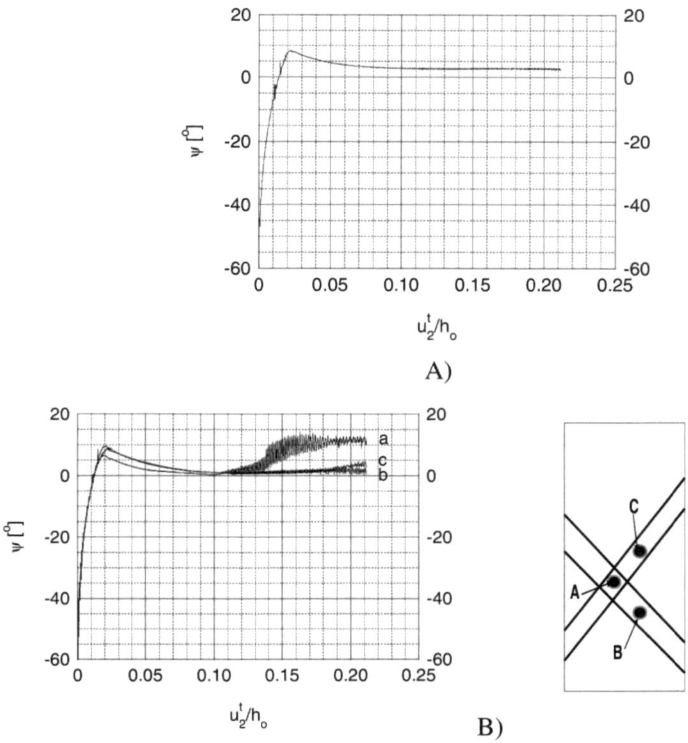

Fig. 5.84. Evolution of mobilized: A) global dilatancy angle and B) mobilized local dilatancy angle (Eq.5.46): a) mid-point of two intersecting shear zones (point A), b) mid-point of one shear zone (point B), c) mid-point of second shear zone (point C) versus normalized vertical displacement of the upper edge u_2^t/h_o

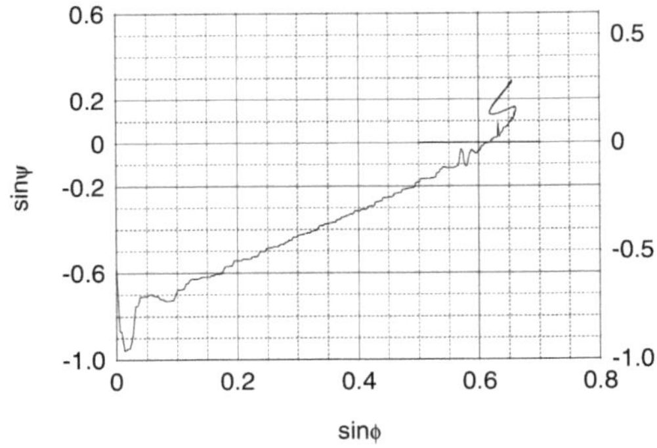

Fig. 5.85. Relationship between the global dilatancy angle ψ (Eq.6.46) and global friction angle ϕ (Eq.5.1) (very rough boundaries)

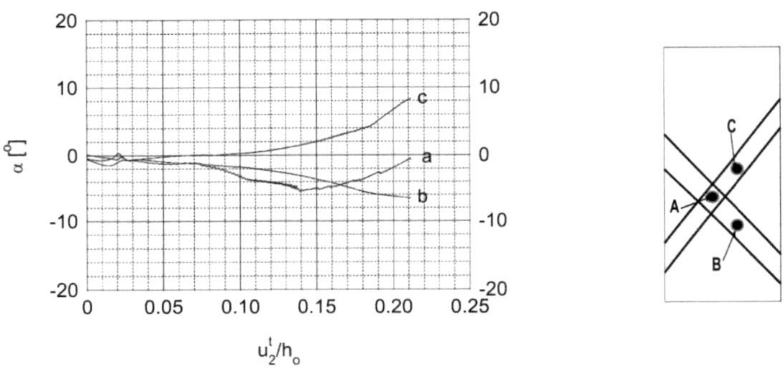

Fig. 5.86. Evolution of orientation of the major principal stress α versus normalized vertical displacement of the upper edge u_2^t/h_o: (a) point A, b) point B, c) point C) (very rough boundaries)

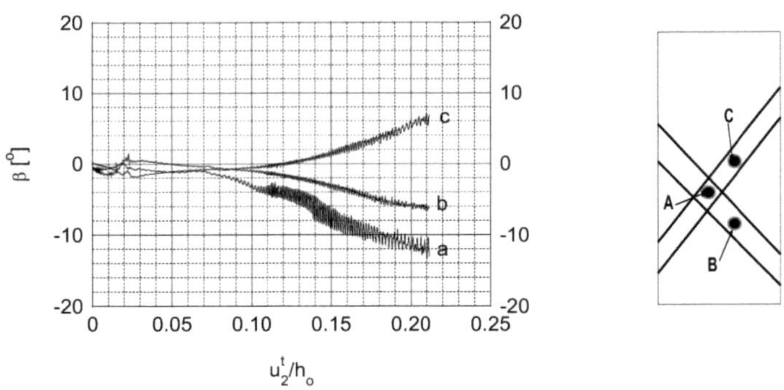

Fig. 5.87. Evolution of orientation of the major principal strain rate β versus normalized vertical displacement of the upper edge u_2^t/h_o: point A, b) point B, c) point C) (very rough boundaries)

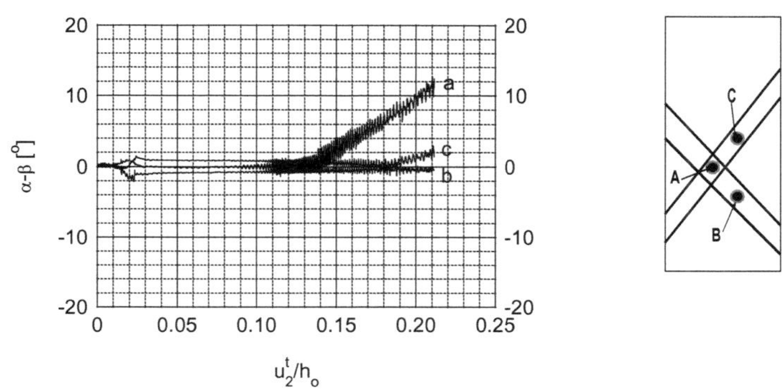

Fig. 5.88. Evolution of orientation of the non-coaxiality angle $\xi=\alpha-\beta$ versus normalized vertical displacement of the upper edge u_2^t/h_o: (a) point A, b) point B, c) point C) (very rough boundaries)

The following conclusions can be drawn from the numerical analyses:

- Non-coaxiality appears from the beginning and increases steadily along with the deformation. For smooth boundaries, a linear increase is observed. At peak, the deviation angle between the principal stress and the principal strain-rate directions is found to be insignificant. At large deformation, the deviation angle increases up to 6°-12° depending on the boundary roughness. This deviation seems to increase with the dilatancy angle. The effect of the non-coaxiality parameter has some pronounced effect only at large deformation.
- The global peak friction angle is found to be independent of the boundary roughness. On the contrary, the global friction angle at large deformation shows strong dependence on the boundary roughness (it is significantly higher for granular specimens with very rough boundaries).
- The local friction angles at large deformation in the shear zone are found to be higher (smooth boundaries) or smaller (very rough boundaries) than the corresponding global friction angles. At peak, the local friction angles remain nearly independent of the boundary roughness.
- The global peak dilatancy angle shows independence on the boundary roughness. The global dilatancy angle at large deformation is higher for very rough boundaries than for smooth boundaries.
- The local dilatancy angles in the shear zones at large deformation differ from the global dilatancy angles. The difference is minimal at the peak.
- The stress-dilatancy plot (according to Eqs.5.46 and 5.1) intercepts the *t/s* axis at about 38° before the peak. After the peak the intercept is observed only for smooth boundaries at about 35°.
- At large deformation, the global behaviour is contractant for smooth boundaries and dilatant for very rough boundaries. For both boundaries, however, the local dilatancy angle in the shear zones has an increasing tendency with increasing deformation.

- The evolutions of the mobilized dilatancy angle suggested by Wan and Guo (2004), Vardoulakis and Georgopoulos (2005) and Gutierrez and Vardoulakis (2007) can approximately be confirmed locally in the shear zones up to u_2^t/h_o=0.10.
- The shear zone emerges always before the peak on the basis of an increase of Cosserat rotations.
- At the peak, some deformation can still be observed outside the shear zone. The assumption that all deformation occurs inside the shear zone cannot be fully justified at peak.

5.6 Textural Anisotropy

Granular materials are heterogeneous and discrete systems composed of grains with different shape, size and roughness. Thus, their behaviour is influenced by the orientation of grains with respect to the direction of sedimentation (Boehler and Sawczuk 1977, Kanatani 1984, Oda 1972, Satake 2005, Khidas and Jia 2005, Al Hattamleh et al. 2007). This inherent anisotropy due to texture (fabric) is called a transverse isotropy since the material has a rotational symmetry with respect to one of the co-ordinates axes. The plane perpendicular to the orientation direction is called the bedding plane and it is a plane of isotropy.

The laboratory experiments in Fig.5.89 show evidently that the orientation of the bedding plane relative to the principal stress directions has a pronounced effect on the stress-strain behaviour (Arthur and Phillips 1975, Lam and Tatsuoka 1988, Tatsuoka et al. 1990, 1991, 1994, 1997, Oda 1972, Abelev and Lade 2003). The shear stiffness, peak friction angle and average volume change are usually larger and strains corresponding to the peak value are smaller for loading perpendicular to the bedding plane than for loading parallel to it. The inclination of the shear zone with respect to the bottom becomes smaller. For large monotonic shearing, the stress ratio approaches a stationary value (Yamada and Ishihara 1979, Tatsuoka et al. 1994), i.e. anisotropy vanishes at residual state (critical state) at large shear deformation due to the so-called SOM-effect (swept out of memory effect) (Gudehus 1997). The DEM simulations seem to confirm also this SOM-effect in granular bodies (Pena et al. 2005).

In enhanced hypoplastic models, two approaches of the inherent textural anisotropy have been already proposed. The first one was suggested by Tejchman et al. (2007) for second-gradient hypoplasticity wherein a structure tensor was included to take into account the spatial orientation of the bedding plane. The model was based an idea of Boehler and Sawczuk (1977). The structural tensor was introduced into the nonlinear part of the hypoplastic equation. In this way, both strength and dilatancy were increased for the major normal stress perpendicular to the bedding plane. However, since the effect of anisotropy diminished in critical states at large shear deformation, this model was unable to describe the anisotropic behaviour if the initial void ratio of granulates was equal to the critical void ratio. The second approach was proposed by Tejchman and Niemunis (2006) within the framework of micro-polar hypoplasticity. They introduced a scalar density function, which depends on the orientation of the major principal stress with reference to the bedding plane. This formulation can be used to describe the anisotropic behaviour at all densities. In both enhanced models, only two additional material parameters were needed.

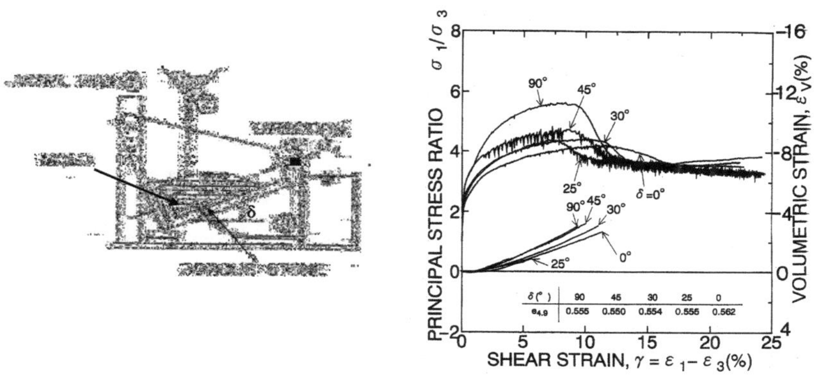

Fig. 5.89. Method used to prepare a granular specimen (SLB sand) and relationships among the stress ratio σ_1/σ_3, the average shear strain $\gamma=\varepsilon_1-\varepsilon_3$ and the average volumetric strain $\varepsilon_v=\varepsilon_1+\varepsilon_3$ for different angles of δ between 90 and 0 degrees at σ_3=80 kPa (Tatsuoka et al. 1994)

Below, a novel approach is attempted to describe textural anisotropy with usual isotropic micro-polar hypoplasticity (Tejchman and Wu 2007). First, the initial void ratio in the granular specimen was distributed stochastically by using a random correlated Gaussian field (Section 5.3). Next, this field was rotated by different angles to simulate the specimen preparation process in laboratory experiments, which was characterized by the angle between the filling and loading direction (Fig.5.89). An isotropic micro-polar hypoplastic constitutive model was used. In addition, the numerical results from the isotropic model were compared with three different anisotropic micro-polar constitutive models for an uniform distribution of the initial void ratio.

FE calculations of plane strain compression tests were performed with a sand specimen with initial dimensions of h_o=0.14 m (height), b=0.04 m (width) and l=1.0 m (depth – due to plane strain calculations).

A quasi-static deformation was imposed through a constant vertical displacement increment Δu prescribed at nodes along the upper edge of the specimen. The boundary conditions of null shear stress are imposed at the top and bottom of the specimen. To preserve the stability of the specimen against horizontal sliding, the node in the middle of the top edge was kept fixed.

As the initial stress state, a K_0-state with $\sigma_{22}=\gamma_d x_2$ and $\sigma_{11}=K_0\gamma_d x_2$ was assumed in the specimen; x_2 is the vertical coordinate measured from the top of the specimen, γ_d=16.5 kN/m^3 denotes the initial volume weight and K_0=0.50 is the earth pressure coefficient at rest. Next, a confining pressure of σ_c=200 kPa was prescribed.

Isotropic model

The numerical results for three different spatially correlated random fields of the initial void ratio (with λ_{x1}=1, λ_{x2}=3 and s_{eo}=0.05, Eq.5.27) rotated by θ=0°, θ=45°and θ=90° (Fig.5.90) are depicted in Fig.5.91. The normalized load-displacement curves and the deformed FE meshes with the distribution of the equivalent total strain $\bar{\varepsilon}=\sqrt{\varepsilon_{ij}\varepsilon_{ij}}$ are shown in Fig.5.91 (P – resultant vertical force on the top, b=0.04

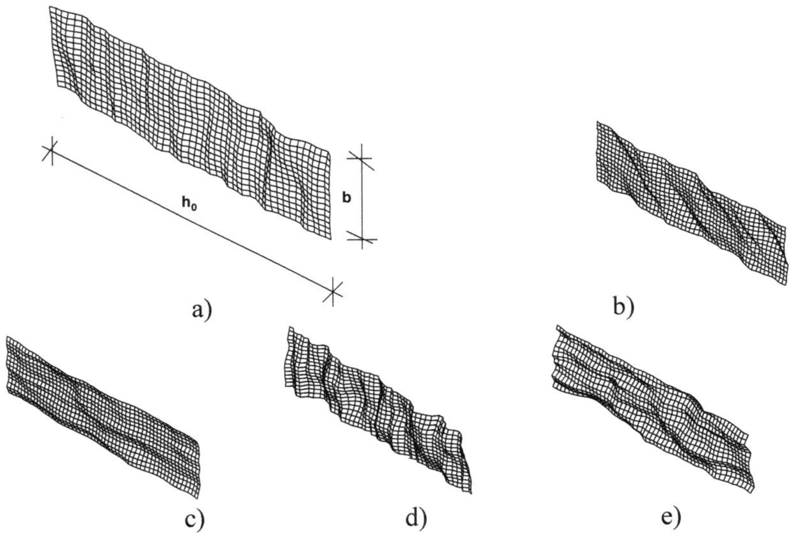

Fig. 5.90. Two-dimensional distribution of initial void ratio e_o in the granular specimen for the following parameters: a) λ_{x1}=1, λ_{x2}=3, s_{eo}=0.05, θ=0°,b) λ_{x1}=1, λ_{x2}=3, s_{eo}=0.05, θ=45°, c) λ_{x1}=1, λ_{x2}=3, s_{eo}=0.05, θ=90°, d) λ_{x1}=1, λ_{x2}=3, s_{eo}=0.10, θ=0°, e) λ_{x1}=1, λ_{x2}=3, s_{eo}=0.10, θ=90° (h_o=14 cm – specimen initial height, b=4 cm – specimen initial width, θ – bedding plane inclination, s_d – standard deviation, λ_{xi} - decay coefficients)

m - specimen width, u_2 – vertical displacement of the top, h=0.14 m – initial height of the sand body, l=1.0 m – specimen depth, ε_{ij} – strain tensor). The darker the region, the higher the equivalent total strain $\bar{\varepsilon}$. The entire range of strain was divided into 20 different grey scales.

The normalized overall vertical force in the granular specimen is the largest for the rotation angle θ==0°, smaller for θ=45° and the smallest for the angle θ=90° (as in the experiments, Fig.5.89). The same normalized vertical force was obtained in the residual state at large deformation for all rotation angles. The friction angle at the peak calculated from the principal stresses $\sigma_1=P/(bl)$ and $\sigma_2=\sigma_c$ (Eq.4.1) decreases from 41.81^o (θ=0°) down to 41.23^o (θ=45°) and to 40.67^o (θ=90°), respectively. The residual internal friction angle is the same, 30°. This corresponds to the assumed critical friction angle of 30° (Eq.3.82). The shear strain corresponding to the peak internal friction angle is u_2^t/h_o=0.024 (θ=0° and θ=45°) and u_2/h=0.022 (θ=90°).

One shear zone occurs inside of the specimen which crosses the specimen and whose location is caused by a stochastic distribution of the initial void ratio. The shear zone thickness is insignificantly influenced by the rotation angle θ (it slightly increases with increasing θ). By making use of Cosserat rotations the shear zone thickness is about 8-9 mm [(16-18)$\times d_{50}$]. The shear zone inclination against the bottom always decreases slightly with increasing angle θ from υ=54° (θ=0°) down to υ=50° (θ =90°).

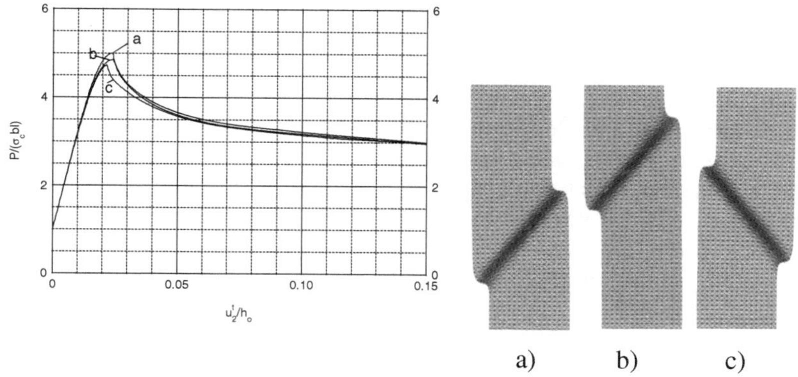

Fig. 5.91. Effect of the rotation angle θ on the normalized load-displacement curve and deformed meshes with the distribution of equivalent total strain $\bar{\varepsilon} = \sqrt{\varepsilon_{ij}\varepsilon_{ij}}$ at residual state with stochastic distribution of e_o using isotropic micro-polar hypoplastic model: a) λ_{x1}=1, λ_{x2}=3, s_{eo}=0.05, θ=0°, b) λ_{x1}=1, λ_{x2}=3, s_{eo}=0.05, θ=45°, c) λ_{x1}=1, λ_{x2}=3, s_{eo}=0.05, θ=90°

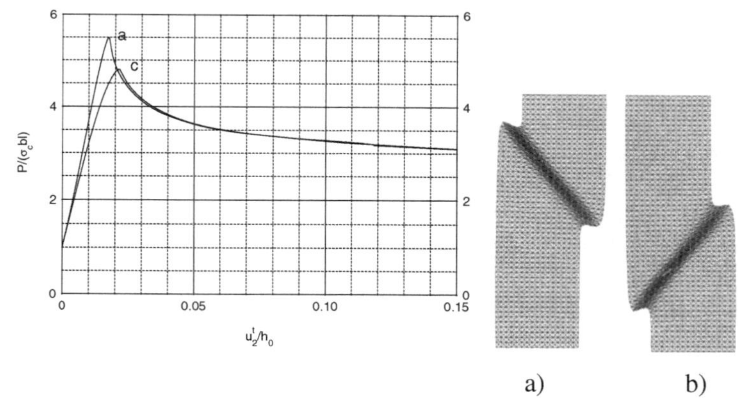

Fig. 5.92. Effect of the rotation angle θ on the normalized load-displacement curve and deformed meshes with the distribution of equivalent total strain $\bar{\varepsilon} = \sqrt{\varepsilon_{ij}\varepsilon_{ij}}$ at residual state with stochastic distribution of e_o using isotropic micro-polar hypoplastic model: a) λ_{x1}=1, λ_{x2}=3, s_{eo}=0.10, θ=0° c) λ_{x1}=1, λ_{x2}=3, s_{eo}=0.10, θ=90°

The effect of the standard deviation s_{eo} is demonstrated in Fig.5.92. The larger is the parameter s_{eo}, the larger the effect of anisotropy. The internal friction angle decreases from 43.81° at u_2/h=0.017 (θ=0°) down to 41.21° (θ=90°) at u_2/h=0.021. The differences between the peak internal friction angles are about 5%. The shear zone thickness decreases by about 1 mm and the its inclination against the bottom increases by about 2° as compared to results with s_d=0.05.

Anisotropic model

As can be seen from the last section, even an isotropic model can describe the effect of anisotropy properly, provided the specimen is generated by random fields and rotated by an angle corresponding to laboratory experiments. In what follows, the numerical results of three anisotropic hypoplastic models are presented (assuming a uniform distribution of the initial void ratio).

In the approach by Tejchman et al. (2007), in order to take into account anisotropic properties, Eq.3.68 was enhanced by a structure tensor B_{kl} (Bauer et al. 2004):

$$\overset{o}{\sigma}_{ij} = f_s [L_{ij}(\hat{\sigma}_{kl}, \hat{m}_k, d^c_{kl}, k_k d_{50}) + f_d B_{ijkl}(s_{kl}) N_{kl}(\hat{\sigma}_{kl}) \sqrt{d^c_{kl} d^c_{kl} + k_k k_k d^2_{50}}], \quad (5.52)$$

where B_{ijkl} is the fourth order tensor:

$$B_{ijkl} = (\eta_1 + \eta_3 - 2\eta_2) s_{ij} s_{kl} + \eta_3 \delta_{ik} \delta_{jl} + (\eta_2 - \eta_3)(s_{ik} \delta_{jl} + s_{jl} \delta_{ik}), \quad (5.53)$$

and the tensor s_{ij} represents the dyadic product of the normal vector s_i (Fig.5.93)

$$s_i = [-\sin\theta, \cos\theta, 0] \quad (5.54)$$

of the bedding plane with a bedding angle θ.

$$s_{ij} = \begin{bmatrix} \sin^2\theta & -\sin\theta\cos\theta & 0 \\ -\sin\theta\cos\theta & \cos^2\theta & 0 \\ 0 & 0 & 0 \end{bmatrix}. \quad (5.55)$$

Herein η_i (i=1, 2, 3) are material parameters related to anisotropy. If η_i =1, the constitutive relation for an initially isotropic material (1) is recovered. To describe diminishing anisotropy in critical state due to the SOM-effect, the additional anisotropic parameters were assumed to depend on the density factor f_d in (Eq.3.79) as follows: $\eta_1=\eta_{10}(f_d-1)$, $\eta_2=\eta_{20}(f_d-1)$, $\eta_3=\eta_{30}(f_d-1)$. The following anisotropic parameters are assumed: η_{10}=0.8, η_{20}=0.25 and η_{30}=0 (Tejchman et al. 2007). Due to the assumption of η_{30}=0, the micro-polar anisotropic hypoplastic model requires 2 additional constants.

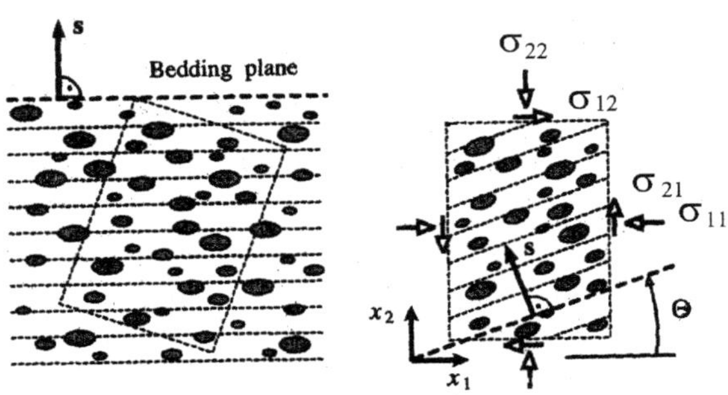

Fig. 5.93. Normal vector s of the bedding plane (a) and its inclination θ (b) with respect to a fixed co-ordinate system (θ - bedding plane inclination, υ - shear zone inclination)

In turn, in the approach by Niemunis (2003a, 2003b) and Tejchman and Niemunis (2006), the critical void ratio in Eq.3.82 is replaced by

$$e_c = \bar{e}_{c0}\, exp[-(-\sigma_{kk} / h_s)^n] \,, \tag{5.56}$$

where

$$\bar{e}_{c0} = e_{c0} + \Delta e_{c0}\, \vec{s}_{ij}\, \hat{\sigma}^*_{ij} \,, \tag{5.57}$$

In the above relationship, the normalized deviator $\hat{\sigma}^*_{ij} = \sigma_{ij}^* / \sigma_{kk}$ is multiplied by the unit dyadic tensor

$$\vec{s}_{ij} = s_{ij} / \sqrt{s_{kl} s_{kl}} \tag{5.58}$$

Eq.5.56 takes into account the inclination of the deviatoric stress with reference to the bedding plane. The parameter Δe_{c0} describes the degree of anisotropy of the critical void ratio e_{c0} from the isotropic reference at $p=0$. Assuming that anisotropic effects diminish at large deformations, we have $\bar{e}_{co} \rightarrow e_{co}$, hence $\Delta e_{c0} \rightarrow 0$ at large deformations. The decay of the rate of Δe_{c0} is assumed to have the following form

$$\Delta \dot{e}_{c0} = -\Delta e_{c0} B \sqrt{d_{ij}^{\,c} d_{ij}^{\,c} + k_i k_i d_{50}^2} \,. \tag{5.59}$$

The initial value of Δe_{c0} at t=0 and the rate of decay B are the only additional material constants. The evolution equation for Δe_{c0} is objective since d_{ij}^c and k_i are objective. The anisotropic parameters were assumed to be Δe_{c0} (t=0)=0.25 and B=5 (Tejchman and Niemunis 2006).

Another possibility is to modify the granular hardness h_s (similarly to $\bar{e}_{co}$) which affects directly the stiffness factor in Eq.3.77:

$$\bar{h}_s = h_s + \Delta h_s\, \vec{s}_{ij}\, \hat{\sigma}^*_{ij} \,. \tag{5.60}$$

After some trial calculations, the following values were found suitable: Δe_{c0} (t=0)=1.0 and B=5.

The FE analyses were carried out with the material constants for so-called Karlsruhe sand: e_{i0}=1.30, e_{d0}=0.51, e_{c0}=0.82, ϕ_c=30°, h_s=190 MPa, β=1, n=0.40, α=0.20, $a_c=a_1^{-1}$ and d_{50}=0.5 mm. The parameters n and α were slightly reduced to obtain a similar shear resistance as in the isotropic micro-polar model.

The numerical results for 3 modified anisotropic micro-polar hypoplastic models are depicted in Figs.5.94 and 5.95 (with λ_{x1}=1, λ_{x2}=1 and s_{eo}=0.05). The results are qualitatively similar and close to those presented by Tejchman et al. (2007) and Tejchman and Niemunis (2006). They are also similar as the results with an isotropic micro-polar constitutive law (Figs.5.91 and 5.92). The shear resistance decreases with increasing bedding angle, and the shear zone width increases and its inclination υ

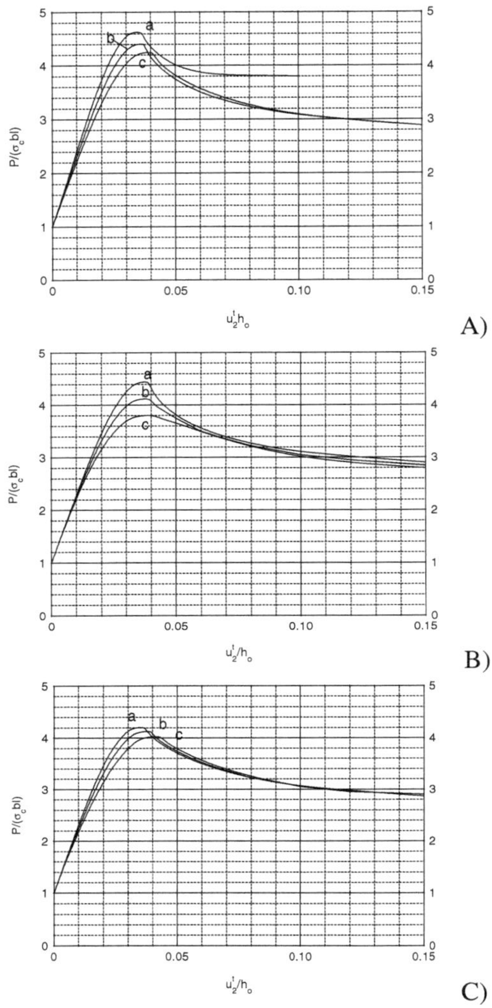

Fig. 5.94. Effect of the bedding plane inclination θ of Fig.5.237 on the normalized load-displacement curve with stochastic distribution of initial void ratio e_o: A) anisotropic model by Tejchman et al. (2007), B): anisotropic model by Tejchman and Niemunis (2006), C) anisotropic model by Tejchman and Niemunis (2006) with modified granular hardness h_s for bedding angle inclinations: a) θ=0°, b) θ=45°, c) θ=90°)

decreases with increasing bedding angle. The shear zone width is slightly different for each model (for the same angle θ) due to the various rate of softening influencing the shear zone formation (the smaller rate of softening, the wider the shear zone width, Tejchman et al. 1999). The difference is about 1-2 mm. The evolution of one load-displacement curve after the peak in the anisotropic model by Tejchman et al. (2007) at θ=90° (curve 'a' in Fig.5.94A) significantly differs from the remaining curves after

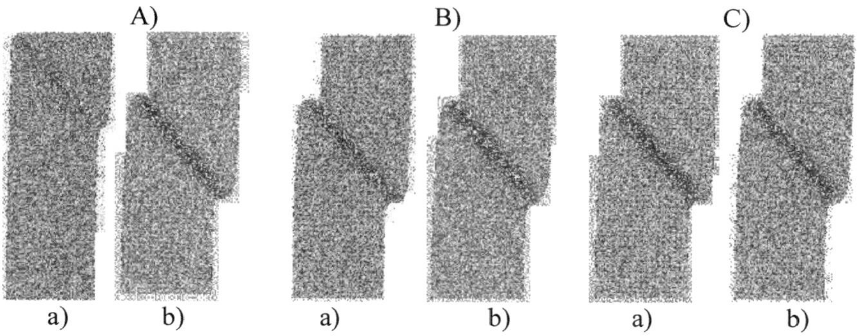

Fig. 5.95. Deformed meshes with the distribution of equivalent total strain $\bar{\varepsilon} = \sqrt{\varepsilon_{ij}\varepsilon_{ij}}$ at residual state with stochastic distribution of e_o: A) anisotropic model by Tejchman et al. (2007), B): anisotropic model by Tejchman and Niemunis (2006), C) anisotropic model by Tejchman and Niemunis (2006) with modified granular hardness h_s for bedding angle inclinations: a) θ=0°, b) θ=90°

the peak (the calculation was broken at u_2^t/h_o=0.10) due to the fact that shear localization bumps into the top boundary at the left corner changing significantly the width of the top boundary (Fig.5.95A, a). Thus, it directly influences the vertical force acting on the top edge. In the remaining cases, shear localization occurs inside of the material far from the horizontal boundaries and the load-displacement diagrams are qualitatively the same.

The following conclusions can be derived on the basis of FE calculations:

- The anisotropic effect is realistically described. The larger is the bedding plane inclination, the smaller the peak internal friction angle. The vertical strain corresponding to the peak increases with increasing bedding plane inclination. The anisotropic effect vanishes in the residual state.
- The shear zone thickness slightly decreases with increasing bedding plane inclination. The shear zone inclination decreases with increasing bedding plane inclination.
- The larger is the scatter of the initial void ratio, the larger the effect of anisotropy. Therein, the shear zone width slightly decreases and its inclination with the base slightly increases.
- The position of the shear zone depends strongly of the stochastic distribution of the initial void ratio.
- The results with isotropic micro-polar constitutive model and stochastic distribution of the initial void ratio are qualitatively similar to those with anisotropic micro-polar constitutive models and uniform distribution of the initial void ratio.

The numerical results based on isotropic and anisotropic models show that anisotropy can be described realistically by both approaches. Although both the isotropic and the anisotropic models are of phenomenological nature, the approach based on the isotropic model is to be favored for its clear physical background. The isotropic approach mimics the sedimentation process. In fact, both approaches represent some homogenization of an anisotropic medium. In the anisotropic models, the homogenization is assumed at the element level via the constitutive equation. In the approach based on the isotropic model, the homogenization is performed numerically at the structure level.

References

Abelev, A.V., Lade, P.V.: Effects of cross anisotropy on 3-dimensional behaviour of sand. II: Stress-strain behaviour and shear banding. J. Eng. Mech. ASCE 129(2), 167–174 (2003)

Airey, D.W., Al-Douri, R.H., Poulos, H.G.: Estimation of pile friction degradation from shear box tests. ASTM Geotech. Test, J. 15(4), 388–392 (1992)

Al Hattamleh, O., Muhunthan, B., Zbib, H.M.: Stress distribution in granular heaps using multi-slip formulation. Int. J. Num. Anal. Meth. Geomech. 29, 713–727 (2005)

Al Hattamleh, O., Muhunthan, B., Shalabi, F.: Numerical simulation of fabric anisotropy and strain localization of sand under simple shear. Int. J. Num. Anal. Meths. Geomech. (submitted for publication) (2007)

Arthur, J.R.F., Menzies, B.K.: Inherent anisotropy in a sand. Geotechnique 22(1), 115–128 (1972)

Arthur, J.R.F., Phillips, A.B.: Homogeneous and layered sand in triaxial compression. Geotechnique 25(4), 799–815 (1975)

Arthur, J.R.F., Assadi, A.: Ruptured sand sheared in plane strain. In: Proc. 9th Int. Conf. Soil Mech. Found. Engng., Tokyo, vol. 1, pp. 19–22 (1977)

Bazant, Z.P., Lin, K.L.: Random creep and shrinkage in structures sampling. J. Struct. Engng. ASCE 111(5), 1113–1134 (1985)

Bazant, Z.P., Chen, E.P.: Scaling of structural failure. Appl. Mech. Rev. 50(10), 593–627 (1997)

Bazant, Z., Planas, J.: Fracture and size effect in concrete and other quasi-brittle materials. CRC Press LLC, Boca Raton (1998)

Bazant, Z.P., Yavari, A.: Is the cause of size effect on structural strength fractal or energetic-statistical? Engng. Fract. Mech. 72, 1–31 (2005)

Bazant, Z., Pang, S.D.: Computational structural reliability – a major challenge and opprtinity for concrete and other quasibrittle structures. In: Meschke, G., de Borst, R., Mang, H., Bicanic, N. (eds.) Computational Modelling of Concrete Structures, EURO-C 2006, pp. 845–856. Taylor and Francis, Abington (2006)

Becker, M., Lippmann, H.: Plane plastic flow of granular model material. Arch. Mech. 29, 829–846 (1977)

Bielewicz, E., Górski, J.: Shell with random geometric imperfections. Simulation-based approach. Int. J. Non-Linear Mech. 37(4–5), 777–784 (2002)

Boulon, M.: Numerical and physical modeling of piles behaviour under monotonous and cyclic loading. In: Modelling soil-water-structure interactions, Balkema, pp. 285–293 (1988)

Brockbank, R., Huntley, J.M., Ball, R.C.: Contact force distribution beneath three-dimensional granular pile. J. Physics II 7, 1521–1532 (1997)

Brumund, W.F., Leonards, G.A.: Experimental study of static and dynamic friction between sand and typical construction materials. J. Test. Eval. 1, 162–165 (1973)

Carpinteri, A., Chiaia, B., Ferro, G.: Multifractal scaling law: an extensive application to nominal strength size effect of concrete structures. In: Mihashi, M., Okamura, H., Bazant, Z.P. (eds.) Size effect of concrete structures, pp. 173–185. E&FN Spon (1994)

De Jong, J.T., Randloph, M.F., White, D.J.: Interface load transfer degradation during cyclic loading: a microscale investigation. Soils Found. 43(4), 81–93 (2003)

De Jong, J.T., White, D.J., Randolph, M.F.: Microscale observation and modeling of soil-structure interface behavior using particle image velocimetry. Soils Found. 46(1), 15–28 (2006)

Desai, C.S., Drumm, E.C., Zaman, M.M.: Cyclic testing and modeling of interfaces. J. Geot. Engng. ASCE 111(6) (1985)

Desrues, J., Viggiani, G.: Strain localization in sand: overview of the experiments in Grenoble using stereo photogrammetry. Int. J. Numer. Anal. Meth. Geomech. 28(4), 279–321 (2004)

Devroye, L.: Non-uniform random variate generation. Springer, Heidelberg (1986)

Fenton, G.A., Griffiths, D.V.: Probabilistic foundation settlement on spatially random soil. J. Geotech. Geoenviron. Eng. 128(5), 381–389 (2002)

Florian, A.: An efficient sampling scheme: Updated latin hypercube sampling. Probab. Engng. Mech. 2, 123–130 (1992)

Górski, J.: Non-linear models of structures with random geometric and material imperfections simulation-based approach. Monograph, Gdansk University of Technology, 68 (2006)

Gudehus, G., Nübel, K.: Evolution of shear bands in sand. Geotechnique 54(3), 187–201 (2004)

Gutierrez, M., Ishihara, K., Towhata, I.: Flow theory for sand during rotation of principal stress direction. Soils Found. 31(4), 121–132 (1993)

Gutierrez, M.A., de Borst, R.: Energy dissipation, internal length scale and localization patterning – a probabilistic approach. In: Idelsohn, S., Onate, E., Dvorkin, E. (eds.) Computational Mechanics, CIMNE, Barcelona, pp. 1–9 (1998)

Gutierrez, M.A.: Size sensitivity for the reliability index in stochastic finite element analysis of damage. Int. J. Fract. 137(1-4), 109–120 (2006)

Gutierrez, M., Ishihara, K.: Non-coaxiality and energy dissipation in granular materials. Soils Found. 40(2), 49–59 (2000)

Gutierrez, M., Vardoulakis, I.: Energy dissipation and post-bifurcation behaviour of granular soils. Int. J. Num. Anal. Meth. Geomech. 31, 435–455 (2007)

Haaker, G.: Measurement of wall friction and wear in bulk solids handling. In: Proc. Int. Conf. On Silos – Forschung und Praxis, Tagung 88, Karlsruhe, pp. 389–405 (1988)

Hassan, A.H.: Etude experimentale et numerique du comportement local et global d'une interface sol granulaire structure. PhD thesis, Grenoble University (1995)

Hill, R.: The mathematical theory of plasticity, Oxford (1950)

Huang, W., Bauer, E., Sloan, S.W.: Behaviour of interfacial layer along granular soil-structure interfaces. Struct. Eng. Mech. 15(3), 315–329 (2003)

Huck, P.J., Saxena, S.K.: Response of soil-concrete interface at high pressure. In: Proc. 10th ICSMFE, vol. 2, pp. 141–144 (1981)

Hummel, F.H., Finnan, E.J.: The distribution of pressure on surfaces supporting a mass of granular materials. Minutes of Proc. Int. Civil Eng. Session 212, 369–392 (1920)

Huntley, J.M.: Force distribution in an inhomogeneous sandpile. The European Physical Journal B 8(3), 389–397 (1999)

Hurtado, J.E., Barbat, A.H.: Monte Carlo techniques in computational stochastic mechanics. Arch. Comput. Meth. in Engng. 5(1), 2–20 (1988)

Khidas, Y., Jia, X.: Acoustic measurements of anisotropic elasticity in glass beads packings. In: Garcia-Rojo, Herrmann, McNamara (eds.) Powders and Grains, pp. 309–312. Taylor and Francis Group, London (2005)

Kishida, H., Uesugi, M.: Tests of the interface between sand and steel in the simple shear apparatus. Geotechnique 37, 45–52 (1987)

Knabe, W., Przewłócki, J., Różyński, G.: Spatial averages for linear elements for two-parameter random field. Prob. Engng. Mech. 13(3), 147–167 (1998)

Lam, W.K., Tatsuoka, F.: Effect of initial anisotropic fabric and on strength and deformation characteristics of sand. Soils and Foundations 28(1), 89–106 (1988)

Le Bellego, C., Dube, J.F., Pijaudier-Cabot, G., Gerard, B.: Calibration of nonlocal damage model from size effect tests. European Journal of Mechanics A/Solids 22, 33–46 (2003)

Liffman, K., Chan, D.Y.C., Hughes, B.D.: On the stress depression under a sandpile. Powder Tech. 78, 263–271 (1994)

Liffman, K., Nguyen, M., Metcalfe, G., Cleary, P.: Forces in piles of granular material: an analytic and 3D DEM study. Granular Matter 3(3), 165–176 (2001)

Löffelmann, F.: Theoretische und experimentelle Untersuchungen zur Schüttgut-Wand-Wechselwirkung und zum Mischen und Entmischen von Granulaten. Publication Series of the Institut für Mechanische Verfahrenstechnik, Universität Karlsruhe (1989)

Michalowski, R.L., Park, N.: Admissible stress fields and arching in piles of sand. Geotechnique 54(8), 529–538 (2004)

Michalowski, R.L., Park, N.: Arching in soils. Numerical Models in Geomechanics. In: Pande, G.N., Pietruszczak, S. (eds.) NUMOG X, pp. 433–438. Taylor and Francis Group, Abington (2007)

Miura, K., Miura, S., Toki, S.: Deformation behaviour of anisotropic sand under principal axes rotation. Soils and Foundations 26(1), 36–52 (1986)

Neuffer, F., Leibnitz, A.: Über den Gleitwiderstand zwischen Erdstoffen und Bauwerksflächen. Bericht aus der Bauforschung 37 (1964)

Niemunis, A.: Extended hypoplastic models for soils. Monograph, Gdansk University of Technology (2003a)

Niemunis, A.: Anisotropic effects in hypoplasticity. In: Proc. Int. Symp. on Deformation Characteristics of Geomaterials, Lyon, France, vol. 1, pp. 1211–1217 (2003b)

Niemunis, A., Wichtmann, T., Petryna, Y., Triantafyllidis, T.: Stochastic modeling of settlements due to cyclic loading for soil-structure interaction. In: Proc. Int. Conf. Structural Damage and Lifetime Assessment, Rome, pp. 1–8 (2005)

Paikowsky, S.G., Player, C.M., Connors, P.J.: A dual interface apparatus for testing unrestricted friction of soil along solid surfaces. ASTM Geotech. Test. J. 18(2), 168–193 (1995)

Pena, A.A., Herrmann, H.J., Lizcano, A., Alonso-Marroquin, F.: Investigation of the asymptotic states of granular materials using a discrete model of anisotropic particles. In: Garcia-Rojo, Herrmann, McNamara (eds.) Powders and Grains, pp. 697–700. Taylor and Francis Group, London (2005)

Potyondy, J.G.: Skin friction between various soils and construction materials. Geotechnique 4, 339–353 (1961)

Poulos, H.G.: Cyclic axial loading analysis of piles in sand. ASCE Geotech. Engrg. J. 115(6), 836–851 (1989)

Pradhan, T.B.S., Tatsuoka, F.: Experimental stress-dilatancy relations of sand subject to cyclic loading. Soils and Foundations 29, 45–64 (1989)

Przewłócki, J., Górski, J.: Stochastic FEM analysis of strip foundation. TASK Quarterly 3(2), 173–181 (1999)

Przewłócki, J., Górski, J.: Strip foundation on 2-D and 3-D random subsoil. Probabilistic Engineering Mechanics 16, 121–136 (2001)

Rubinstain, R.Y.: Simulation and the Monte Carlo method, New York. J. Wiley & Sons, Chichester (1981)

Regueiro, R.A., Borja, R.I.: Plane strain finite element analysis of pressure sensitive plasticity with strong discontinuity. Int. J. Solids and Structures 38(21), 3647–3672 (2001)

Satake, M.: A theory on the stress-induced anisotropy. In: Garcia-Rojo, Herrmann, McNamara (eds.) Powders and Grains, pp. 155–158. Taylor and Francis Group, London (2005)
Savage, S.B.: Problems in the statics and dynamics of granular materials. In: Behringer, R.P., Jenkins, J.T. (eds.) Powders and Grains, Balkema, pp. 185–194 (1997)
Savage, S.B.: Modeling and granular material boundary value problems. In: Herrmann, H., Hovi, J.P., Luding, S. (eds.) Physics of Dry Granular Media, pp. 25–95. Kluwer, Dordrecht (1998)
Smid, J., Novosad, J.: Particle technology. In: Proc. of the Conference POWTECH 1981, IChE Symposium, vol. 63, D3/V/1-D3/V/12 (1981)
Sondermann, W.: Spannungen und Verformungen bei bewehrter Erde. Mitteilungen des Institutes für Grundbau, Universität Braunschweig (1983)
Tatsuoka, F., Nakamura, S., Huang, C., Tani, K.: Strength anisotropy and shear band direction in plane strain tests of sand. Soils Found. 30(1), 35–54 (1990)
Tatsuoka, F., Okahara, M., Tanaka, T., Tani, K., Morimoto, T., Siddiquee, M.S.A.: Progressive failure and particle size effect in bearing capacity of footing on sand. In: Proc. of the ASCE Geotechnical Engineering Congress, vol. 27(2), pp. 788–802 (1991)
Tatsuoka, F., Siddiquee, M.S.A., Yoshida, T., Park, C.S., Kamegai, Y., Goto, S., Kohata, Y.: Testing methods and results of element tests and testing conditions of plane strain model bearing capacity tests using air-dried dense Silver Leighton Buzzard sand. Internal Report of the Tokyo University, pp. 1-120 (1994)
Tatsuoka, F., Goto, S., Tanaka, T., Tani, K., Kimura, Y.: Particle size effects on bearing capacity of footing on granular material. In: Asaoka, A., Adachi, T., Oka, F. (eds.) Deformation and Progressive Failure in Geomechanics, pp. 133–138. Pergamon, Oxford (1997)
Tejchman, J.: Scherzonenbildung und Verspannungseffekte in Granulaten unter Berücksichtigung von Korndrehungen. Publication Series of the Institute of Soil and Rock Mechanics, University Karlsruhe, vol. 117, pp. 1–236 (1989)
Tejchman, J., Wu, W.: Experimental and numerical study of sand-steel interfaces. Int. J. Num. Anal. Meth. Geomech. 19(8), 513–537 (1995)
Tejchman, J., Bauer, E.: Numerical simulation of shear band formation with a polar hypoplastic model. Comp. Geotech. 19(3), 221–244 (1996)
Tejchman, J.: Modelling of shear localisation and autogeneous dynamic effects in granular bodies. Publication Series of the Institute of Soil and Rock Mechanics, University Karlsruhe, vol. 140, pp. 1–353 (1997)
Tejchman, J.: Shearing of an infinite narrow granular layer between two boundaries. In: Mühlhaus, H.-B. (ed.) Bifurcation and Localisation Theory in Geomechanics, Swets, Zeitlinger, Lisse, pp. 95–103 (2001)
Tejchman, J.: Influence of a characteristic length on shear zone formation in hypoplasticity with different enhancements. Comp. Geotech. 31(8), 595–611 (2004)
Tejchman, J., Bauer, E.: FE simulations of a direct and a true simple shear test within a polar hypoplasticity. Comp. Geotech. 21(1), 1–16 (2005)
Tejchman, J., Niemunis, A.: FE studies on shear localization in an anistropic micro-polar hypoplastic granular material. Granular Matter 8(3-4), 205–220 (2006)
Tejchman, J., Bauer, E., Wu, W.: Effect of fabric anisotropy on shear localization in sand during plane strain compression. Acta Mech. 189(1-2), 23–53 (2007)
Tejchman, J., Gorski, J.: FE investigations of a deterministic and statistical size effect in granular bodies within a micro-polar hypoplasticity. Granular Matter 9(6), 439–453 (2007)
Tejchman, J., Wu, W.: Modeling of of textural anisotropy in granular materialbodies within micro-polar hypoplasticity. International Journal of Non-Linear Mechanics 42, 882–894 (2007)

Tejchman, J., Górski, J.: Deterministic and statistical size effect during shearing of granular layer within a micro-polar hypoplasticity. Int. J. Num. Anal. Meth. Geomech. 32(1), 81–107 (2008a)

Tejchman, J., Górski, J.: Computations of size effects in granular bodies within micro-polar hypoplasticity during plane strain compression. Int. J. for Solids and Structures 45(6), 1546–1569 (2008b)

Tejchman, J., Górski, J.: Effect of initial void ratio and boundary roughness on shear localization in granular bodies during plane strain compression. Int. J. Non-linear Mechanics (submitted, 2008c)

Tejchman, J., Bauer, E.: FE modeling of shear resistance degradation in granular materials during cyclic shearing under CNS condition. Comp. Geotech. (2008), doi:10.1016/j.compgeo.2008.02.006

Tejchman, J., Wu, W.: FE calculations of stress distribution under prismatic and conical sandpiles within hypoplasticity. Granular Matter 10(5), 399–405 (2008)

Tejchman, J., Wu, W.: FE investigations of non-coaxiality and stress-dilatancy rule in dilatant granular bodies within micro-polar hypoplasticity. Int. J. Num. Anal. Meth. Geomech. (2008b), doi:10.1002/nag715

Tejchman, J.: FE investigations of granular matrial behaviour during cyclic wall shearing under CNS condition. Can. Geotech. J. (2008)(under review)

Uesugi, M., Kishida, H.: Influential factors of friction between steel and dry sand. Soils Found. 26(2), 33–46 (1986)

Uesugi, M.: Friction between dry sand and construction. Ph.D thesis, Tokyo Institute of Technology (1987)

Unterreiner, P., Vardoulakis, I., Boulon, M., Sulem, J.: Essential features of a Cosserat continuum in interfacial localisation. In: Chambon, R., Desrues, J., Vardoulakis, I. (eds.) Localisation and Bifurcation Theory for Soils and Rocks, Balkema, pp. 141–155 (1994)

Unterreiner, P.: Contribution a l'etude et al la modelisation numerique des sols cloues: application au calcul en deformation des ouvrages de soutenement. Ph.D thesis, Ecole Nationale des Ponts et Chaussees (1994)

Walukiewicz, H., Bielewicz, E., Górski, J.: Simulation of nonhomogeneous random fields for structural applications. Computers and Structures 64(1-4), 491–498 (1997)

Wan, G.R., Guo, P.J.: Stress dilatancy and fabric dependencies on sand behavior. Journal of Engineering Mechanics 130(6), 635–645 (2004)

Watson, A.: Searching for the sand-pile pressure dip. Science 273, 579–580 (1996)

Weibull, W.A.: Statistical theory of the strength of materials. Journal of Applied Mechanics 18(9), 293–297 (1951)

Wernick, E.: Zusammenhänge zwischen der Mantelreibung von zylindrischen Ankern und Ergebnissen von direkten Scherversuchen. Bautechnik 8, 263–267 (1977)

Wernick, E.: Tragfähigkeit zylindrischer Anker in Sand unter besonderer Berücksichtigung des Dilatanzverhaltens. Publication Series of the Institute for Rock and Soil Mechanics, University Karlsruhe 75 (1978)

Wittmer, J., Claudin, P., Cates, M.E., Bouchaud, J.P.: In explanation for the central stress minimum in sand piles. Nature 382, 336–338 (1996)

Wood, D.M., Budhu, M.: The behaviour of Leighton Buzzard sand in cyclic simple shear tests. In: Int. Symp. on Soils and Cyclic and Transient Loading, Swansea, pp. 9–21 (1980)

Wu, W.: Rational approach to anisotropy of sand. Intern. Journal for Numerical and Analytical Methods in Geomechanics 22, 921–940 (1998)

Vanel, L., Howell, D., Clarck, D., Behringer, R.P., Clement, E.: Memories in sand: experimental test of construction history on stress distribution under sandpiles. Physical Review E 60(5), R5040–R5043 (1999)

Vanmarcke, E.H.: Random Fields: Analysis and Synthesis. MIT Press, Cambridge (1983)

Vardoulakis, I.: Scherfugenbildung in Sandkörpern als Verzweigungsproblem. Dissertation, Institute for Soil and Rock Mechanics, University of Karlsruhe, 70 (1977)

Vardoulakis, I.: Shear band inclination and shear modulus in biaxial tests. Int. J. Num. Anal. Meth. Geomech. 4, 103–119 (1980)

Vardoulakis, I., Shah, K.R., Papanastasiou, P.: Modelling of tool-rock shear interfaces using gradient-dependent flow theory of plasticity. Int. J. Rock Mech. Min. Sci. Geomech. 29(6), 573–582 (1992)

Vardoulakis, I., Georgopoulos, I.O.: The stress-dilatancy hypothesis revisited: shear-banding related instabilities. Soils and Foundations 45(2), 61–76 (2005)

Vesic, A.S.: Bearing capacity theory from experiments. Journal of Soil Mechanics and Foundation Engineering ASCE, Discussion 99, 575–577 (1973)

van Vliet, M.R.A.: Size effect in tensile fracture of concrete and rock. PhD thesis, University of Delft (2000)

Vorechovsky, M., Matesova, D.: Size effect in concrete specimens under tension: interplay of sources. In: Meschke, G., de Borst, R., Mang, H., Bicanic, N. (eds.) Computational Modelling of Concrete Structures, EURO-C 2006, pp. 905–914. Taylor and Francis, Abington (2006)

Yamada, Y., Ishihara, K.: Anisotropic deformation characteristics of sand under three dimensional stress conditions. Soils and Foundations 19(2), 79–94 (1979)

Yoshimi, Y., Kishida, T.: Friction between sand and metal surface. In: Proc. 10th ICSMFE, pp. 831–834 (1981)

6 Epilogue

The presented FE results of different boundary value problems for dry cohesionless granular materials in a quasi-static regime show that shear localization is a fundamental phenomenon. It can occur in the interior domain in the form of a spontaneous shear zone as a single shear zone, a multiple or a regular pattern of zones or it can be also created in the form of an induced single shear zone along walls of structures. Thus, it has to be taken into when modeling the behaviour of granular bodies.

A hypoplastic constitutive model enhanced by a characteristic length of microstructure by means of a micro-polar theory can realistically describe the behaviour of of cohesionless granular materials including shear localization. The model is, in particular, effective for monotonic deformation paths. Due to the presence of a characteristic length, the boundary value problems are well-posed and numerical results (load-displacement diagrams, spacing and thickness of shear zones) are mesh-independent. The characteristic length allows also to capture a deterministic size effect in granular specimens related to the ratio between a mean grain diameter and specimen size.

A micro-polar model has good physical grounds to describe the behaviour of granulates since it takes into account rotations and couple stresses which are observed during shearing but remain negligible during homogeneous deformation. Its other advantages are: the characteristic length is directly related to the mean grain diameter, realistic wall boundary conditions at the interface of granulate with a structure can be derived and material parameters can be easily identified. The Cosserat rotation, couple stresses and asymmetry of the stress tensor are the best indicator to detect shear zones.

A satisfactory agreement between FE results obtained with a micro-polar hypoplastic model and laboratory experiments was achieved on condition that the element size was not greater than five mean grain diameters.

The shear zone thickness has not a constant value; it depends strongly on the boundary value problem considered. It increases mainly with increasing mean grain diameter, initial void ratio and wall roughness. It depends also on pressure level, direction of the deformation rate, wall stiffness and distribution of initial void ratio. The inclination of shear zones from the minor principal stress decreases with increasing initial void ratio, pressure level and mean grain diameter.

The load–displacement curves reach a residual state. The shear resistance grows with decreasing initial void ratio and pressure level, increasing mean grain diameter

J. Tejchman: Shear Localiz. in Gran. Bodies, SSGG, pp. 313–314, 2008.
springerlink.com

and increasing ratio between a characteristic length and specimen size. The behaviour of the material becomes more ductile with increasing ratio between a characteristic length and specimen size. The mobilized friction angle can significantly vary along shear zones up to the residual state. The influence of a statistical size effect becomes more pronounced with increasing specimen size.

Cyclic shearing with large symmetric cyclic amplitudes influences the evolution of the thickness of a localised zone (in contrast to the results for small cyclic shear amplitudes). With an increase of the number of cycles, the thickness of the localised zone increases with an initially dense specimen, and it decreases with an initially loose specimen. Within the localized zone, the void ratio always decreases after changing the shear direction and then increases tending to the critical void value (large and medium shear amplitude), and always decreases after changing the shear direction (small shear amplitude). The degradation of shear resistance during cyclic shearing increases with increasing initial void ratio and vertical normal stiffness.

Shear zones have a tendency for reflection only from fixed or moving rigid boundaries.

Shear localization with a continuous densification can be also created in initially loose granular materials. The global material softening is not necessary to obtain shear localization whose formation mainly depends on the boundary conditions of the entire system.

The interface behaviour within a micro-polar continuum can be described among others by a kinematic condition including the ratio between the Cosserat rotation and slip. The ratio can be connected approximately with the wall roughness. When using micro-polar boundary conditions, the mobilised wall friction angle is obtained as natural outcome rather than prescribed.

The statistical size effect is significantly weaker than the deterministic one. The mean shear resistance at peak with a stochastic distribution of the initial void ratio is always smaller than with a uniform distribution of the initial void ratio. It diminishes with increasing specimen height and decreasing mean grain diameter.

The calculations of shear localization with a micro-polar hypoplastic model will be continued. To simulate large geotechnical problems, the constitutive model will be used together with a remeshing technique or interface elements. To take into account large mesh deformation, an Arbitrary Lagrangian-Eulerian formulation or a particle-in-cell approach will be used. The statistical size effects will be investigated using different non-symmetric random distributions of the initial void ratio and grain diameter (mutually correlated) in problems wherein varying pressure is of importance. To describe rapid granular flow, the model will be enhanced by viscous terms. In addition, the FE calculations will be performed with water-saturated granular specimens.

List of Symbols

a_c	micro-polar parameter
A [m^2]	cross-section area
b [m]	width
b_o [m]	initial width
c	non-coaxiality parameter
d_{50} [m]	mean grain diameter
d_{ij}	rate-of-deformation tensor
e	void ratio
e_c	critical void ratio
e_d	minimum void ratio
e_i	maximum void ratio
e_o	initial void ratio
E [kPa]	modulus of elasticity
f_d	density factor
f_i^B [kN/m^3]	volume body forces
f_s [kN/m^2]	stiffness factor
g [m/s^2]	gravitational acceleration
h [m]	specimen height
h_o [m]	initial height
h_s [kPa]	granulate hardness
K	pressure coefficient
K_o	earth pressure at rest
l [m]	specimen length
m_i [kN/m]	Cauchy couple stress vector
$\hat{m}_i$ [kN/m]	normalized couple stress vector
$\overset{o}{m}_i$ [kN/m]	objective couple stress rate vector
$\overset{\square}{m}_i$ [kN/m]	time derivative couple stress vector
m^B [kNm/m^3]	volume body moment
n	coefficient of compression
N [kN]	normal force

p [MPa]	pressure
r_w [m]	wall roughness
s_{eo}	standard deviation
s_{ij} [kPa]	deviatoric stress tensor
t [s]	time
T [kN]	shear force
u_1 [m]	horizontal displacement
u_2 [m]	vertical displacement
W_{ij}	non-polar spin tensor
w^c	micro-polar spin
w	weighting function
v [m/s]	velocity
$v_{i,j}$ [1/s]	velocity gradient
V [m^3]	volume
x_1 [m]	rectangular horizontal co-ordinate
x_2 [m]	rectangular vertical co-ordinate
α	coefficient of pycnotropy
β	coefficient of barotropy
δ_{ij}	Kronecker delta
ε_{ij}	deformation tensor
γ, γ_d [kN/m^3]	volume weight
κ_I	curvature vector
λ	decay coefficient
ϕ [°]	internal friction angle
ϕ_c [°]	critical angle of internal friction
ϕ_w [°]	wall friction angle
ρ [kg/m^3]	mass density
θ [°]	shear zone inclination
σ_c [kPa]	confining pressure
σ_i [kPa]	principle stresses
σ_{ij} [kPa]	Cauchy stress tensor
σ^*_{ij} [kPa]	deviatoric part of Cauchy stress tensor
$\hat{\sigma}_{ij}$ [kPa]	normalized stress tensor
$\overset{o}{\sigma}_{ij}$ [kPa]	objective stress rate tensor
$\dot{\sigma}_{ij}$ [kPa]	time derivative stress tensor
σ_n [kPa]	normal stress
τ [kPa]	shear stresses
υ	Poisson's ratio
ω^c	Cosserat rotation
ξ	non-coaxiality angle
ψ	dilatancy angle

Springer Series in Geomechanics and Geoengineering

Edited by W. Wu, R.I. Borja

Tejchman, J.
Shear Localization in
Granular Bodies with Micro-Polar
Hypoplasticity
316 p. 2008 [978-3-540-70554-3]

MIX
Papier aus verantwortungsvollen Quellen
Paper from responsible sources
FSC® C105338

If you have any concerns about our products,
you can contact us on
ProductSafety@springernature.com

In case Publisher is established outside the EU,
the EU authorized representative is:
Springer Nature Customer Service Center GmbH
Europaplatz 3, 69115 Heidelberg, Germany

Printed by Libri Plureos GmbH
in Hamburg, Germany